New Techniques for
Future Accelerators

ETTORE MAJORANA
INTERNATIONAL SCIENCE SERIES
Series Editor:
Antonino Zichichi
European Physical Society
Geneva, Switzerland

(PHYSICAL SCIENCES)

Recent volumes in the series:

Volume 21   **ELECTROWEAK EFFECTS AT HIGH ENERGIES**
             Edited by Harvey B. Newman

Volume 22   **LASER PHOTOBIOLOGY AND PHOTOMEDICINE**
             Edited by S. Martellucci and A. N. Chester

Volume 23   **FUNDAMENTAL INTERACTIONS IN LOW-ENERGY
             SYSTEMS**
             Edited by P. Dalpiaz, G. Fiorentini, and G. Torelli

Volume 24   **DATA ANALYSIS IN ASTRONOMY**
             Edited by V. Di Gesù, L. Scarsi, P. Crane,
             J. H. Friedman , and S. Levialdi

Volume 25   **FRONTIERS IN NUCLEAR DYNAMICS**
             Edited by R. A. Broglia and C. H. Dasso

Volume 26   **TOKAMAK START-UP: Problems and Scenarios
             Related to the Transient Phases of a
             Thermonuclear Fusion Reactor**
             Edited by Heinz Knoepfel

Volume 27   **DATA ANALYSIS IN ASTRONOMY II**
             Edited by V. Di Gesù, L. Scarsi, P. Crane,
             J. H. Friedman, and S. Levialdi

Volume 28   **THE RESPONSE OF NUCLEI UNDER EXTREME
             CONDITIONS**
             Edited by R. A. Broglia

Volume 29   **NEW TECHNIQUES FOR FUTURE ACCELERATORS**
             Edited by M. Puglisi, S. Stipcich, and G. Torelli

Volume 30   **SPECTROSCOPY OF SOLID-STATE LASER-TYPE
             MATERIALS**
             Edited by Baldassare Di Bartolo

Volume 31   **FUNDAMENTAL SYMMETRIES**
             Edited by P. Bloch, P. Pavlopoulos, and R. Klapisch

---

A Continuation Order Plan is available for this series. A continuation order will bring delivery of each new volume immediately upon publication. Volumes are billed only upon actual shipment. For further information please contact the publisher.

New Techniques for Future Accelerators

Edited by

Mario Puglisi

Pavia University
Pavia, Italy

Stanislao Stipcich

INFN, Frascati National Laboratory
Frascati, Italy

and

Gabriele Torelli

Pisa University
Pisa, Italy

Plenum Press • New York and London

Library of Congress Cataloging in Publication Data

Seminar on New Techniques for Future Accelerators (1986: Erice, Sicily)
New techniques for future accelerators.

(Ettore Majorana international science series. Physical sciences)
Proceedings of the first workshop of the INFN eloisatron project: Seminar on New
Techniques for Future Accelerators, held May 11–17, 1986, in Erice, Trapani, Sicily.
Bibliography: p.
Includes index.
1. Particle accelerators—Congresses. I. Puglisi, Mario. II. Stipcich, Stanislao. III.
Torelli, Gabriele. IV. Instituto nazionale di fisica nucleare. V. Title. VI. Series.
QC787.P3S46 1986   539.7′3   87-20248
ISBN 978-1-4684-9116-6   ISBN 978-1-4684-9114-2 (eBook)
DOI 10.1007/978-1-4684-9114-2

Proceedings of the First Workshop of the INFN Eloisatron Project: Seminar on New
Techniques for Future Accelerators, held May 11–17, 1986, in Erice, Trapani, Sicily.

© 1987 Plenum Press, New York
Softcover reprint of the hardcover 1st edition 1987
A Division of Plenum Publishing Corporation
233 Spring Street, New York, N.Y. 10013

PREFACE

    This Seminar has been organized in Erice, in the frame of the
Eloisatron project activities, with the special purpose of bringing
together an interdisciplinary group of distinguished physicists
with prominent interest in the development of the accelerators.

    Listening to the invited lectures, examining the new topics and
reviewing ideas for the acceleration of particles to energies beyond
those attainable in machines whose construction is under way or is
now contemplated are all important moments of this Seminar that will
offer to the Italian Physicists a very important opening over the
scenario of the accelerators.

    In connection with the Eloisatron project developments future
Workshop-Seminars are now envisioned, each one aimed to a very
specific topic in the field of the particle accelerators.

The Editors

CONTENTS

Overview of Linear Collider Studies 1
  K. Johnsen

Principles of Beat-Wave Accelerators 15
  U. de Angelis, R. Fedele and V.G. Vaccaro

Wake Field Acceleration 29
  W. Bialowons, H.D. Bremer, F.-J. Decker, M. v. Hartrott,
    H.C. Lewin, G.-A. Voss, T. Weiland, P. Wilhelm, Xiao
    Chengde and K. Yokoya

Energy Efficiency and Choice of Parameters for Linear Colliders ... 45
  J. Clauss

A Two-Stage RF Linear Colliders using a Superconducting Drive
    Linac 67
  W. Schnell

The Micro Lasertron. An Efficient Switched-Power Source of mm
    Wavelength Radiation 89
  R.B. Palmer

Collider Scaling and Cost Estimation 105
  R.B. Palmer

Cooling Rings for TeV Colliders 121
  R.B. Palmer

Linear Colliders Driven by a Superconducting Linac-FEL System ... 139
  U. Amaldi and C. Pellegrini

Low Technology FELs and IFELs: How can a small Laboratory
    contribute 163
  G. Dattoli and E. Sabia

Developments in the Physics of High Current Linear Ion
    Accelerators 181
  T.P. Wangler

RFQ Application 201
   A. Schempp

High Power 35 GHz Testing of a Free-Electron Laser and Two-Beam
      Accelerator Structures 219
   D.B. Hopkins, R.W. Kuenning, F.B. Selph, A.M. Sessler,
   J.C. Clark, W.M. Fawley, T.J. Orzechowski, A.C. Paul,
   D.Prosnitz, E.T. Scharlemann, S.M. Yarema and
   B.R. Anderson

Acceleration of Electrons by the Wake Field of Proton Bunches 241
   A.G. Ruggiero

Analytic Tools for solving Nonlinear Problems in Particle
      Accelerators: A Review and an Example 249
   S.P. Petracca and I.M. Pinto

Approximate Solutions of the Equation for the Longitudinal
      Motion of Particles in an RFQ Accelerator 259
   M. Leo, R.A. Leo, M. Puglisi, C. Rossi, G. Soliani and
   G. Torelli

A Model of Four Rods RFQ 265
   A. Fabris, A. Massarotti and M. Vretenar

RFQ Field Stabilization using Dipole Suppressors 271
   M. Vretenar

Participants 277

Subject Index 281

OVERVIEW OF LINEAR COLLIDER STUDIES

K. Johnsen
CERN
Geneva, Switzerland

ABSTRACT

Against the background of the well-known Livingston diagram, the current status and complementarity of the hadron collider projects SSC, LHC and Eloisatron, the lepton collider projects LEP and SLC, and the hybrid collider HERA are reviewed. Whereas future hadron colliders have no clear technological limit, lepton colliders are at an important turning point and the search for new techniques in the direction of linear colliders is in full flood. A natural approach is to extrapolate parameters for normal conducting structures to fit the special requirements of linear colliders. Superconducting accelerating structures, the use of pulsed rather than harmonic RF power with optoelectric switches, wake fields and the plasma beatwave scheme are some of the new ideas which are reviewed in this paper. Much of what is included comes from the findings of the Long Range Planning Committee established by the CERN Council under the chairmanship of C. Rubbia.

1. INTRODUCTION

Particle accelerators became an important tool for experimental nuclear physics from around 1930. Since then there has been tremendous progress in the construction of such accelerators with an increase of about one and a half orders of magnitude in beam energy per decade, as illustrated by the Livingston plot shown in Fig. 1. In spite of this energy increase of eight orders of magnitude the cost of a typical accelerator installation has only gone up by one order of magnitude if inflation is corrected for.

It is interesting to look a little more closely at how this has happened. The progress of each type of accelerator has saturated fairly quickly, whereas new ideas have been proposed regularly and have been the main contributors to the rapid advance. Two startling examples can be pointed out: the invention of the alternating gradient focusing in the early fifties and the application of colliding beams in the sixties and onwards. For the future the technological development within such

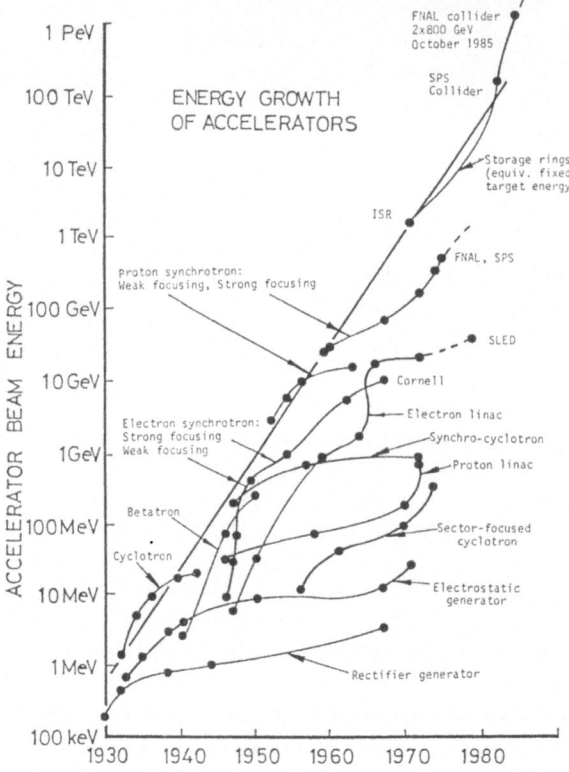

Fig. 1 Livingston plot showing the maximum energy as a function of year
for various accelerator techniques

fields as superconductivity will be of great significance. And to satisfy the far-reaching visions
of high-energy physics in the more distant future, we may have to rely upon new inventions and
developments. This talk will be speculative on how this development will continue in the future in
order to satisfy the desires of the high-energy physics community for ever higher energy and at the
same time, increased luminosity. Another interesting observation can be made with reference to
Fig. 1. The highest energies attainable at present are already far beyond what can ever be reached
by fixed-target machines. In other words, when we talk about future facilities we are only talking
about accelerators that can function as colliding beam machines.

For hadron colliders there is no clear technological limit. For this reason future plans for
such colliders are based on known principles, but extended in energy and luminosity as far as one can
possibly hope to reach within assumed financial limits.

For electron colliders the situation is very different. The difficulties created by the radiation from electrons in circular machines severely limits the further development of circular electron colliders to such an extent that it is considered unlikely that such colliders will be constructed beyond the LEP energies. New ways must therefore be sought.

2. STUDIES OF FUTURE HADRON COLLIDERS

The highest energy hadron colliders are the proton-antiproton collider (Sp$\bar{\mathrm{p}}$S) at CERN of about 2 × 300 GeV energy and the proton-antiproton collider (Tevatron I) at Fermilab of about 2 × 800 GeV. The former has been operational since 1982, the latter had a trial run in October, 1985, and will operate for physics experimentation from late 1986. Both will always have inherently modest luminosities, although these are incredibly high when we take into account how they are achieved.

The next natural step is to put proton rings with the highest possible magnetic field into the LEP tunnel (see Fig. 2). Indeed, such considerations formed part of the arguments that determined the radius and size of the LEP tunnel. A range of possibilities exists for such a collider. The conceptually simplest option would be a p$\bar{\mathrm{p}}$ ring with a single beam channel, preferably with the highest possible guide field. The luminosity would, however, be relatively low because antiproton sources are not very intense. Furthermore, in order to make provision for bunch separation at unwanted crossings, the aperture must be somewhat enlarged compared with a single-beam machine.

Another, and probably more interesting, option would be two beam-channels side by side. For space reasons, it is, in that case, an advantage to use the so-called "2 in 1" solution to the magnet design, i.e. the two channels within the same magnet yoke and the same cryostat. The two-channel approach will allow high-luminosity pp collisions with many bunches. For this reason this approach has formed the basis of some studies within ECFA and CERN of such a project. The main parameters that have come out of these studies are listed in Tables 1 and 2.

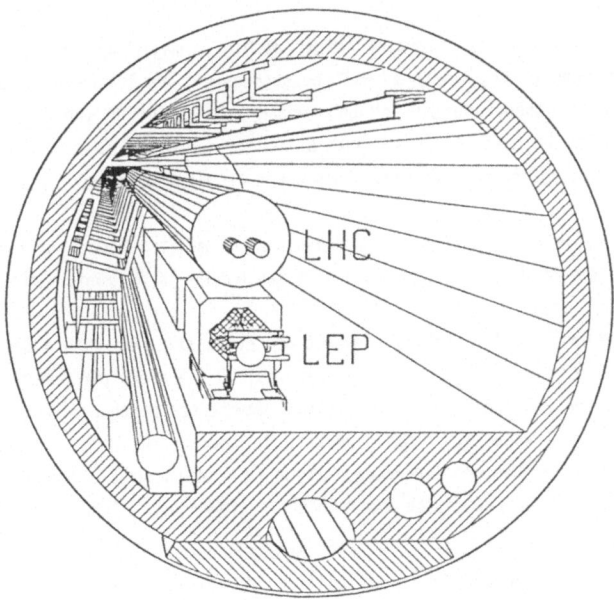

Fig. 2 Large hadron collider in the LEP tunnel

Table 1

General parameters

Collider in LEP tunnel	Proton-proton
Dipole magnetic field (T)	10
Operating beam energy (TeV)	8-9
Separation between orbits (mm)	165-180
Number of bunches	3564
Bunch spacing (ns)	25
Number of crossing points	8
Beta value at crossing point (m)	1
Full crossing angle (μrad)	96

Table 2

Performance

$\langle n \rangle$ at $\sum$ = 100 (mb)	1	4
Luminosity (cm^{-2} s^{-1})	4×10^{32}	1.5×10^{33}
Beam-beam tune shift	0.0013	0.0025
Beam stored energy (MJ)	63	121
RMS beam radius (μ)m*	12	
Beam life-time (h)**	42	21

* at interaction point for β* = 1 m
** particle loss due to beam-beam collisions

A few clarifying comments to these tables may be useful.

Since the tunnel is given it is important to get as high a magnetic field as possible. It is hoped that an R & D programme on Nb_3Sn magnets will be successful, which should result in about 10 T bending field. This should then give 8-9 TeV beam energy, depending on the final lattice design, where a few options are still open. Another possibility is to follow the more conventional NbTi approach, but with 1.8 K operating temperature. This would result in about 8 T field.

The shortest acceptable bunch spacing (and therefore the maximum number of bunches) is determined by the response times of detector elements, which can hardly be assumed shorter than 25 ns.

Another uncertain element is the number of events per bunch crossing that a detector system can accept. It is believed to be of the order of one. In Table 2 one event per bunch crossing has been assumed for the first column. A second column where four events per bunch crossing have been assumed is also presented. It seems clear that for a machine to operate at 10^{33} cm^{-2} s^{-1} some detector development is needed, whereas the corresponding machine parameters seem alright. If higher luminosities are needed both machine performance and detector performance will have to be stretched.

This study has demonstrated that a hadron collider in the LEP tunnel would technically and scientifically be a very sound proposition. Centre-of-mass energies of 10-12 TeV would be attainable with present-day superconducting technology, with the prospect of 15-20 TeV with successful Nb_3Sn

development. Luminosities of 10^{32}–10^{33} cm^{-2} s^{-1} seem entirely feasible in each of the eight inter-action regions.

Whether or not such a machine will be built depends, however, on the development over the next few years of high-energy physics in the world in general and in Europe more specifically. It also depends on how new ideas develop during the rest of this century.

In the USA a vigorous effort goes into the study of a 2 × 20 TeV facility called the SSC (for Super Superconducting Collider). For this project there is no existing tunnel to put constraints on the magnitude of the guide field. They have therefore considered many field options between 3 T and 6.5 T, with little emphasis on the 10 T option since the time element is so important in their case. Otherwise the considerations are very similar to the ones described in the previous paragraph. A summary of the present results for the preliminary study is given in Table 3 and Fig. 3.

Table 3

Primary SSC design objectives

Maximum beam energy (TeV)	20		
Injection energy (TeV)	1		
Maximum luminosity (cm^{-2} s^{-1})	10^{33}		
Maximum number of interactions per bunch crossing (at max. luminosity)	10		
Number of interaction regions	6 (4 initially developed)		
Field (T)	6.5	5	3
Circumference (km)	90	113	164

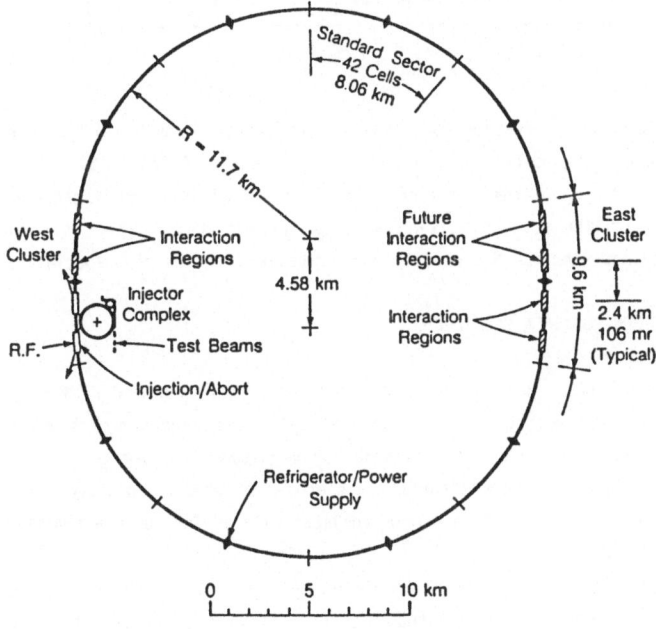

Fig. 3 The SSC project layout

The studies of this project are far enough advanced that technically the project is ripe for a decision. The fate of the project is therefore now largely in the hands of the politicians. With the options chosen by the American physicists the construction of this project, or an approximately equivalent one, has become very urgent as the HEP community will be very short of experimental facilities in some years time.

Another European hadron collider study should be mentioned: the Eloisatron project in Italy. Its aim is an energy of 2 × 50 TeV with 10 T magnets on a circumference of about 150 km. The Italian involvement in HERA, by the supply of 50% of the bending magnets for the superconducting proton ring of this facility, is considered as a kind of preparatory work for the Eloisatron project.

Before we turn to e^+e^- colliders, a few words should be said about the possibility of electron-proton colliders, which falls somewhat in between the pure hadron colliders and the electron colliders. The first and only such facility is under construction at DESY, with the name of HERA, already mentioned above. The main parameters for HERA are listed in Table 4.

This project will be operational in 1989, and will then constitute an experimental facility very complementary to other facilities in the world.

Table 4

Main parameters for HERA

	p-ring	e-ring
Energy (GeV)	820	30
Circumference (km)	6.3	
Dipole strength (T)	4.5	0.18
Luminosity (cm^{-2} s^{-1})	3 × 10^{31}	
Number of interaction points	4	

Assuming this becomes as interesting for physics experimentation as hoped for, one would want to think of even higher energy e-p collisions in the future. Again LHC lends itself naturally to this possibility, as it is sitting in the LEP tunnel, and relatively minor modifications are needed to collide the electrons with the protons, giving an order of magnitude higher collision energy than HERA. To do similar things with the SSC would require the construction of a special electron ring.

3. PROSPECTS FOR ELECTRON-POSITRON COLLIDERS

Electron-positron colliders have a few clear advantages over hadron colliders. The main ones are: the e^+e^- collisions are much cleaner than hadron collisions, and a much higher luminosity can be exploited, if achievable. In addition, an order of magnitude less energy is required since the colliding particles are themselves "constituents". Energy-wise an e^+e^- collider with centre-of-mass energy of 2 TeV is roughly equivalent to a hadron collider with 20 TeV centre-of-mass energy.

However, the disadvantage is that the radiation makes it so much harder to make very high energy e^+e^- colliders. It is therefore believed that LEP presently under construction at CERN, with a circumference of 27 km, will be the largest circular e^+e^- collider ever to be built.

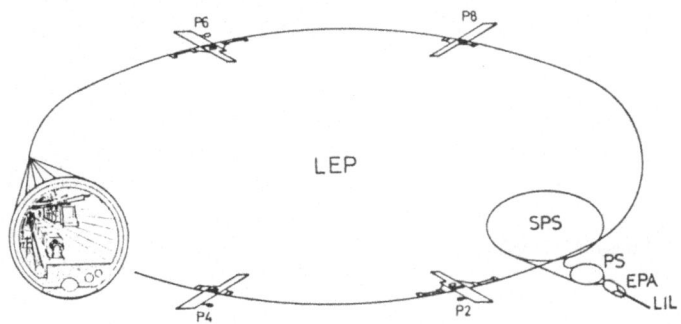

Fig. 4 The LEP machine

A layout of LEP is shown in Fig. 4.   As presently being built, the beam energy of LEP will be 50 GeV. However, it is foreseen to extend the project to higher energies, as illustrated in Table 5.   The luminosity is expected to be a little above 10^{31} cm^{-2} s^{-1} for Phase 1, increasing with increasing energy to perhaps 10^{32} cm^{-2} s^{-1} for the 'ultimate' phase.   The project is under construction and is planned to have beams in the spring of 1989.

Table 5

LEP stages

	Beam energy (GeV)	Power Dissipation (MW)	Installed RF power (MW)	Expt. areas at points
Phase 1	50	75	16 (copper cav.)	2,4,6,8
Planning permission	100 a)	150	48 (supercond.cav.)	1,2,4,5,6,8
'Ultimate'	125 a)	(a)	96 (supercond.cav.)	all 8

The radiation problem in a circular e^+e^- collider is fundamental and can only be counteracted or solved by increasing the circumference of the machine, hence the very large circumference of LEP. The ultimate limit of this approach is to go to infinity with the bending radius, in other words to consider colliding beams from linear accelerators.   The first speculations on this kind of approach, with a host of new problems, started about 10 years ago and has recently gained considerable momentum. Because of the importance attached to this possibility, it has been made the subject of one of the Advisory Pannels organised by C. Rubbia for the Long Range Planning Committee (LRPC) set up by the CERN Council.

The first attempt at colliding beams from a linear accelerator will be done with a facility now under construction at SLAC, the so-called SLAC Linear Collider (SLC), with a beam energy of 50 GeV and luminosity of 6×10^{30} cm^{-2} s^{-1}. The layout and main parameters are shown in Fig. 5. Contrary to circular colliders the beams pass through each other only once, which means that beam parameters must be optimised such as to make maximum use of this one crossing.   One consequence of this is illustrated by the very small beam sizes that have to be achieved in the collision region with correspondingly difficult requirements on the quality of the system for the final focus. The project will provide crucial information in such areas when it starts its operation in late 1986 or early 1987.

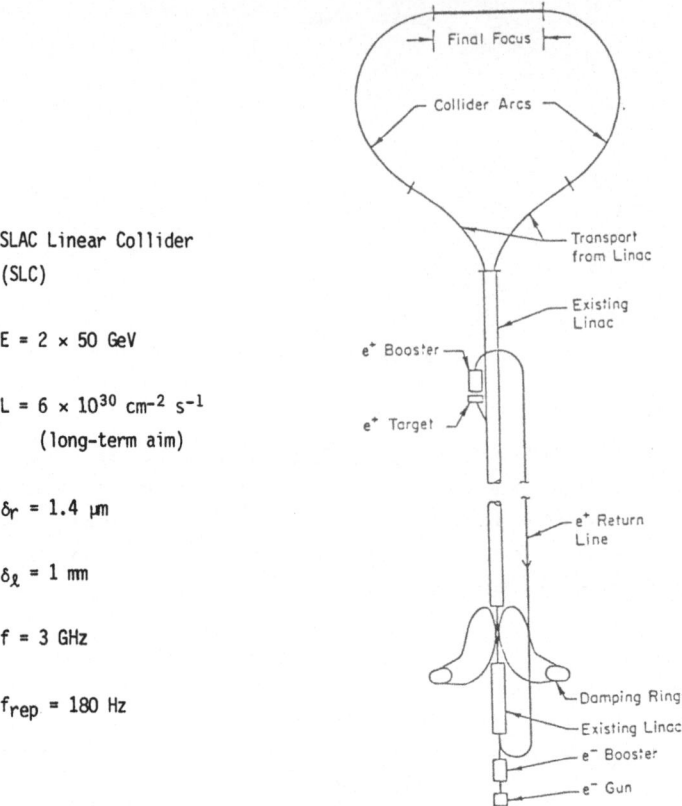

SLAC Linear Collider
(SLC)

$E = 2 \times 50$ GeV

$L = 6 \times 10^{30}$ cm^{-2} s^{-1}
 (long-term aim)

$\delta_r = 1.4$ μm

$\delta_\ell = 1$ mm

$f = 3$ GHz

$f_{rep} = 180$ Hz

Fig. 5 SLC

At CERN we are considering the possibilities of a linear collider (CLIC) with energies in the TeV range, and luminosities up to 10^{34} cm^{-2} s^{-1}, i.e. more than an order of magnitude higher energy than SLC, and about three orders of magnitude higher luminosity. This requires more than a simple extrapolation of present-day techniques.

Below we list the most important approaches that have been proposed for acceleration in linear colliders, with some comments on their main characteristics, and the need for further detailed studies before reliable comparisons can be made and firm conclusions drawn.

3.1 Normal-conducting accelerating structures

The classical travelling wave linac appears to be capable of accelerating gradients well above 100 MV/m, especially at high microwave frequencies. A high frequency is also desirable in order to reduce stored energy. Longitudinal and transverse wake fields as well as construction problems may impose a limit on the choice of frequency and this remains to be studied in detail.

The Q factor is too small to permit conservation of stored energy over a realistic repetition period. In spite of this, RF to beam efficiencies above 5% can be obtained. The price for this is a very short power pulse and an appreciable but probably correctible energy spread of the beam, owing to the need for extracting about 10% of stored electromagnetic energy from the accelerating structure at each beam pulse. The RF to beam efficiency could be raised above 10% if the electromagnetic energy remaining at the output of the travelling wave structure could be recovered. Two two-stage schemes resulting from the CERN study seem to permit that. Multibunch operation with bunch-to-bunch compensation of energy deviation may yield 30% efficiency provided higher-order wakefield problems are solved and a suitable final focus scheme is found.

The worst remaining problem seems to be the economic and efficient generation of peak RF power. Microwave power sources are being developed which may permit feeding individual linac sections with tens of megawatts peak power at wavelengths down to the order of 1 cm, but an economic problem remains. Alternatively a two-beam scheme employing microwave free electron lasers, induction linacs and a low energy drive beam has been proposed and is being developed (Sessler, 1982).

Two proposals of two-stage schemes, both employing a superconducting RF drive linac and a drive beam have emanated from our group (Schnell, 1986; U. Amaldi and C. Pellegrini, 1986). (See contribution by W. Schnell to this seminar.)

A few tentative parameters of a 1 TeV main linac are given in Table 6. Energy recovery at the output of the travelling wave structure, would raise the luminosity (for given input power) by a factor of two. Compensated multibunch operation may raise the luminosity to 0.6×10^{34} at only 25% extension of length and average power.

Table 6

Tentative parameters of a 1 TeV main linac

Energy	1	TeV
Frequency	29	GHz
Accelerating gradient	80	MV/m
Active length	12,5	km
Peak power per metre of section length	96	MW/m
Total average RF power	80	MW
Beam power	5	MW
Repetition frequency	5,8	kHz
Bunch population	5.4×10^9	
Beam radius at collision	77	nm
Luminosity (single bunch without energy recovery)	10^{33}	$cm^{-2} s^{-1}$

3.2 Superconducting accelerating structures

Linear colliders based on superconducting cavities (SC) were proposed and studied at CERN more than ten years ago (U. Amaldi, 1976). In recent years studies of thermal breakdowns and electron loading in SC cavities have led to the construction of 350 MHz cavities having a gradient of 7 MV/m and a Q-value of 3×10^9. For small 3 GHz cavities gradients of 18 MV/m and $Q = 10^{10}$ have been obtained. A few years of technological developments on the preparation of clean and defect-free surfaces may allow us to reach of 25 MV/m and $Q = 5 \times 10^{10}$. The development of type II SC, like Nb_3Sn and NbN, possibly sputtered on a copper substrate, offers the promise of reaching even higher gradients and quality factors. For linear colliders economic fabrication and treatment will also become of paramount importance. Examples have been studied by U. Amaldi, H. Lengeler and A. Piel (1986) assuming the following parameters: $\nu = 1$ GHz, G = 25 MeV/m, $Q = 5 \times 10^{10}$, temperature = 1.8 K.

Table 7 lists the main parameters of two reference designs of a high luminosity (10^{34} cm^{-2} s^{-1}) (1+1) TeV collider for which some cost optimisations have been performed. The first design uses the principle of energy recovery to save on the total power and has parameters which are extrapolations by a factor of 10 only with respect to the SLC design. U. Amaldi (1986) and others have demonstrated that the damping ring system is a very important and expensive part of the complex. The second reference design has no energy recovery and requires an emittance 10 times smaller, at the limit of what can be achieved today with the best storage rings for synchrotron radiation.

The cost optimisation procedure, which takes into account klystron replacement and electricity consumption over 10 years, leads to the conclusion that the accelerator has to be run with a macroscopic duty factor C = 0.15, and that both designs need about 350 MW for running.

Table 7

The two reference designs of a superconducting (1$^+$1) TeV collider
with L = 10^{34} cm^{-2} s^{-1} (D = 2)

Accelerator	Energy recovery	No recovery
Beam power P(MW)	100	10
Recovery efficiency η	0.90	0.0
Invariant emittance ε (m)	2.5×10^{-6}	3.1×10^{-7}
Bunch length σ_z (mm)	3.6	0.36
β-value at crossing (mm)	10.7	1.07
Bunch radius $\sigma_x = \sigma_y$ (μm)	0.12	0.013
Particles/bunch N	5.4×10^9	6.5×10^8
Macroscopic duty factor C	0.15	0.15
Bunch peak frequency f (kHz)	670	820
Train peak frequency	67	82
Gradient (MeV/m)	25	25
Q-value	5×10^{10}	5×10^{10}

3.3 Structures excited by optoelectric switches

3.3.1 Conventional structures driven by optoelectric RF generators

The first such device to be suggested was the laser-klystron, where a train of laser pulses at 3 or 6 GHz striking a photocathode switches on bunches of electrons, which are accelerated by a static field and passed through a resonant cavity to extract the energy. It is hoped that the efficiency can be high and the cost lower than for conventional klystrons. Another approach uses the wire photocathodes discussed in the next section with laser pulse trains at 30 GHz, the "microlasertron" (Palmer, 1986).

3.3.2 Radial pulse line structure driven by optoelectric switched power

This proposal (Willis, 1984) uses very short pulses instead of RF power, increasing the breakdown limit on the accelerating gradient. It may prove that the cost of switches may be relatively low and their efficiency high, despite the fact that the peak power is quite high in these non-resonant structures.

Tests are underway on two types of photodiodes, vacuum and semiconductor. For the former, measurements are being made on wire photocathodes, charges close to the field emission limit by nanosecond electrical pulses and then illuminated by a 20 ps pulse of 0.35 nm light. It is known that in this "field assisted photoemission" regime, the electron yield from rugged photocathodes is considerably increased. The efficiency of these photodiodes mounted in a radial line have been studied by the numerical simulation program "MASK". Using a sequence of four suitably timed laser pulses to discharge the wire, an overall efficiency of 15% has been predicted.

Studies have been started on the volume photoconductivity in GaAs (Cr) induced by 1.06 μm laser pulses. Results indicate that 1-5 mJ of light can switch one GW of electrical power.

The performance of the radial lines, consisting of copper discs spaced by a small gap and driven at the outer radius, has been studied in scaled up models, with measurements being made on the effect of misalignments.

A set of tentative parameters for a typical example with a beam energy of 1 TeV and a luminosity of 10^{33} cm^{-2} s^{-1} are given in Table 8.

Table 8

Outer radius of radial line	120 mm
Inner radius (beam hole)	1 mm
Gap between disks	2 mm
Photodiode capacitance	16 pF
Pulse length	25 ps
Charging voltage	80 kV
Stored energy	48 MJ
Switched power	2 GW
Optical energy	3 mJ
Gradient on beam, in gap	0.6 GV/m
Average gradient, for 2 mm disc	0.3 GV/m
Stored energy, electrical	24 KJ/km
Beam energy, for 5% loading	1.2 KJ/km
Number of electrons per pulse	10^{10}
Optical energy	1.5 KJ/km
Bunch radius σ_r*	0.1 μm
Bunch length σ_z	1 mm
Disruption parameter D	2
Beam radiation parameter δ	0.3
Repetition frequency f	2 kHz

3.4 Wake-field acceleration

High accelerating gradients can be produced by the passage of bunches of charged particles through structures such as cavities or disk-loaded wave-guides. A number of accelerators based on this principle have been proposed and models have been or are being constructed in several laboratories around the world.

Very high transformer ratios, i.e. ratios of the maximum accelerating field behind to the decelerating field inside the driver, are hoped to be obtainable with a "wake-field transformer", in which the driver passes along a different trajectory from the beam to be accelerated (Voss and Weiland,

1982). In the coaxial type, the driver passes through an annular slot on the outside of a series of circular cylindrical cavities. The induced fields converge towards the beam to be accelerated on the axis. A model of such an accelerator is under construction at DESY and will be tested shortly. More limited transformer ratios can be obtained in elliptical cavities which have been tested recently in Japan. (See contribution by T. Weiland to this seminar.)

3.5 Plasma wakefields

A bunch of electrons passing through a plasma will excite plasma oscillations travelling with a phase velocity equal to the bunch velocity. Gradients of several GeV/m seem possible. The fields behind the driving beam are similar to the wakefields in a linac structure, with the difference that only a single mode is present and the group velocity is zero so that the field pattern does not spread out laterally. A following bunch can be accelerated by this field.

The principle was proposed by Chen, Huff and Dawson (1984). It seemed at first that the driven beam cannot gain more than twice the energy than the driving beam loses on average. However, it was found by Chen & Dawson (1985) that a driving beam with a sawtooth-shaped longitudinal density profile will permit larger transformer ratios. Another way to obtain this is suggested by van der Meer (1986).

Some of the properties of the scheme important to accelerator application were described by Ruth, Chao, Morton and Wilson (1984). It appears that the transverse focusing by the wakefield tends to be very strong; with the low-emittance beams needed for a narrow final focus this results in beam diameters of the order of a micron. The driving beam, however, must have a diameter of the order of a plasma wavelength (0.1 - 1 mm for typical plasma densities) to prevent excessive radiation loss by particles with large transverse amplitudes (Zotter, 1986). It is not evident that the energy transfer between the beams will be efficient enough with parameters suitable for TeV accelerators. The inherent simplicity of the scheme seems, however, to justify further effort.

Theoretical work on the scheme is at present going on at the University of California, at SLAC and at some other places, including CERN. Much more of this will certainly be needed. Experimental tests on a small scale, to be performed at ANL, have been proposed by a Wisconsin group (Rosenzweig et al., 1985). An electron linac capable of producing dense, short bunches is essential for such tests. The LIL electron linac at CERN has also been suggested as a possible source for tests (E.J.N. Wilson, 1985).

3.6 Plasma beat-wave acceleration

Two collinear laser beams propagating in a plasma with frequencies differing by the plasma frequency can drive a Langmuir wave whose phase velocity is close to the velocity of light. Resonant build-up of plasma-electron oscillations into the non-linear regime generates a ponderomotive force leading to charge separation and a travelling electrostatic wave. With plasma densities in the range of $10^{16} - 10^{17}$ cm^{-3} accelerating fields of several GeV/m can be produced.

Initial experiments (UCLA and Quebec, CO_2 lasers), together with computer simulations, have shown that the basic principle is valid. A Rutherford Lab./Imperial College experiment, in which CERN is participating, is underway; it uses the RAL Nd-glass laser (1.06 micron) and is thus complementary to experiments with CO_2 lasers (10.6 microns). Other groups, both in the USA and in Europe, intend to start experimental work in this field.

The plasma beat-wave accelerator (BWA) and the plasma wakefield accelerator are unique in being the only accelerating methods so far proposed that could plausibly offer energy gains of several GeV/m. These high fields result from charge separation in a medium which, being fully ionised, cannot break down, and offer a particular attraction for the very long term. However, plasmas are notoriously difficult to control and are prone to many instabilities.

The plasma beat-wave principle could be applied to the generation of very strong focusing fields for the interaction region. In this application the problems of tolerances and stability over long distances would be much less severe than for a TeV accelerator. There is therefore a considerable incentive to pursue plasma studies in this regime for possible use with accelerators of a more conventional type in the not-so-far future. (See also contribution of V. Vaccaro to this seminar.)

4. CONCLUDING REMARKS

When demands for e^+e^- collisions first moved into the TeV energy region, with corresponding increases in desired luminosity, conventional approaches to acceleration seemed unpromising and this gave a strong incentive to consider more exotic ideas. In particular it seemed desirable to create extremely high accelerating fields to reduce the physical size of the accelerator. A large number of ideas were born, much larger than those listed earlier. Some were rather short-lived, but quite a few are sufficiently promising that it is well worth putting effort into further studies and development. Nevertheless, no obvious snag-free solution has yet been found, and this is the main reason why new emphasis has recently been put on more conventional approaches. By being open-minded towards the possibilities of very different frequencies, very different power sources, very different beam structures etc., optimism has increased that quasi-conventional solutions maybe found, so that linear colliders need not wait for more exotic ideas to mature. However, several important problems still need attention before one can have sufficient confidence to start detailed design and preparation for construction.

In short, for linear e^+e^- colliders extensive accelerator R & D is needed, but they offer great promises for the somewhat far future. Hadron colliders, on the other hand, can be built for $E_{cm} >$ 10 TeV any time, with requirements to surveying techniques not substantially more difficult than on present day projects.

The CERN Council established in June 1985, a Long Range Planning Committee with Rubbia as Chairman to evaluate the problems related to the many different options for future accelerator facilities and to propose a balanced long-term programme for Europe, also taking into account complementarity in the world. Much of what I have reported is a result of studies initiated by this Rubbia Committee, whose final report is due in June, 1987. Although it is perhaps hoped that this report will crystalise clearly which path to follow into the future, it will also have to contain flexibility and options to allow for desirable changes as programmes develop and become more concrete in other parts of the world, in particular in the USA, where considerable uncertainty reigns at the moment.

DISCUSSION

COURANT:
It seems that in going from conventional to superconducting technique in RF, one loses maximum attainable accelerating gradient, i.e. requires increased length. This is in contrast to the situation in magnets

for storage rings, whose superconductivity improves compacteness. Can you comment on this?

KLEIN and SCHNELL:
Thermal problems can be overcome by slowly pulsing (intervals of seconds) the accelerating field.

PALMER:
I insist that the main problem is on ultrafocusing. The dominant cast factor for linear colliders is not the machine length but the achievement of the desired luminosity. Therefore the lower accelerating gradients of superconducting machines may still be of interest because they promise lower required power, i.e. greater efficiency.
The achievement of maximum accelerating gradient is certainly not the only concern, or even the dominant one.

COURANT and JOHNSEN:
Because of luminosity requirements, required damping rings and their associated problems are common to all recently proposed accelerator types.

PALMER:
The requirement for many cooling rings refered to was formulated by Amaldi on certain specific assumption. He concluded that cooling times were long and that this together with the high required rep rate meant that a large number of bunches would be damping at any one time. Then given normal kicker speeds he derived the requirement for many rings. As I will show on Wednesday morning I have found that with wiggler damping rings the cooling times are quite fast and the requirements are not so severe.
Typical maximum allowable vibrations, e.g. by quadrupole magnets, is likely to be in the 10 angstrom range. This sounds terribly restrictive but is not really so bad considering the technology of lasers and optical bench builders. It is probably achievable but will undoubtedly require new or different techniques.

JOHO:
What are the limitations on the final beam size due to the quantum fluctuations coming from beam deflections in the quadrupoles?

JOHNSEN:
The largest possible benefit to machine builders would result from physicists stating that a luminosity of $L=10^{33}$ was not really necessary.

PRINCIPLES OF BEAT-WAVE ACCELERATORS

U. de Angelis, R. Fedele, and V.G. Vaccaro

Dipartimento di Fisica dell'Università di Napoli
and INFN - Sezione di Napoli, Napoli, Italy

1. INTRODUCTION

One of the most efficient ways of accelerating charged particles is to exploit the very large electric field associated with a longitudinal electrostatic wave.

Having the phase velocity and the accelerating field the same direction, it is possible at least in the principle, to "phase-lock" a beam of injected particles and to accelerate them to very high energies.

To sustain a longitudinal electrostatic wave a "plasma" is necessary, so that the crucial point is how to produce large amplitude electrostatic waves in a plasma. The large electric fields associated with laser beams offer an immediate candidate but there are two problems to be considered.

1) Electromagnetic waves are transverse;

2) A plasma is not transparent to signals below the plasma frequency ω_p: while this is a natural resonance frequency, it is also a cut off frequency for any transverse electromagnetic perturbation. In nature ionosphere offers an example of this phenomenon. It is known that the index of refraction $N(\omega)$ is given by (unmagnetized, isotropic, cold electron plasma)

$$N(\omega) = \left(1 - \frac{\omega_p^2}{\omega^2}\right)^{1/2} \quad ; \quad \omega_p^2 = \frac{4\pi e^2 n_o}{m_e} \tag{1}$$

where n_o is the density of free electrons and e, m_e the electron charge and mass. For the ionosphere $n_o \simeq 10^4 - 10^6$ cm^{-3} corrisponding to $\omega_p \simeq 6 \cdot 10^6 - 6 \cdot 10^7$ rad $\cdot s^{-1}$. The presence in the ionosphere of several layers of plasma with varying density makes also N dependent on height. We can study the electron density at various heights by means of the pulses of radiation transmitted vertically upwards. According to the dependence of electron density on height, a pulse of a given frequency ω_1 enters the layer without reflection if $\omega_1 > \omega_p(h_L)$. When the density is large enough to make $\omega_1 \simeq \omega_p(h_L)$, the index of reflection vanishes and the pulse is reflected.

As a conclusion, in order to achieve a longitudinal wave excitation in plasmas we must first of all penetrate the plasma with e.m. transversal waves and subsequently transform them into longitudinal wave. This is achieved in the Beat Wave Accelerator.

At this point the following question rises spontaneously: why a plasma accelerator? Well, let us immagine to create an electric field by means of a dielectric filled plane capacitor to accelerate charged particles. The acceleration increases as the potential between plates increases. But an intrinsic limit exists because of the dielectric rigidity.

If the electric field overcomes dielectric rigidity breakdown and "perforation" of dielectric occurs. On the contrary, breakdown can not occur in a fully ionized plasma: it can support ultra-high fields. For a plasma a limit to the field also exists, as shown by Dawson (1959). Its expression follows from Poisson's equation:

$$\overline{\nabla} \cdot \overline{E} \sim kE = 4\pi e n_1 \tag{2}$$

where n_1 is the density perturbation.

The maximum field in a cold plasma obtains when $n_1 \sim n_o$, so that: $(k \sim \omega_p/c)$

$$eE^{max} \sim \frac{4\pi n_o e^2}{\omega_p/c} = m\omega_p c \simeq \sqrt{n_o} \;\; (eV/cm) \tag{3}$$

For $n_o \simeq 10^{18} cm^{-3}$, we have $eE^{max} \sim 1 GeV/cm$. In conventional accelerators we have lower acceleration gratients ($eE^{max} \sim 20 MeV/m$) so that with plasma based accelerators high energies can be reached in much shorter distances. In the following we shall give a fluid description (forced oscillator equations) of the plasma accelerator based on the concept of "ponderomotive force" which we introduce first for completeness.

2. THE PONDEROMOTIVE FORCE

A charged particle in an electromagnetic wave essentially oscillates in the direction of the $\overline{E}$ field (figure 8 motion). If however the field envelope has a spatial gradient it can exert a longitudinal (direction of wave propagation) force on the particle known as the ponderomotive force.

Consider a cold, collisionless electron-ion plasma. For this system we write the following Lorentz-Maxwell equations:

$$m_\alpha \frac{d\overline{v}_\alpha}{dt} = q_\alpha \overline{E} + \frac{q_\alpha}{c} \overline{v}_\alpha x \overline{B} \quad \text{(motion)}$$

$$\frac{\partial n_\alpha}{\partial t} + \overline{\nabla} \cdot (n_\alpha \overline{v}_\alpha) = 0 \quad \text{(continuity)}$$

$$\overline{\nabla} \cdot \overline{E} = 4\pi \sum_\alpha q_\alpha n_\alpha \;\;, \;\; \overline{\nabla} \cdot \overline{B} = 0 \tag{4}$$

$$\overline{\nabla} \cdot \overline{E} = -\frac{1}{c} \frac{\partial \overline{B}}{\partial t}$$

$$\overline{\nabla} x \overline{B} = \frac{4\pi}{c} \sum_\alpha q_\alpha n_\alpha v_\alpha + \frac{1}{c} \frac{\partial \overline{E}}{\partial t}$$

where the α index varies on the fluid species (α=e,i) and $\overline{v}_\alpha(\overline{r}, t)$ is the velocity field, $\overline{E}$ and $\overline{B}$ are electric and magnetic fields, n_α is the number density field, q_α is the electric charge. Here we used the fluid derivative $\frac{d}{dt} = \frac{\partial}{\partial t} + \overline{v} \cdot \overline{\nabla}$.

Consider the propagation in the plasma of an electromagnetic wave with a space- and time-dependent amplitude. Assuming z as the direction of propagation we consider the one-dimensional case, that is all quantities only depend on z, t and not on x, y (i.e. $\overline{\nabla} = \hat{z}\frac{\partial}{\partial z}$). Then, writing the vector potential of the wave as:

$$\overline{A}(z,t) = \overline{A}_s(z,t)exp[i(kz - \omega t)] \tag{5}$$

where $\overline{A} = (A_x, A_y, 0)$ and $\overline{A}_s(z,t)$ is a complex amplitude and separating the equation of motion into longitudinal (z) and transverse components we have:

$$m_\alpha \frac{dv_{\alpha z}}{dt} = -q_\alpha \frac{\partial \phi}{\partial z} + \frac{q_\alpha}{c}[\overline{v}_\alpha \text{x}(\overline{\nabla}\text{x}\overline{A})]_z \tag{6}$$

$$m_\alpha \frac{d\overline{v}_{\alpha\perp}}{dt} = -\frac{q_\alpha}{c}\frac{\partial \overline{A}}{\partial t} - \frac{q_\alpha}{c}[\overline{v}_\alpha\text{x}(\overline{\nabla}\text{x}\overline{A})]_\perp \tag{7}$$

In this geometry it is:

$$[\overline{v}_\alpha\text{x}(\overline{\nabla}\text{x}\overline{A})]_z = \overline{v}_{\alpha\perp} \cdot \frac{\partial \overline{A}}{\partial z} \quad , \quad [\overline{v}_\alpha\text{x}(\overline{\nabla}\text{x}\overline{A})]_\perp = -(\overline{v}_\alpha \cdot \overline{\nabla})\overline{A}$$

so that eq.n(7) gives:

$$\frac{d\overline{v}_{\alpha\perp}}{dt} = -\frac{q_\alpha}{m_\alpha c}\frac{d\overline{A}}{dt} \tag{8}$$

which is immediately integrated into:

$$\overline{v}_{\alpha\perp} = -\frac{q_\alpha}{m_\alpha c}\overline{A} \tag{9}$$

so that the longitudinal equation becomes:

$$m_\alpha \frac{dv_{\alpha z}}{dt} = -q_\alpha \frac{\partial \phi}{\partial z} - \frac{q_\alpha^2}{2m_\alpha c^2}\frac{\partial}{\partial z}|\overline{A}|^2 \equiv -q_\alpha \frac{\partial \phi}{\partial z} + |\overline{f}_\alpha| \tag{10}$$

where

$$\overline{f}_\alpha = -\hat{z}\frac{q_\alpha^2}{2m_\alpha c^2}\frac{\partial}{\partial z}|\overline{A}|^2$$

The concept of ponderomotive force is conveniently introduced for cases when the wave amplitude $\overline{A}_s$ is "slowly varying", that is when

$$\frac{1}{|A_s|}\frac{\partial |A_s|}{\partial t} << \omega$$

so that the wave field is a low frequency envelope modulating a high frequency (ω) oscillation. Then the ponderomotive force $\overline{f}_{NL}$ is defined as the average value of $\overline{f}$ in a wave period $T = 2\pi/\omega$:

$$\overline{f}_{NL}^{(\alpha)} = <\overline{f}_\alpha> = \frac{1}{T}\int_0^T \overline{f}_\alpha(z,t)dt = -\frac{q_\alpha^2}{4m_\alpha\omega^2 c^2}\frac{\partial}{\partial z}|\overline{A}_s|^2 z \tag{11}$$

Only for the case of a circularly polarized wave it is $\overline{f}_{NL} = \overline{f}$. Some considerations follow from (11).

1) Given the inverse dependence on the particle mass the ponderomotive force on ions is much smaller than on electrons. Consequently on the electron timescale (ω_{pe}^{-1}) the ions can be considered immobile (cold plasma): only the electrons are initially

displaced by $\overline{f}_{NL}$ and a longitudinal electric field is consequently generated by the charge separation. At later times $(\sim \omega_{pi}^{-1})$ ion motion has also to be taken into account.

2) We remark that we used the following method to find an expression of the ponderomotive force. Initially, a transverse e.m. wave field only exists: the plasma density is unperturbed $(n = n_o)$ and the z-component of both electric field and fluid velocity are zero. At this order of approximation the net force on the plasma is absent: in fact, the oscillating factor in eq.(5) gives a zero-average force on the plasma particles. At the next order the effect related to the slow-varying term $A_s(z, t)$ is introduced, so that a radiation pressure acts on the plasma. As a consequence, the plasma is perturbed. That this nonlinear force has a non- zero-average is easily seen from (11) where the square of A_s is present. Thus, we note that there are two time-scales in this description: a fast-time scale $(t \sim \omega^{-1})$ and a slow-time scale $(t > \omega^{-1})$. Evidently, ponderomotive force, longitudinal plasma motion and scalar potential appear on the slow-time scale. This is referred to as "low-frequency plasma response".

3. It is possible to write the ponderomotive force in terms of a "ponderomotive potential" ϕ_{NL} as $\overline{f}_{NL} = \overline{\nabla} \phi_{NL}$ with

$$\phi_{NL} = -\frac{\omega_p^2}{\omega^2} \frac{< E^2 >}{8\pi} \tag{12}$$

where E is the transverse wave electric field $(E \sim \frac{\omega}{c} A)$.

3. BEAT WAVE MECHANISM

In the Beat Wave Accelerator (BWA), as proposed by Tajima and Dawson (1979), the longitudinal field associated with the plasma wave is generated "beating" two e.m. waves (hereafter referred to as pumps) of frequencies and wave-vectors

$$\omega_1 - \omega_2 = \omega_p \quad , \quad k_1 - k_2 = k_p = \frac{\omega_p}{c} \tag{13}$$

(resonance conditions)

where (ω_p, k_p) are the frequency and wave-number of the excited plasma wave.

This resonant excitation of a plasma wave is a three-wave process (Raman forward decay) where eq.s(13) can be considered (multiplied by ƫ) as the laws of energy and momentum conservation for the two incoming photons and the generated plasmon. This process is best understood in terms of the ponderomotive force associated with the beat wave and a simple fluid model for the plasma.

Introducing the isotropic plasma dielectric constant (from (1))

$$\varepsilon = 1 - \frac{\omega_p^2}{\omega^2} \tag{14}$$

and taking into account (11) and (12) we get:

$$\overline{f}_{NL} = (\varepsilon - 1) \overline{\nabla} \frac{< E^2 >}{8\pi} = (\varepsilon - 1) \hat{z} \frac{\partial}{\partial z} \frac{< E^2 >}{8\pi} \tag{15}$$

This (nonlinear) force moves the electron in the longitudinal (beat propagation) direction opposite to the gradient $(\varepsilon < 1)$ until the restoring force due to the immobile $(\omega_{pi} >> \omega_{pe})$ ions equals the ponderomotive force, thus establishing plasma oscillations, i.e. density perturbations δn with an associated longitudinal electric field E_p. The maximum amplitude of such a field in a cold plasma is given by (section 1):

$$E_p^{max} = \frac{m_e c}{e} \omega_p \tag{16}$$

and wave-breaking occurs for $E_p > E_p^{max}$.

The phase velocity v_ϕ of the plasma wave is:

$$v_\phi \equiv \frac{\omega_p}{k_p} = \frac{\omega_1 - \omega_2}{k_1 - k_2} \simeq \frac{d\omega}{dk} \equiv v_g \tag{17}$$

where v_g is the group velocity of the pump and we used the conditions $\omega_1 \simeq \omega_2$, $k_1 \simeq k_2$ (i.e.: $\omega_p << \omega_1, \omega_2$; $k_p << k_1, k_2$). For propagation in an unmagnetized plasma it is:

$$v_g = c(1 - \omega_p^2/\omega_o^2)^{1/2} \quad , \quad \omega_o \equiv \frac{\omega_1 + \omega_2}{2} \tag{18}$$

and then

$$v_\phi \simeq v_g \simeq c \quad \text{for} \quad \omega_p/\omega_o << 1 \tag{19}$$

The condition (19) has to be associated to the resonance conditions (13) for efficient acceleration: in this case in fact the beat wave and the plasma wave move in synchronism through the plasma and the plasma field E_p can thus trap and accelerate a beam of particles injected (at the end of the beat pulse) with velocity v_ϕ i.e. with an initial energy:

$$m_e c^2 \gamma_{particle} = m_e c^2 (1 - v_\phi^2/c^2)^{-1/2} = \frac{1}{2} \frac{\omega_o}{\omega_p} (MeV) \tag{20}$$

4. BWA THEORY IN FLUID DESCRIPTION

a). *Equations and solutions (linear treatment).*

We assume a cold isotropic plasma of classical electrons and an immobile, neutralizing ion background and we look for the plasma response to an incoming wave (pump) on a timescale that ignores ion motion. Assume two linearly polarized pumps of amplitude E_{o1} and E_{o2} propagating in the z-direction:

$$\overline{E}_j(z,t) = \overline{E}_{oj} \cos(k_j z - \omega_j t) \quad (j = 1, 2) \tag{21}$$

The sum of the two is the "beat wave":

$$\overline{E}(z,t) = \overline{E}_{o1} \cos(k_1 z - \omega_1 t) + \overline{E}_{o2} \cos(k_2 z - \omega_2 t)$$

Consequently:

$$< \overline{E}^2 > = \frac{E_{o1}^2 + E_{o2}^2}{2} + E_{o1} E_{o2} \cos(k_s z - \omega_s t) \tag{22}$$

and

$$\phi_{NL} = -\frac{\omega_p^2}{\omega^2} \left[\frac{E_{oT}^2}{16\pi} + \frac{E_{o1} E_{o2}}{8\pi} \cos(k_s z - \omega_s t) \right] \tag{23}$$

where the average was made on $T = 2\pi/\omega$ with $\omega = (\omega_1 + \omega_2)/2$, $E_{OT}^2 \equiv E_{o1}^2 + E_{o2}^2$, and $\omega_s = \omega_1 - \omega_2$, $k_s = k_1 - k_2$, with $\omega_s << \omega_1, \omega_2$ and $k_s << k_1, k_2$.

From (15) we obtain:

$$F_{NL} = \frac{\nu_{o1}\nu_{o2}}{2} e E_p^{max} \sin(k_s z - \omega_s t) \tag{24}$$

where we have used $\omega_1 \omega_2 = \omega^2 - \omega_s^2 \simeq \omega^2$ and we have introduced the quiver velocity:

$$\nu_{oj} = \frac{e E_{oj}}{m_e \omega c} \quad (j = 1, 2) \tag{25}$$

Denoting with ϕ the ambipolar potential associated with the charge separation due to the ponderomotive effect on electrons, i.e. $\overline{E}_p = -\overline{\nabla}\phi$, the fluid model equations are (low-frequency plasma response):

$$m_e \left(\frac{\partial}{\partial t} + v_e \frac{\partial}{\partial z} \right) v_e = -e E_p + F_{NL} \quad \text{"motion"} \tag{26}$$

$$\frac{\partial n_e}{\partial t} + \frac{\partial}{\partial z}(n_e v_e) = 0 \qquad \text{"continuity"} \tag{27}$$

$$\overline{\nabla} \cdot \overline{E}_p = 4\pi e(n_o - n_e) \qquad \text{"Poisson"} \tag{28}$$

where v_e is the longitudinal electron velocity and n_e the perturbed electron density. Linearizing the equations for:

$$n_e(z,t) = n_o + n(z,t) \quad ; \quad v_e(z,t) = v(z,t) \tag{29}$$

we have the system:

$$m_e \frac{\partial v}{\partial t} = -e E_p + F_{NL} \tag{30}$$

$$\frac{\partial n}{\partial t} + n_o \frac{\partial v}{\partial z} = 0 \tag{31}$$

$$\frac{\partial E_p}{\partial z} = -4\pi e n \tag{32}$$

Making use of (24) this system can be written as:

$$\frac{\partial^2 v}{\partial t^2} + \omega_p^2 v = \frac{-\nu_{o1}\nu_{o2}}{2} (e E_p^{max}) \frac{\omega_s}{m_e} \cos(k_s z - \omega_s t) \tag{33}$$

$$\frac{\partial^2 n}{\partial t^2} + \omega_p^2 n = -\frac{\nu_{o1}\nu_{o2}}{2} (e E_p^{max}) \frac{k_s n_o}{m_e} \cos(k_s z - \omega_s t) \tag{34}$$

$$\frac{\partial^2 \varepsilon}{\partial t^2} + \omega_p^2 \varepsilon = \frac{\nu_{o1}\nu_{o2}}{2} \omega_p^2 \sin(k_s z - \omega_s t) \tag{35}$$

where:

$$\varepsilon = E_p / E_p^{max} \tag{36}$$

The resonant condition of these "forced harmonic oscillator" equations are exactly the conditions (13) $(\omega_s = \omega_p, k_s = k_p)$ corrisponding to the solution:

$$v(z,t) = -\frac{\nu_{o1}\nu_{o2}}{4}\frac{eE_p^{max}}{m_e\omega_p}\sin k_p z \sin\omega_p t - \frac{\nu_{o1}\nu_{o2}}{4}\frac{eE_p^{max}}{m_e}t\sin(k_p z - \omega_p t) \tag{37}$$

$$n(z,t) = -\frac{\nu_{o1}\nu_{o2}}{4}\frac{eE_p^{max}}{m_e}\frac{k_p n_o}{\omega_p^2}\sin k_p z \sin\omega_p t - \frac{\nu_{o1}\nu_{o2}}{4}\frac{eE_p^{max}}{m_e}\frac{k_p n_o}{\omega_p}t\sin(k_p z - \omega_p t) \tag{38}$$

$$\varepsilon(z,t) = \frac{\nu_{o1}\nu_{o2}}{4}\cos k_p z \sin\omega_p t + \frac{\nu_{o1}\nu_{o2}}{4}\omega_p t \cos(k_p z - \omega_p t) \tag{39}$$

These represent the sum of a stationary wave and of a plasma wave of growing amplitude. The ratio of the stationary to the growing wave amplitude is $(\omega_p t)^{-1}$ and the plasma wave can therefore be taken as the "secular" solution to the problem for $\omega_p t >> 1$.

b). *Relativistic saturation.*

A nonlinear calculation of the growth rate including the effect of relativistic mass variation (Rosembluth and Liu, 1972) shows that ε cannot increase indefinitely as shown by eq.(39) but, because of relativistic detuning (ω_p changes) of the harmonic oscillator equation (35), saturates at

$$\varepsilon_{max} = \left(\frac{16}{3}\nu_{o1}\nu_{o2}\right)^{1/3} \tag{40}$$

The plasma wave growth occurs in the pulse of the beat wave: the breaking limit (40) combined with eq.(39) gives therefore a limit for the pulse duration, that is:

$$\tau_{pump} \le (0.32\omega_p)^{-1}\left(\frac{\nu_{o1}\nu_{o2}}{4}\right)^{-2/3}\,\sec \tag{41}$$

which is in the nanosecond region per laser pumps. For $t \sim \tau_{pump}$ the longitudinal field has reached its maximum amplitude and particles injected at this time with the right energy (eq.(20)) can be accelerated. The maximum energy gain ($\varepsilon \simeq \varepsilon_{max}$) in the BWA scheme is (Tajima and Dawson, 1979):

$$W_{max} = 2\varepsilon_{max}\left(\frac{\omega_o}{\omega_p}\right)^2\,(MeV) \tag{42}$$

with the time and length to reach this energy given by:

$$t_{eff} \simeq \frac{W_{max}}{ecE_p} \simeq \frac{4}{\omega_p}(\omega_o/\omega_p)^2 \tag{43}$$

$$l_{eff} \simeq ct_{eff} = \frac{4c}{\omega_p}(\omega_o/\omega_p)^2 \tag{44}$$

Since $W_{max} \sim n_o^{-1}$ and $W_{max}/l_{eff} \sim \sqrt{n_o}$, the higher the final energy desired, the slower the gradient in the simple BWA Scheme (Beat Wave Dilemma).
[This intrinsic difficulty can be solved in the surfatron scheme (Katsouleas and Dawson, 1983)]

As a conclusion we give the estimates for Conventional, BWA and Surfatron Accelerators ($\lambda = 1$ micron, $n_o = 10^{18}$ cm^{-3}, $B = 50$ KG, $\varepsilon = 0.5$):

CONVENTIONAL   1 TeV in 50 km
BWA   1 TeV in 600 m
SURFATRON   1 TeV in 60 cm x 20 m

5. SOME IMPORTANT PARAMETRIC PROCESSES RELATED TO BWA (INSTA-BILITIES)

One of the approximations in the model presented so far has been to consider the transverse field (beating "pumps") as unchanged as it travels through the plasma. A more complete theory would have to consider the effects of plasma motion on the field such as Raman cascading on plasma density perturbations: put in other words in a plasma the fields have to be self- consistent solutions of the Lorentz-Maxwell system and a "given" pump can only be taken as an initial or boundary value for the final solution.

The excited plasma wave, when it reaches large enough amplitudes, can also drive other plasma instabilities. The model presented can therefore only be considered as a zero-order solution to the problem: other effects, such as the three-dimensional nature of the beam entering the plasma (Fedele et al. 1986), the optimization of the frequency mismatch $\omega_s = \delta\omega_p$ (Tang et al. 1985, Noble 1985), the role of plasma noise and ion motion on wave saturation (Horton and Tajima 1985) as well as the role of pump-driven and plasma-driven parametric instabilities (Fried et al. 1976) have to be considered for a better understanding of the BWA scheme.

Here we shall give a brief summary of some parametric instabilities thought to be relevant to the BWA.

Consider a large amplitude wave (electromagnetic or electrostatic) of wavelength λ_o, incoming in the plasma, as a pump, toward a density perturbation. For instance, suppose that the last has wavelength $\lambda_1 = \frac{1}{2}\lambda_o$. This perturbation will be a unidimensional lattice for the pump. So it will produce a partial reflection and the incoming and reflected waves will generate a stationary wave with maxima corrisponding to minima of perturbation. Consequently a ponderomotive force will appear because of the interference envelope. It will push the electrons toward lower intensity field regions, corrisponding to higher densities. Thus the density perturbation will be enhanced where it was already higher and this fact will produce, in turn, an increasing backscattering which, in turn, will intensificate more and more the density perturbation and so forth. This phenomenon is enhanced when the density perturbation corrisponds to a plasma eigenmode (plasma oscillations).

We will give some examples about these kind of processes:

a). *Stimulated Brillouin Backscattering Instability (SBBI).*

In the SBBI processes a large amplitude e.m. wave (ω_o, k_o) decays in another e.m. wave (ω_2, k_2) and a iono-acoustic wave (ω_1, k_1). This instability has the following growth rate:

$$\gamma_g \sim 1.6\text{x}10^{14} \left(\frac{2\alpha_o}{\gamma_\phi}\right)^{2/3} s^{-1} \tag{45}$$

where α_o is the normalized pump amplitude (quiver velocity) and $\gamma_\phi = \omega_o/\omega_p$. For $\alpha_o \sim 0.1$ and $\gamma_\phi \simeq 100$:

$$\gamma_g = 2.5\text{x}10^{12} s^{-1}.$$

The influence on the beat wave mechanism is not yet well known.

b). *Stimulated Raman Backscattering Instability (SRBI).*

In the SRBI processes the e.m. pump (ω_o, k_o) decays in another e.m. wave (ω_2, k_2) and an electron plasma wave (ω_1, k_1). The growth rate is:

$$\gamma_g \sim \frac{10^{15}}{\lambda_o} \frac{\alpha_o}{\gamma_\phi^{1/2}} \tag{46}$$

For $\alpha_o \sim 0.1$, $\gamma_\phi \sim 100$:

$$\gamma_g \sim 10^{13} s^{-1}.$$

c). *Plasma Decay Instability (PDI).*

In the PDI processes a large amplitude electron plasma wave (ω_o, k_o) decays in another electron wave (ω_2, k_2) and a iono-acoustic wave (ω_1, k_1). These processes represent excitation mechanisms for iono-acoustic waves. The growth rate is:

$$\gamma_g = \frac{\alpha_o^{2/3}}{\gamma_\phi} \omega_o \tag{47}$$

For $\alpha_o \sim 0.1$, $\gamma_\phi \sim 100$, $\lambda_o \sim 1\mu m$:

$$\gamma_g \sim 10^{13} s^{-1}$$

d). *Filamentation Instability (FI).*

To describe this process it will be useful to refer to an example. Consider in a plasma an ion density perturbation perpendicular to the direction of propagation of an e.m. wave incoming in the plasma. The latter will propagate along constant density lines. But the index of refraction is larger in the depressions of density and smaller where the density is higher. On the other hand, it is known that the e.m. radiation prefers to propagate in the regions with high index of refraction. Thus it will concentrate greatly in the density-depressed regions (filamentary structure). The filamentation leads to an instability. In fact, the filaments introduce a ponderomotive force (like in the self-focussing of beam laser of finite diameter) that acts in such a way that there is a self-focussing for each of them. So that a further new radiation remains trapped which in turn produces an additional ponderomotive effect and so forth.

The growth rate is given by:

$$\gamma_g \simeq \frac{\omega_{pi}}{\sqrt{2}} \alpha_o \tag{48}$$

For $\gamma_\phi = 100$, $\lambda_o = 1\mu m$:

$$\gamma_g \sim 5 \times 10^{10} s^{-1}$$

6. MICROWAVE DRIVEN RESONANT EXCITATION OF PLASMA WAVES

The difficulties associated with the use of high intensity laser pumps (in terms of cost, time, technology and competing instabilities) have stimulated the proposal for different schemes for plasma accelerators. Chen et al. (1985), for instance, suggest the use of electron beams to excite the plasma wave and Formisano et al. (1986) suggest the use of microwave (μw) pumps instead of lasers for scaled experiments based on recent results by de Angelis et al. (1986) who have shown that resonant excitation of large amplitude plasma waves $(\geq 10^4 Volt/cm)$ is possible through the beating of two microwaves of nearly equal frequency in an "open resonator" filled with plasma of subcritical density.

Given the lower power ($\sim MW$) and larger pulse widths ($\sim \mu sec$) of the present generation of μw generators with respect to lasers an immediate consequence of the scaling will be a less efficient acceleration; the advantage on the other hand is to make experiments more affordable (in terms of cost, time and technology) so that, while the ultimate goal for media accelerators rests with high intensity laser beams, it could be worth to test the basic theories with scaled experiments. It should also be mentioned however, that the μw range can also offer possibilities that are out of the laser range and that μw technology is in rapid expansion so that a full investigation of schemes in the millimetric wavelength could prove useful not only for scaled experiments but also for alternative solutions to some of the problems in media accelerators.

As an example we mention the possibility of power flux confinement. Consider the pump intensity $I_o = P/A$ where P is the power in the pump and $A = \pi R^2$ is the focussing area: for lasers typically it is $R \sim 10\lambda_p$ where λ_p is the plasmon wavelength whereas for μw it is possible to scale at $R < \lambda_p$ and regain part of the loss due to the lower power. In fact, if λ_o is the pump wavelength (~ 1 mm), it is possible to take

$$R \sim \lambda_o = \lambda_p \frac{\omega_o}{\omega_p} \tag{49}$$

and since $\omega_o/\omega_p >> 1$ is a necessary condition for efficient acceleration we can have $R << \lambda_p$ with a corrisponding gain in pump intensity. All the effects of strong focussing can be generated and studied in the μw regime much better than in the laser regime. One of the effects is the generation of radial fields: a recent calculation (Fedele et al., 1986) in cylindrical coordinates for a gaussian pump profile gives:

$$E_r(r, z, t) \sim \delta \frac{r}{R^2} \exp(-2r^2/R^2)[1 - \omega_p t sin(k_p z - \omega_p t)] \tag{50}$$

$$E_z(r, z, t) \sim \exp(-2r^2/R^2)\delta\omega_p t \cos(k_p z - \omega_p t) \tag{51}$$

where:

$$\delta = \frac{1}{4}\nu_{01}\nu_{02} \tag{52}$$

for the radial and longitudinal electric field components.

Notice that as R is made smaller the radial field increases: this could be advantageous as particles are pushed towards the axis where the accelerating field E_z is higher. The problem of betatron oscillations and of strong focussing in general can be carefully investigated in the μw range.

In conclusion, we report for completeness the main results from the cited work of de Angelis et al. (1986) on microwave driven plasma waves.

Consider (Fig.1) an open resonator of length $2l$ with spherical copper mirrors of radius R_o, enclosed in a plasma chamber transparent to microwaves with plasma of sub-critical density for the incoming radiation: microwaves from a generator coupled to the left mirror (grid). The field distribution in the cavity has the typical caustic shape with maximum focusing at the center

$$\pi W_o^2 = \lambda[l(R_o - l)]^{1/2} \quad \text{"focusing area"} \tag{53}$$

The field is fairly uniform between $-z_R$ and $+z_R$ (Raileigh distance) where

$$z_R = \pi w_o^2/\lambda \quad \text{"Raileigh distance"} \tag{54}$$

In (53) and (54) λ is the pump wavelength, $\lambda = \frac{1}{2}(\lambda_1 + \lambda_2) \equiv \lambda_f$ for the two beating pumps case.

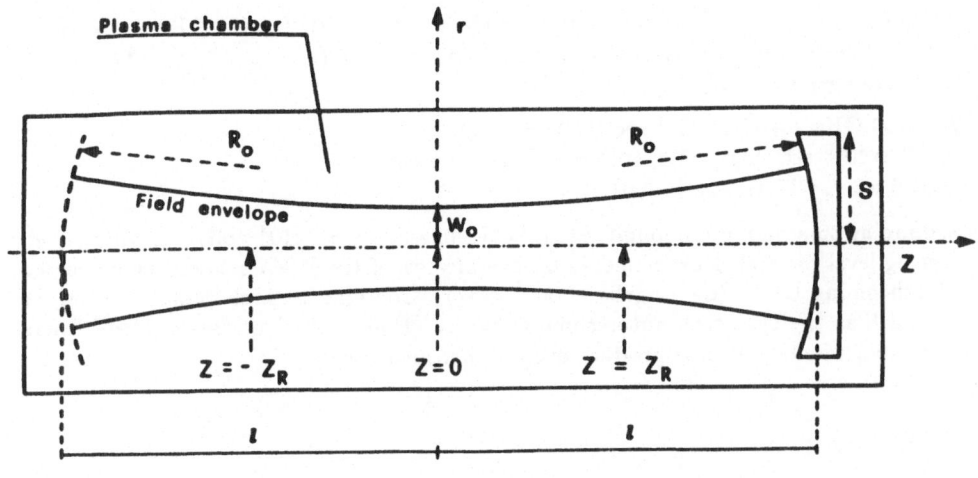

Fig. 1

The beat amplitude grows (because of multiple reflections in the cavity) and so does the resulting ponderomotive force on electrons so that the driven plasma wave will also grow until relativistic saturation or collisional saturation, depending on the pump amplitude.

There are two relevant timescales in the problem: the resonator saturation time τ_R (time for the pump field to reach its maximum amplitude) and the plasma wave saturation time τ_s.

These depend on the resonator dimensions, on the generator characteristics and coupling and on the plasma parameters and are determined as:

$$\tau_R = \frac{8\pi^2\mu^2\nu}{\alpha\omega_f} \quad , \quad \tau_s = \tau_R(63/K)^{1/7} \tag{55}$$

where

$$\mu = w_o/\lambda_f \quad , \quad \nu = l/z_R \quad , \quad \omega_f = \frac{1}{2}(\omega_1 + \omega_2) \tag{56}$$

$$\alpha = \frac{1}{2}[0.2(\omega_f/10^{16})^{1/2} + |T|^2] \tag{57}$$

where $|T|^2$ is the transmission coefficient of the mirror grid and

$$K = 1.3\text{x}10^{-11}(P_{KW}|T|^2)^2\nu^3\mu^2\alpha^{-7} \tag{58}$$

where P_{KW} is the generator power in kilowatts. The above results are given for a choice $\Delta\lambda/\lambda_1 = 0.1$, that is $\omega_f/\omega_p = 10$.

The generated plasma field has a radial and a longitudinal (resonator axis) component and its maximum amplitude will be $\varepsilon(\tau_s)$.

An estimate of the maximum field on axis ($r = 0$) gives:

$$\varepsilon_{max} = 4.7\frac{\nu}{\alpha^2}\frac{P_{KW}}{10^6}|T|^2[x + \frac{1}{2} - 2(1 - \frac{1}{2}e^{-x})^2] \tag{59}$$

where $x = \tau_s/\tau_R$.

As an example consider the NRL gyrotron (Read 1985) with $f_1=120$ GHz and $P_{KW} = 10^2$ coupled to a resonator with $l=10$ cm and $w_o = 2\lambda_f$ (i.e. $\mu =2$, $\nu =2.8$) and a transmission coefficient $|T|^2=0.1$. This gives:

. $f_2= 109$ GHz, $\omega_f=7.2\text{x}10^{11}$ sec^{-1}, $\lambda_f=2.6$ mm,
$\omega_p=6.9\text{x}10^{10}$ sec^{-1}, $n_o=1.4\text{x}10^{12}$ cm^{-3}, $z_R=3.6$ cm,
$\tau_R=26$ ns, $\tau_s=27$ ns, $\varepsilon_{max}=0.01$

corresponding to a maximum longitudinal field (on axis) of $1.3\text{x}10^4$ volt/cm, which is an interesting level for scaled experiments on the physics of the BWA scheme as suggested in a forthcoming paper (de Angelis et al. 1987). Coupling to high power microwave devices such as the cyclotron autoresonance maser ($P_{KW} = 10^4$) or free electron lasers ($P_{KW} \simeq 10^5$) would of course produce even higher field levels.

REFERENCES

1) Chen P., Dawson J.M., Huff R.W., Katsouleas T.: Ph. Rev. Lett. 54, 693 (1985)
2) Dawson J.M., Ph. Rev. 113, 383 (1959)
3) de Angelis U., Fedele R., Miano G., Nappi C.: Plasma Phys. and Contr. Fusion (1986, submitted)
4) de Angelis U., De Menna L., Fedele R., Miano G., Nappi C., Vaccaro V.: to appear in IEEE Trans. on Plasma Science, Special Issue on Plasma Based High Energy Accelerators (April 1987)
5) Fedele R., de Angelis U., Katsouleas T.: Ph. Rev. A 33, 4412, (1986)
6) Formisano V., de Angelis U., Vaccaro V.: Europhys. Letters (1986), to be published
7) Fried B.D., Ikemura T., Nishikawa N., Schmitt G.: Phys. Fl. 19, 1975 (1976)
8) Katsouleas T., Dawson J.M.: Ph. Rev. Lett. 51, 392 (1983)
9) Noble R.J.: Ph. Rev. A 32, 450 (1985)
10) Read M.E.: "Millimetric wave sources for fusion applications", NRL International Note (1985)
11) Rosenbluth M.N., Lin C.S.: Ph. Rev. Lett. 29, 70 (1972)
12) Tang C.M., Sprangle P., Sudan R.N.: Phys. Fl. 28, 1974 (1985)
13) Tajima T., Dawson J.M.: Ph. Rev. Lett. 43, 267 (1979)

DISCUSSION

CAVENAGO:
The surfatron concept, while solving the problem of longitudinal synchro-nization of the particle beam with the electrostatic wave, introduces an unacceptable angle between laser and particles path. On the contrary, there is some hope that the lasers selfconfine in a rippled fiber, changing the phase velocity of the electrostatic wave to the beam velocity ($\sim$ c) Tajuma 85, but computations are needed to support this very non-trivial effect.

PINTO:
Saturation will ultimately limit the performance of the BWA. Accordingly, one should adopt a more refined mathematical technique than standard per-turbation theory.
Transverse dynamics can be relevant, and has been neglected in your ap-proach.

BOSCOLO:

The problem that worries me is that in a beat experiment two quasi mono-chromatic radiation beams are required. The goal of high power and very narrow bandwidth is very difficult to get. I know it is reached in a CO TEA lasers driven by a single mode - single frequency oscillator. The difficulty of using a FEL consists in the fact that these lasers have normally a quite large band width.

WAKE FIELD ACCELERATION

W. Bialowons, H.D. Bremer, F.-J. Decker, M.v. Hartrott, H.C. Lewin, G.-A. Voss, T. Weiland, P. Wilhelm, Xiao Chengde[x] and K. Yokoya[o]

Deutsches Elektronen Synchrotron DESY
Notkestrasse 85, 2000 Hamburg 52, West Germany

ABSTRACT

We are investigating the possibility of accelerating particles with high gradients in a "Wake Field Transformer"[1,2]. The progress of this experiment will be described. The developement of the high current hollow beam electron gun was continued. In the conventional linac, the hollow beam was accelerated to about 6 MeV. Beam monitors came into operation, two gap monitors, two fluorescent monitors and a Cerenkov monitor. Calculations with the computer code WAKTRACK[3] gave the final details for the high energy section of the accelerator that will be installed during 1986.

INTRODUCTION

Circular accelerators for electrons are operating very successfully in the GeV range. As the synchrotron radiatioin increases with the fourth power of the energy, linear accelerators become attractive at very high energies, above 100 GeV. In the TeV region they are the only feasible choice.

To become achievable with respect to cost, they have to operate with high acceleration gradients. The total length of a linac is given approximately by

$$L \sim \frac{\text{center of mass energy}}{\text{accelerating gradient}} \cdot$$

The common technology allows gradients up to 17 MeV/m[4], with superconductive cavities one reaches up to 20 MeV/m[5]. Laser accelerators[6], plasma

---

(x) On leave from Tsinghua University, Beijing, People's Republik of China.
(o) On leave from KEK National Lab. for High Energy Physics, Japan.

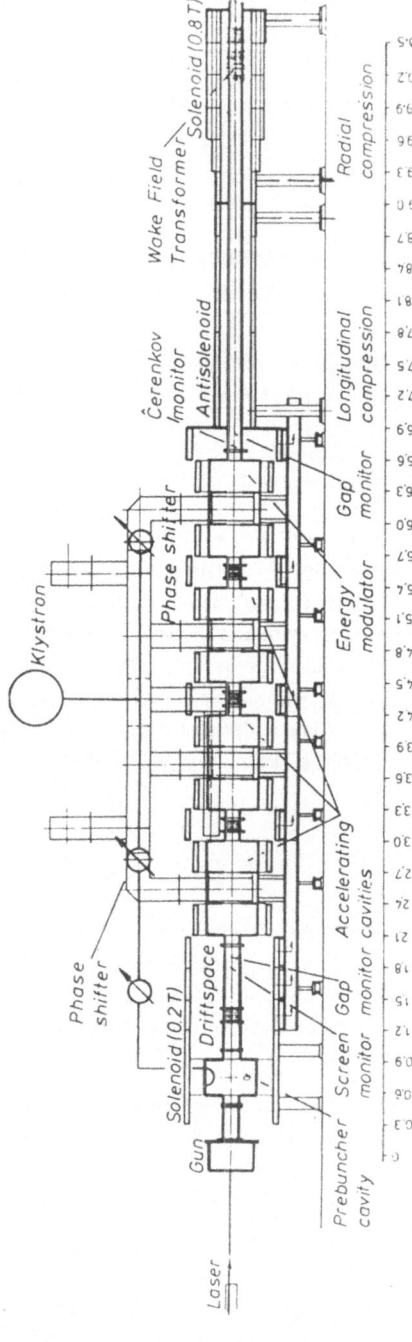

Fig. 1. Overall layout of the Wake Field Transformer experiment at DESY shown from left to right: The infrared light beam produced by a Q-switched Nd-Yag laser (peak power > 100 MW) is focused on a ring shaped tantalum cathode. The emitted thermoionic- and photoelectrons are extracted by a voltage of about 100 kV and guided by solenoid fields of 0.2 T into the linac. First the hollow beam becomes compressed longitudinally in a prebuncher. Then four 3-cell cavities (500 MHz) accelerate it to 8 MeV. A pulsed klystron (peak power 1 MW) feeds the cavities. In the antisolenoid further longitudinal compression is achieved. By increasing solenoidal fields at the end radial compression takes place too. After that the electron ring enters the actual Wake Field Transformer, consisting of 80 cylindrical cells arranged one after another. For monitoring and adjustment of the ring beam, several beam monitors have been set up.

accelerators[9,10] promise much higher gradients (about 1 order of magnitude). Several methods have been discussed[6,7,8]. The DESY Wake Field Transformer experiment is carried out in order to recognize the inherent physical and technical problems in more detail and possibly to overcome them. As the principle has been described in detail elsewhere[1,2], we only recall the basics: A driving beam with high charge (≈ 1 μC) passes the "Wake Field Transformer" (WFT), where it excites wake fields that lead to deceleration. By proper shaping of the WFT, the wave packet is subsequently spatially focussed. Thus the field strength increases proportionally to the inverse of the square root of the volume containing the wake fields. A second beam with lower charge (≈ 0.01 μC) passes the transformer with a suitable delay and experiences a much greater acceleration than the deceleration of the driving beam.

Among different geometrical arrangements of the driving and driven beam[1,2], a hollow beam driving a thin beam on its center may be achieved most easily. Furthermore for this case the calculated magnitude of the transformation ratio is 10, which is large compared to other geometries. Assuming a total charge of 1 μC and a bunch length of 5 mm, we find values for the accelerating gradient above 100 MeV/m for the inner beam.

In the experiment (Fig. 1), a laser driven gun produces a high current (1 kA, 150 keV) hollow ring beam of 10 cm diameter. In the prebuncher the hollow beam is bunched to a length of about 5 cm. The following linac consists of four 3-cell-cavities (500 MHz) powered by a 1 MW klystron. At the end of the linac, the bunch has a length of about 3 cm (rms) and an energy of 8 MeV. The whole equipment is put into a solenoid guiding field of 0.2T. The final longitudinal compression is achieved in the high energy buncher, after which the ring beam is fed into the Wake Field Transformer.

So far, the experimental set up has been installed up to the end of the linac.

THE GUN

Fig. 2 shows the design of the gun. An infrared light pulse of a Nd-YAG laser, consisting of a 3 ns main pulse with two side maxima, is focused on a ring at the cathode. There electrons are extracted from the metal via heating and photo effect[10]. These are accelerated by the high voltage and follow the field lines of the surrounding solenoid magnets through a slot hole in the anode.

High voltage

Using an available ceramic of the DESY linacs the high voltage is limited to about 50 kV. In order to reach the design value of 150 kV the gun is surrounded by an insulating material for which SF_6 gas is used. Since the laser light must not penetrate the SF_6 gas, because otherwise the glass surfaces of the windows would be etched, it travels in air inside a tube (see Fig. 2). The HV feed through has also been improved. With these modifications, the high voltage was limited by sparking in the vacuum chamber.

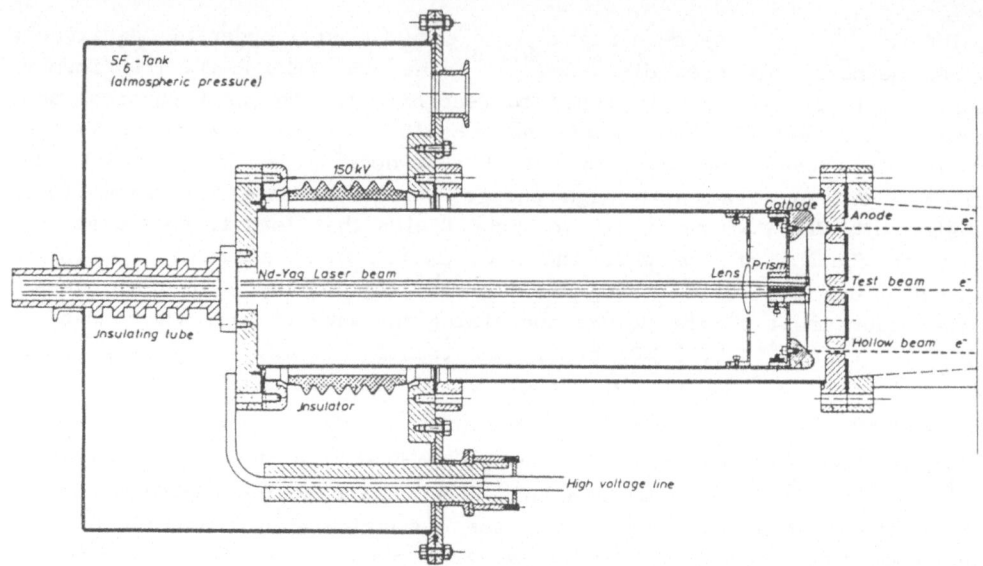

Fig. 2. Cross section of the laser driven hollow beam gun. The infrared laser light is focussed on a ring at the tantalum cathode. The op
tical train consists of a viewing port, a focusing lens and a conical
mirror. The conical mirror is a quartz cylinder with a polished inver
se cone at the top, making use of the total reflection at the glass
vacuum surface. The emission mechanism is explained by the generalized Richardson effect (pure thermionic effect and thermionic support
ed multiphotoneffect). The emitted electrons are extracted by a high
voltage and guided by a solenoid field through a slot hole in the ano
de. In order to increase the break down voltage the insulating   ceramic is surrounded by SF_6-gas at atmospheric pressure. The optical
train is separated from the gas by an insulating tube.

This was improved by increasing the distance between anode and cathode and
installing an additional ion sputter pump (400 l/s). Finally the voltage
has been raised to 130 kV. However at the required magnetic field strength
(~ 0.2 T) the maximum value is reduced to about 80 kV. Some possible ex-
planations could be a surface current on the glass prism or Penning effect.

Current

The peak current measured with a gap monitor was about 15 A at 50 kV.
This is one order of magnitude lower than the expected value for a space
charge limited current. If the limitation of the current is not due to
space charge it can be enhanced by increasing the laser power density at
the cathode surface. However, it may also be possible that it is limited
by space charge or parts of the beam are lost by mechanical obstacles. In
the next step an anode with a wider ring slot will be used.

Pulse length

The laser produces a light pulse of about 10 ns total length. The main
intensity is concentrated in about 3 ns; the two side maxima contain less

then 10% of the light intensity. This profile was measured with a fast vacuum photo diode. The current response lasts about 9 ns (FWHM), measured with a gap monitor.

Azimuthal homogeneity of the ring beam

It is important that the azimuthal charge distribution in the ring be as homogeneous as possible. Any inhomogeneity causes transverse wake fields which lead to transverse instabilities of the inner beam. However it seems very difficult to create a homogeneous electron ring beam.

Fig. 4 shows the upper and lower part of the electron ring detected by the two cameras on the movable fluorescent screen monitor. Until now two tantalum cathodes have been tested.

Several causes are considered for this inhomogeneity, e.g. an assymmetry in the laser profile or in the light path, a surface effect of the cathode or a mechanical obstacle for the electrons. The laser profile has been measured (see Fig. 3) and it seems not to be the main cause for the inhomogeneity.

The four tags holding the inner part of the anode cause holes in the hollow beam (see Fig. 4). However these holes do not effect the inner beam very much due to their fourfold symmetry.

THE LINAC

Prebuncher cavity

So far, the prebuncher cavity has concentrated a low current electron beam to 5 short bunches of about 3 cm length each. With higher currents (higher laser power) the bunches are longer. This is interpreted as a space charge effect.

The prebuncher and the accelerating cavities are powered by one klystron which is pulsed at 25 Hz with 100 μs duration. As the buncher cavity is located in the magnetic field of the solenoid, multipactoring is much more severe than usual. The field strength breaks down after about 20 μs. Thus satble conditions cannot be achieved easily. In order to avoid this instability, the laser is triggered in the first rise before multipactoring takes place, about 10 μs after the rf has been turned on. Due to filling time ($\approx$ 20 μs), however, at this point of time the rf level in the accelerating cavities has reached only a small fraction of its final value, so that no proper acceleration of the electrons takes place. Thus a separate rf amplifier for the prebuncher cavity, that can be triggered separately, is being installed, so that we can use the first stable region.

Accelerating cavities

In the four 3-cell cavities the electrons are accelerated to about 6 MeV. A first phase shifter between the prebuncher cavity and the first ac-

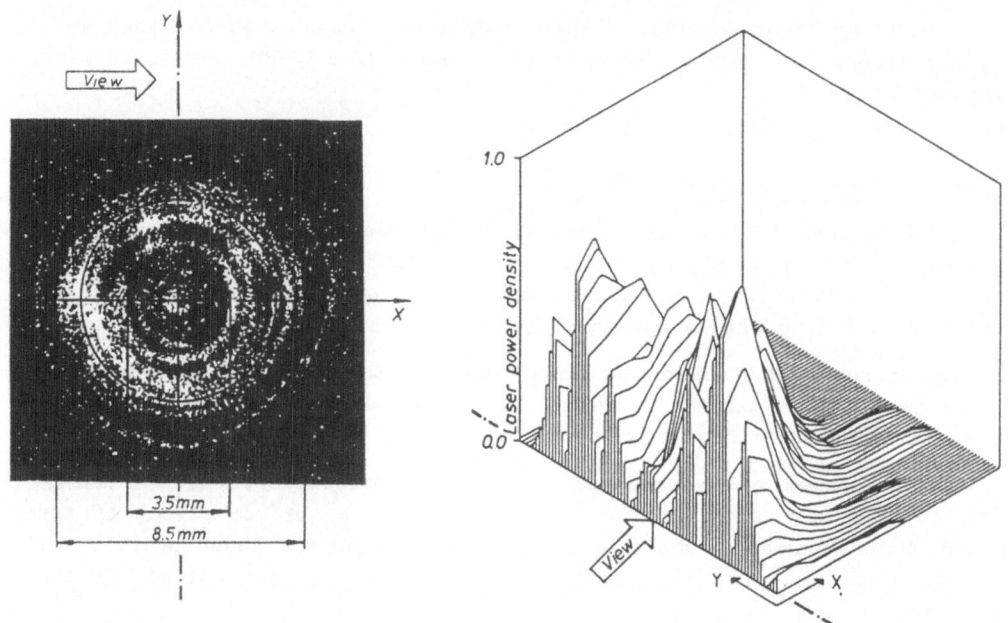

Fig. 3. Spatial intensity profile of the Q-switched Nd-Yag laser beam
(λ=1.064 m). On the left side the profile is recorded on a "burn paper" and on the right side the profile is scanned with a pin hole pow
er meter. A burn paper is an exposed and developed black and white
photo paper. A single shot evaporates the silver coating on the paper
proportional to the laser power density. This pattern gives a first
impression of the spatial intensity profile. A pin hole power meter
is a calorimeter where the sensitive area is reduced by a pin hole
(here with a diameter of 1 mm). The plot shows a vertical center cut
through the profile. In both pictures a diffraction pattern is seen
which is typical for this laser.

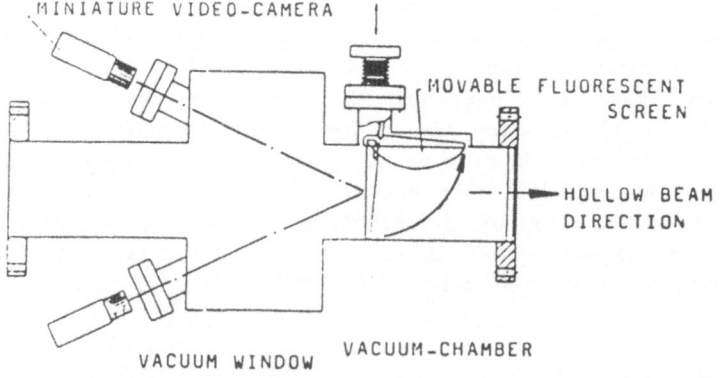

Fig. 4. Movable screen monitor for observation of the ring beam
before acceleration (left at the top).

celerating cavity adjust the injection phase. A second phase shifter between the first and the second cavity corrects for the delay in phase of the not yet relativistic electrons. The third phase shifter at the fourth cavity allows a variable energy spread for the high energy bunching in the antisolenoid.

Correction coils

The movable screen in the low energy part and a fixed, but transparent fluorescent grid at the end of the linac allow the observation of the position of the hollow beam, its actual diameter, its radial thickness and the azimuthal charge distribution.

We found a tendency of the ring to be slightly oval instead of round. This is caused by aberrations from the cylindrical symmetry of the solenoid coils and can be corrected by quadrupole correction coils. With two pairs of correction coils behind the first screen monitor a superposition of a dipole and quadrupole magnetic field can be produced, by adjusting the currents in the four coils individually. Thus both the shape and the position of the hollow be can be controlled.

BEAM DIAGNOSTICS

In the drift space of the prebuncher, a gap monitor and a movable fluorescent screen monitor were installed. For beam diagnostics after the acceleration a Cerenkov monitor, a gap monitor, a fluorescent grid and a special hollow beam spectrometer were added.

Gap monitors

For the current measurement of the long bunches, gap monitors are used which simply interrupt the beam pipe by a ceramic ring. The image current on the wall induced by the beam current flows through some resistors bridging the gap (Fig. 6). The voltage signals appearing across these resistors are picked up. They contain information about the charge distribution longitudinally and transversly.

The first gap monitor in the low energy part consists of 64 resistors, 10 Ω each. Sixteen 50 Ω cables pick up the signals over an additional 50 Ω resistor avoiding the reflection of backwards running signals. Eight of the 16 signals are summed up and can be observed either after a long cable with an oscilloscope or directly near the monitor with a sampling head to avoid damping and distorsion of the signal.

Fig. 5 shows two of five bunches separated by 2 ns. The oscillation of about 3 GHz is caused by a resonant circuit consisting of the capacity of the ceramic ring (C $\approx$ 50 pF) and the inductivity of the resistors and their connections (L $\approx$ 50 pH). This LC resonant circuit is damped by the low resistance of the 64 gap resistors (together 0.156 Ω). Thus the signals produced by short bunches are hidden by oscillations. Fortunately

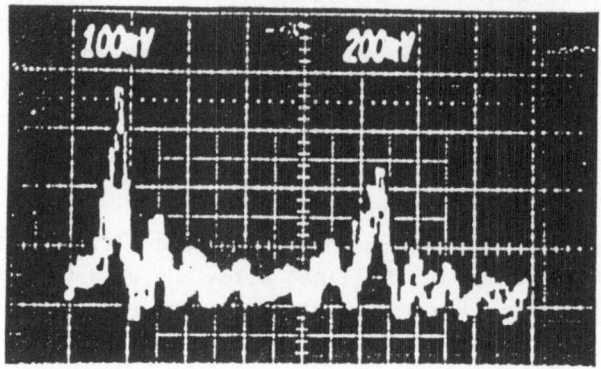

Fig. 5. Two of 5 bunches produced by a single laser shot. The separation between them is 2 ns. The bunch length is about 150 ps (FWHM). The signals are produced by a gap monitor in front of the linac behind a drift space of 1.2 m after the end of the prebuncher cavity. They are detected by a sampling oscilloscope.

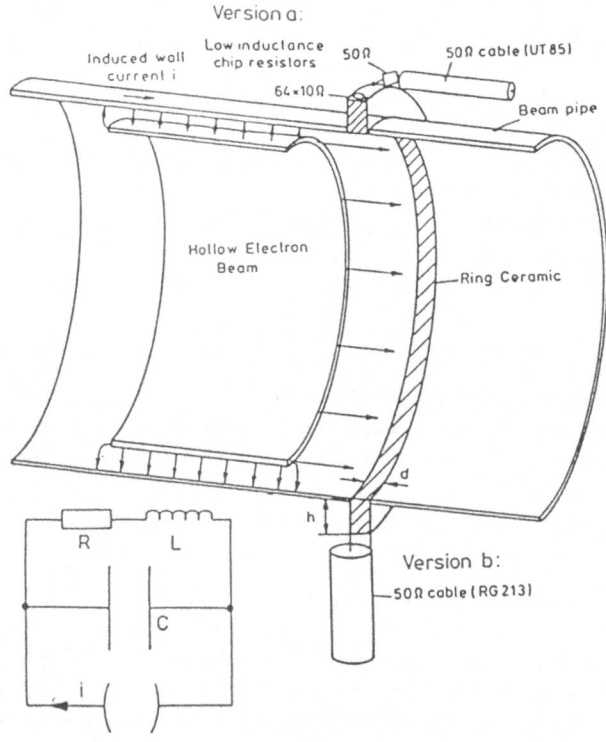

Fig. 6. Physical layout and wiring of the gap monitor at the low energy end of the linac (Version a) and its equivalent circuit (L=50 pH, C=50 pF and R=0.156 Ω), which can explain the observed oscillations of about 3 GHz between the bunches (see Fig.5) (Version b). Short connection from the ceramic (d=4 mm, h=7 mm) to one of the 32 big 50Ω cables damping the oscillations occuring at the energy end periodically.

longer bunches do not excite oscillations with such large amplitudes and thus can be observed reasonably well.

In the gap monitor at the end of the linac very short bunches would excite large oscillations. Therefore here the ohmic load is adjusted to aperiodic damping (R=1.56 Ω) in order to get fast signal response while the oscillations are damped sufficiently. This is achieved without any resistors but only by bridging the gap with 32 pickup coax cables of 50 impedance.

If the ring beam is centered to the axis of the beam pipe the azimuthal charge distribution can be determined from the different amplitudes of the signals around the pipe. On the other hand the absolute value of the total beam is obtained by summing all the pickups. This yields a signal which is approximatelly independent of the azimuthal beam distribution. Those signals have been compared with current measurements performed using a pulse transformer on the high voltage line feeding the cathode. Also here short pulses (10 ns) excite oscillations so that higher amplitudes are simulated. However the corrected values are consistent with the measurements of the gap monitors. The summed signals of the peak current derived from the both gap monitors agree within 15%.

Cerenkov monitor

To measure the length of short bunches (≤ 5 cm), the time resolution of 150 ps of the gap monitors is not sufficient. In order to resolve the expected bunch length of less than 1 cm properly, a time resolution of about 10 ps is necessary. This resolution can be achieved by a commercial streak camera. The light pulse is created by the Cerenkov effect in glass. A similar arrangement has recently been used at Slac[11]. A small part of the ring beam penetrates a quartz wedge and excites Cerenkov radiation. This light is then guided out by total reflections and leaves the beam pipe nearly perpendicular to the beam axis. An optical system (see Fig. 7), consisting of 3 lenses and 2 mirrors, guides the light over 5 m to the streak camera, which is located far from the disturbing solenoid end field.

Fig. 8 shows the streak camera signal of a bunch with 1 cm FWHM. So far, such short bunches can be achieved only at low current.

Hollow beam spectrometer

In order to analyze the energy of the hollow beam, a special hollow beam spectrometer (see Fig. 9) has been developed. In the present stage, it is mounted at the end of the linac.

In the short solenoid end field the particles experience an azimuthal kick and thus change their azimuthal velocity by $\Delta v_\varphi = (erB)/(2\gamma m_o)$ (v_φ: azimuthal velocity after crossing the end field, γm_o: energy, r: radius, B: solenoid field strength). Particles of different energies hit a surrounding fluorescent pipe at different longitudinal positions. The fluorescent light is observed by two television cameras viewing from the end of the pipe. From the longitudinal position, v_φ can be calculated and thus, knowing B and measuring r with a screen monitor,we can determine the energy.

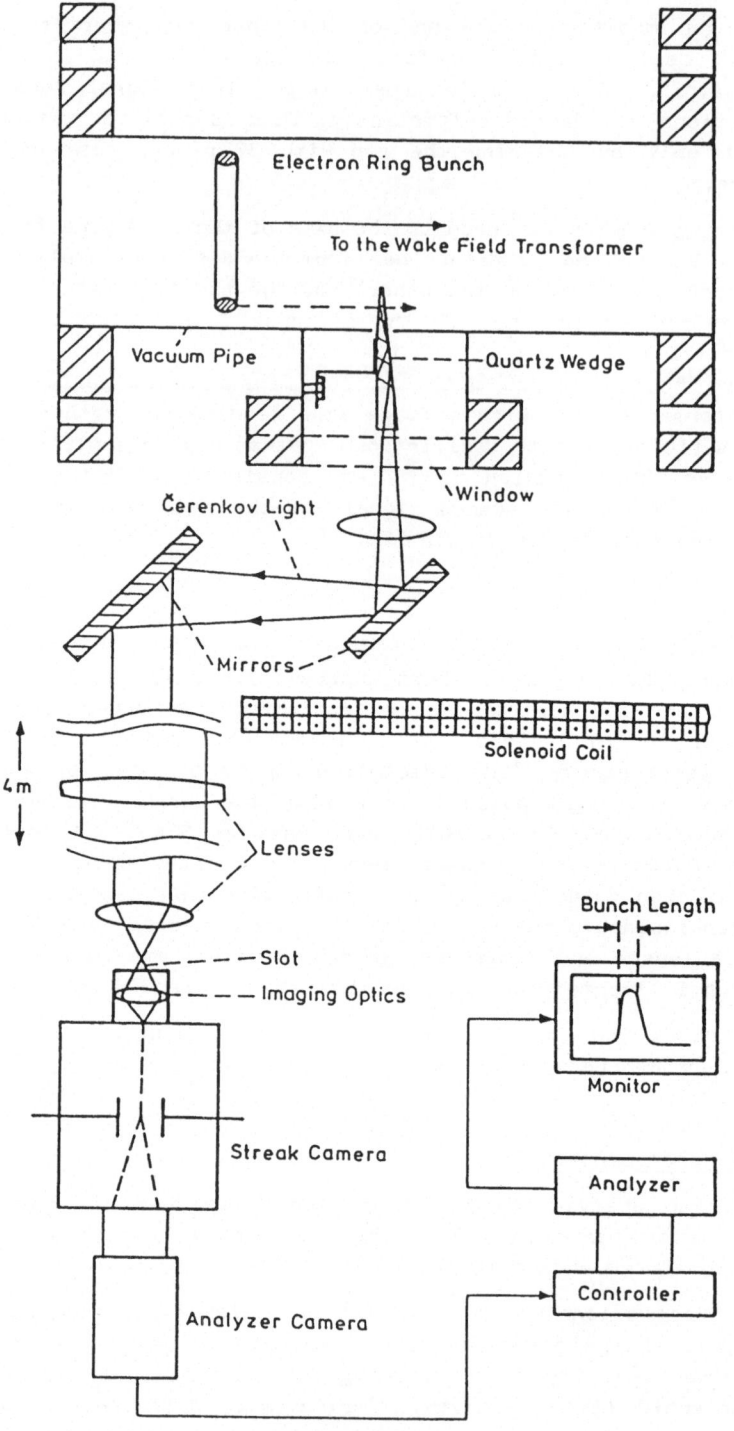

Fig. 7. Setup of the Cerenkov light monitor for high resolution longitudinal beam size measurement of the hollow beam.

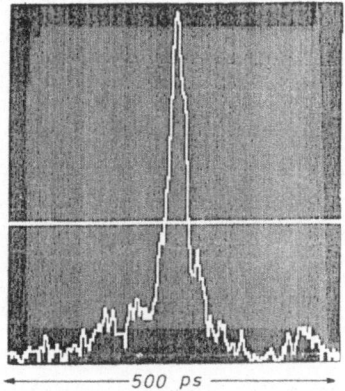

Fig. 8. Single bunch signal taken with the streak camera. The
width (FWHM) corresponds to a length of 1 cm.

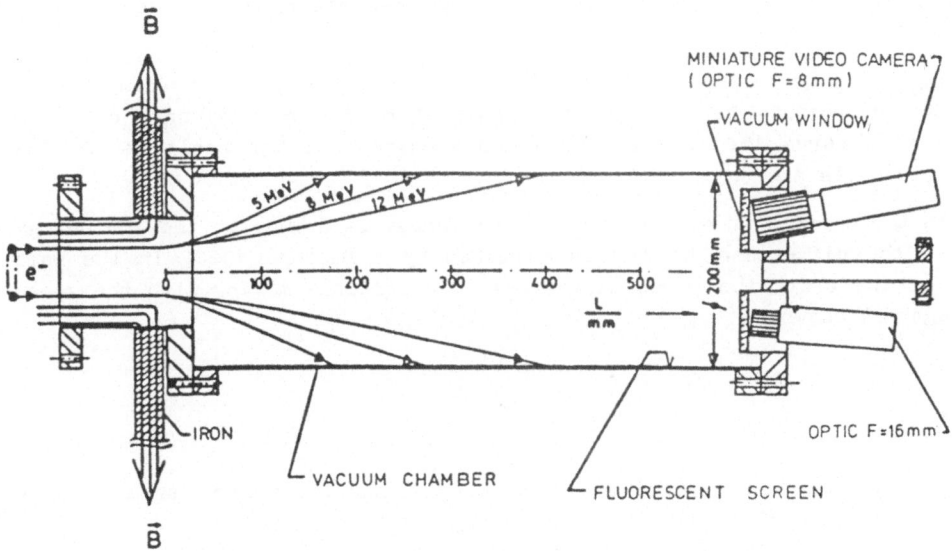

Fig. 9. Hollow beam spectrometer making use of the end field of
the solenoid on the left side (not shown). By an iron sheet (3
cm thick) the inner longitudinal field is bent outwards within
a short range. Here the passing electrons experience a transverse momentum and hit the fluorescent screen at the inner wall
of the spectrometer tube at a longitudinal position corresponding to their energy.

For several reasons (space charge, transverse electric field components in the cavities, gaps between the solenoids) the electrons experience transverse forces and therefore oscillate around the lines of the guiding magnetic field. Presently it is not powwible to determine the initial transverse velocity due to these effects when the electrons enter the end field at the spectrometer. This leads to an uncertainty in the energy measurement. In order to improve the accuracy a longitudinally movable screen has been installed by which the transverse electron velocity can be determined just before the spectrometer.

Furthermore, according to calculations, the oscillations can be suppressed by proper adjustment of the solenoid currents. Thus the movable screen will also be used in order to monitor appropriate adjustments.

COMPUTER SIMULATION

The computer code WAKTRACK has been extended to take space charge into account. Furthermore external data for cavity fields obtained from numerical solutions (URMEL) of Maxwell equations may be handled and wake field effects can be included (TBCI). Static electric and magnetic fields as calculated by the codes PROFI and MAFIA are also accepted and processed by WAKTRACK[12].

Fig. 10 shows a typical plot output. In the lower part, the ring radius is plotted versus the longitudinal position of the ring along the linac. Above this curve the cavities with their phases and the positions of the solenoid coils are drawn.

In the upper part of the figure, the energy is drawn, in the low energy range, the difference to $\gamma=1$ is streched by a factor of 30. In the same part of the plot, the phase difference to a particle moving with the speed of light is shown.

Linac

For the tracking, initially a number of particles are assumed to be distributed on a ring, all having an energy of 75 keV.

The bunching effect of the prebuncher and the acceleration in each cavity cell can be observed.

Oscillations of the electrons around the magnetic field lines are excited by the radial cavity fields and radial magnetic field components, which occur at the gaps between the coils. By varying the current in some of the coils, the collective oscillation can be suppressed. However at the transition of the particles into the antisolenoid of the high energy buncher additional oscillations are excited due to the finite length of solenoid end fields. Investigations have been started to develop hollow beam focusing system to counteract this effect and other non-collective effects such as radial space charge.

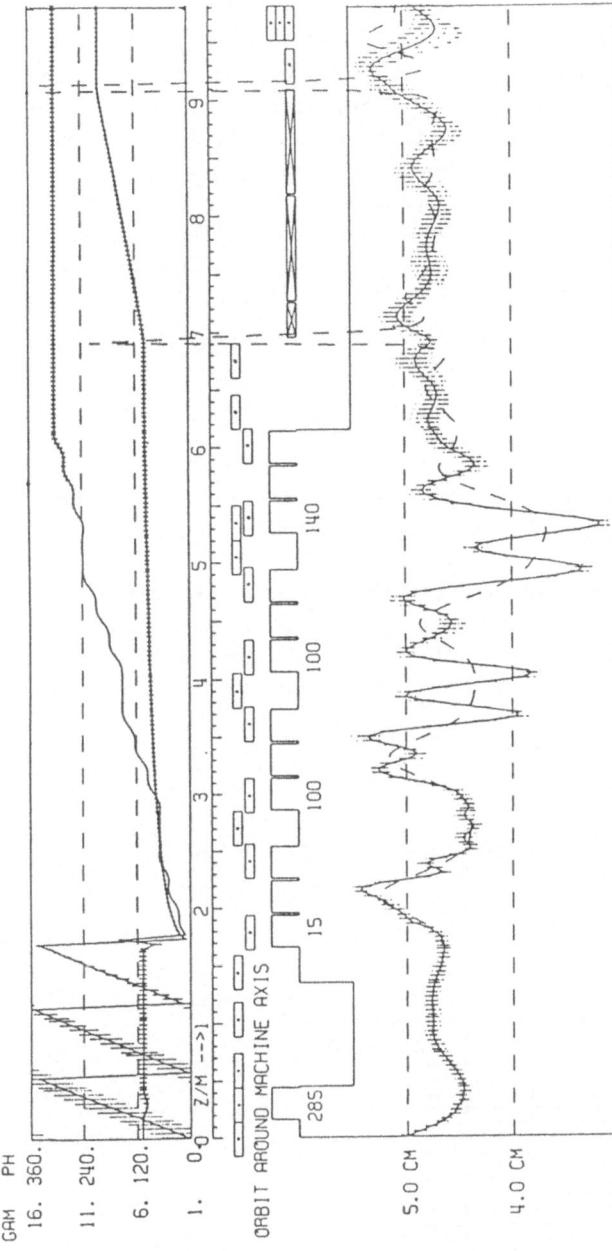

Fig. 10. Typical plot output of the WAKTRACK tracking code, showing several particle properties in the DESY experiment. In the middle between the two frames, a scheme of the experimental set up is drawn, together with a scale, which can be compared to Fig.1. The coils are sketched and near the cavity contours their phases are written in degree. Below this, the orbit of the particles is plotted: the ring radius varies between 3 and 5 cm due to transverse forces at gap between coils and in cavities. The collective oscillation around the dashed reference orbit is reduced in its amplitude by adjusting the current in the coils between the third and the fourth cavity (4.6 m < z < 5.6 m). Above the set up sketch two curves are plotted in one frame, the relativistic γ factor and the phase difference compared to a particle moving with the speed of light. In front of the first accelerating cavity (z<1.7 m), where the particles are nonrelativistic, (γ−1) is streched by a factor of 30. Behind that, the increasing energy in the linac can be seen. The phase difference of nonrelativistic particles changes with a certain slope depending on the particle velocity. When the particle speed approaches c, the phase becomes constant. It only changes again in the high energy buncher (7 m ≤ z < 9 m), where the velocity parallel to the z-axes is decreased by rotating the gollow beam.

High energy buncher

The final longitudinal compression of the hollow beam is achieved in an antisolenoid. The field strength of the solenoid and of the antisolenoid are equal, the region where the field is radial must be very short. Therefore iron plates are inserted between the coils. The resulting fields has been calculated with PROFI and included to the tracking code.

When the ring passes the radial field, the particles experience an azimuthal kick, which is just twice the kick of a solenoid end field. Thus the path of one particle in the antisolenoid is a spiral on a cylinder that has the same diameter as the ring beam, i.e. the ring is rotating, the radius does not change (see Fig. 10). As some of the energy goes into the circular motion, the longitudinal velocity decreases i.e. the phase changes. The phase of the fourth cavity is adjusted such that the earlier particles are accelerated less than the later ones. Thus the ring can be bunched even at high energies, where classical rf bunching mechanisms fail.

At the end of the antisolenoid, the rotation is stopped by an inverse kick.

Wake field transformer

The first model built of the transformer is shown in Fig. 11. For preliminary investigations independent of the available current we intend to create a low energy (50 keV) test beam just in front of the transformer. With this set up it will be possible to detect even small accelerating gradients due to the Wake Field Transformation. Of course 50 keV electrons will not be captured and accelerated optimally as they are nonrelativistic. However computer simulations show that a significant change in energy can be expected.

Acknowledgement

The authors wish to thank all the DESY staff for their help.

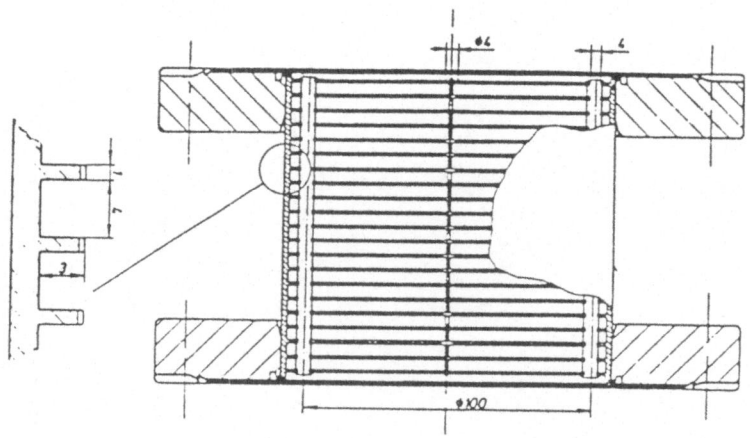

Fig. 11. Sectional drawing of a part of the wake field transformer.

References

1. G.A.Vos and Th. Weiland, Particle acceleration by wake fields, DESY M-82-10 (1982).
2. G.A.Voss and Th. Weiland, Wake Field Acceleration Mechanism, Proceedings of the ECFA Conference "The Challenge of Ultra High Energies", Oxford, September 1982.
3. Th.Weiland and F.Willeke, "Particle Tracking with Collective Effects in Wake Field Accelerators", Proceedings of the 12th Intern. Conf. on High Energy Accelerators, Chicago, 1983, pp. 457-459 and following improved versions by Gary Rodens (Los Alamos) and Kaoru Yokova (on leave from KEK).
4. SLC-design report, SLAC report 229 (1980).
5. H.Piel, Recent Progress in RF Superconducting, IEEE Trans. Nucl. Sci. NS-32:3565 (1985).
6. Proceedings of the Workshop on Laser Acceleration of Particles, AIP Conf. Proc. No. 91 (1982).
7. "The Challenge of Ultra-High Energies", Proceedings of the ECFA-RAL Workshop, Oxford, September 1982, ECFA 83/68 (1983).
8. "The Generation of High Fields for Particle Acceleration to Very High Energies", Proceedings of the CAS-ECFA-INFN Workshop, Frascati, September 1984, ECFA 85/91 CERN 85-07 (1985).
9. W.Bialowons, H.Dehne, A.Febel, M.Leneke, H.Musfeldt, J.Rossbach, R.Rossmanith, G.A.Voss, Th.Weiland and F.Willeke, "A Wake Fiels Transformer Experiment", Proceedings of the 12th Intern. Conf. on High Energy Accelerators, Chicago, 1983.
10. W.Bialowons, H.D.Bremer, F.J.Decker, R.Klatt, H.C.Lewin, S.Ohsawa, G.A. Voss and TH.Weiland, "Wake Field Work at DESY", IEEE TRans. Nucl. Sci. NS-32:3471 (1985).
11. J.C.Sheppard et al., "Real Time Bunch Length Measurements in the SLC Linac", SLAC-PUB-3584 (1985).
12. Th.Weiland, On the numerical solution of Maxwell's equations and applications in the field of accelerator physics, Particle Accelerators 15:245 (1984), and references therein.

Discussion

WILSON
Question on the current used in the hollow beams.

WEILAND
Roughly 10% of the design value and in terms of charge 1/6. The main problem is not the wake transformer but the gun. We are elaborating a new model which will improve these results.

VACCARO

Why do you not use a set of filamentary beams (6, say) distributed around the circumference?

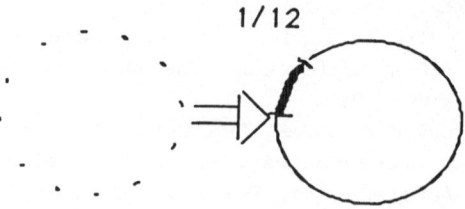

1/12

WEILAND

Because each filament would have to carry a higher charge and consequently would cause higher space forces.

JOHO

This suggestion is analogous to the Megalack where a system using minibeams like the one suggested by Vaccaro, was helpful.

WEILAND

Mentions that no better focusing can be achieved since the driving beam will have 100% energy spread at the end of the transformer.

RUGGIERO

Supposing one was to design a linear collider with the power of 1 MW, how much power is required for the driver and for the system as a whole?

WEILAND

I can only give orientative values. The efficiency of transformation is roughly one over the transformation ratio, about 16%, but it also depends on single or multi bunch operation.

SCHNELL

Even with infinite gradients in the wake field device the average accelerating gradient cannot be larger than the drive linac gradient times the transformation ratio.

ENERGY EFFICIENCY AND CHOICE OF PARAMETERS FOR LINEAR COLLIDERS(*)

J.Clauss

Brookhaven National Laboratory
Upton, NY 11973, USA

Introduction

The cost of large high energy facilities is a matter of strong concern. Such facilities use very large amounts of electric power and a sizeable part of the total cost is directly related to that fact. It is therefore important to have a clear understanding of the factors that influence the power efficiency, i.e., the amount of power used per unit product, where product is the luminosity generated at a specified energy. The overall efficiency may be regarded as the product of the effciencies with which the parts of the system perform their functions: the efficiency with which raw grid power is converted to r.f. power, the efficiency with which r.f. power can be transmitted from source to load, the efficiency of conversion of r.f. power into particle beam power and finally the efficiency with which beam power is converted to luminosity.

In this paper we address some factors that are relevant to the latter two. In the first part we investigate three possible ways of converting beam power into luminosity: two short bunches colliding with each other, two long ones doing so, and two pulses of bunch trains which interact. The last mode appears to be preferable for obtaining high efficiencies, but its possibilities are restricted by the limits imposed on the length of the interaction region by the users. These restrictions can be met if the wave length λ at which the accelerators work can be chosen small in comparison with that length, this leads to $\lambda < 1$ mm. We consider therefore in a second part some of the implications of linacs for very high frequencies, emphasizing the factors that influence the efficiency of converting r.f. powere into luminosity and assume that suitable power sources are or will be available. It appears that a resonant linac accelerates single bunches rather less efficiently than pulses of many bunches with the same total charge. The pulse requires a higher operating frequency however, this re-

(*) Work performed under Contract No. DE-AC02-76CH00016 with the United States Department of Energy.

stricts the charge a single pulse of acceptable length can carry. This can be compensated for by increasing the pulse repetition rate. Operation at higher frequencies may, or may not, permit higher accelerating gradients. This would be an advantage in nearly all respects. Two processes pose limits to the gradient that can be achieved: the occurrence of field emission of electrons from the metal surfaces exposed to the field, and the possibility of physical damage due to mechanical stresses and shock waves generated by differential thermal expansion due to dissipation.

No suitable accelerator structures for very high frequencies are available at present. A third section describes some characteristics of structures that seem feasible.

1.1. LUMINOSITY, BEAM POWER AND MODE OF OPERATION

Single pass colliders can be operated in various modes. In one mode a single bunch of electrons is made to interact with a single counter streaming bunch of positrons at a time. The process is repeated periodically with a, generally low, pulse repetition frequency. The expression for the luminosity produced in this mode is well known[1] and may be written as

$$\mathscr{L} = f \, \frac{N^2}{4\pi\sigma^2} = f \, \frac{N^2\gamma}{4\pi\varepsilon\beta^*} \quad . \tag{1.1}$$

Here is: $\mathscr{L}$ = luminosity;
   f = number of bunch interactions per unit time;
   N = number of particles per bunch (assumed equal in each
      bunch of an interacting pair;
   $\sigma^2 = \varepsilon\beta^*/\gamma$ square of r.m.s. bunch radius;
   ε = invariant emittance;
   $\gamma = E/E_0$ = energy parameter;
   β^* = amplitude function in the interaction point.

The expression is valid if the bunch length σ is much smaller than β^*. The energy E_{sb} stored in each bunch is

$$E_{sb} = N\gamma E_0 \tag{1.2}$$

thus the power in the two beams is

$$P_b = 2 \, fN\gamma E_0 \quad . \tag{1.3}$$

Calling the ratio of the luminosity and the beam power necessary to produce it η_B, one obtains

$$\eta_B = \frac{\mathscr{L}}{P_b} = \frac{N}{8\pi E_0 \, \varepsilon\beta^*} \quad . \tag{1.4}$$

The number of particles per bunch is limited either by the number of particles that the source can produce per unit time, i.e., by the source current i_b, or by the accelerators between sources and interaction point. In the former case:

$$N = i_b \, T/e = \frac{i_b}{e} \frac{\lambda}{c} \tag{1.5}$$

where: e = charge per particle;

$T = \frac{1}{\nu} = \frac{\lambda}{c}$ period, frequency and wavelength in free space of accelerating field.

Using this one finds for the efficiency

$$\eta_B = \frac{1}{8\pi e E_o c} \cdot \frac{i_b}{\varepsilon} \frac{\lambda}{\beta^*} \; . \tag{1.6}$$

The first term in this expression is inviolate since it consists of constants of nature, the second one is a source parameter. Only the third term, λ/β^*, is available for adjustment. The requirement that the bunches be short compared to β^* implies a coupling between λ and β^*: $\lambda/\beta^* = (\lambda/\sigma_z)(\sigma_z/\beta^*) = \bar{B}\,\sigma_z/\beta^*$, where $\bar{B} = \lambda/\sigma_z$ is the bunching factor. If σ_z/β^* is regarded as restricted $\eta_B \propto \bar{B}$. The accelerator restricts the beam current to a value that is proportional to λG, where G represents the accelerating gradient. We conclude therefore that efficient single bunch operation requires bright souces, large bunching factors $\bar{B}$, large accelerating gradients G and a long wavelength, thus a low operating frequency.

1.2. SINGLE LONG BUNCH

If the colliding bunches are not short compared to β^* their local cross sections at any instant will be functions of position along the axis, because β varies with position. Assuming symmetry relative to the interaction point and disregarding the effects of the beam self fields, one has for $\beta = \beta(z)$:

$$\beta = \beta^* (1 + (z/\beta^*)^2) \tag{1.7}$$

where z is the distance to the interaction point.

The contribution to the instantaneous luminosity by a section of length dz at location z is then:

$$d\mathscr{L} = \frac{2\, i_b/e \; i_b/ec}{4\pi} \; \frac{\gamma}{2\beta^*} \; \frac{dz}{1 + (z/\beta^*)^2} \tag{1.8}$$

where i_b represents the instantaneous beam current. The contribution lasts as long as both colliding bunches are present at location z. Consider first bunch 1, which has length σ_z and moves in the direction of positive z. It will be present at location z during the interval $z/c \leq t \leq (z+\sigma_z)c$. Bunch 2, also of length σ_z, but traveling in the opposite direction will be there while $-z/c \leq t \leq (\sigma_z - z)c$; $t=0$ represents the time at which each bunch reaches the interaction point. It follows that both bunches are present while

$$(\sigma_2 - z)/c \leq t \leq (\sigma_z + z)/c. \tag{1.9}$$

The length of the interval of exposure is

$$\Delta t = (\sigma_z - 2|z|)/c$$

regardless of the sign of z because of the symmetry of the system. $\Delta t=0$ for $|z| \geq \sigma_z/2$ because there the counter streaming bunches do not appear simultaneously. Using this and assuming for convenience that i_b is constant, i.e., that the bunch charge is distributed uniformly along its length, one obtains for the contribution to the time integrated luminosity:

$$\int \delta \mathscr{L} dt = \frac{2i_b^2}{4\pi e^2 c^2} \frac{\gamma}{\varepsilon \beta^*} \frac{\sigma_z - 2|z|}{1+(\frac{z}{\beta^*})^2} dz \quad .$$

Integration over the full length $-\tfrac{1}{2}\sigma_z \leq z \leq \tfrac{1}{2}\sigma_z$ yields

$$\int \mathscr{L} dt = 2 \frac{2i_b^2}{4\pi e^2 c^2} \frac{\gamma}{\varepsilon \beta^*} \sigma_z \beta^* \left[\arctan(\sigma_z/2\beta^*) - \frac{\beta^*}{\sigma_z} \ln(1+(\sigma_z/2\beta^*)^2) \right]. \tag{1.10}$$

Repeating this process at a rate of f times per unit time one finds that the time averaged luminosity is

$$\mathscr{L} = f \frac{i_b^2}{\pi e^2 c^2} \frac{\gamma}{\varepsilon} \sigma_z F(\sigma_z/2\beta^*) \tag{1.11}$$

where

$$F(x) = \arctan(x) - (1/2x) \ln(1+x^2) \quad .$$

This luminosity requires a beam power P_b:

$$P_b = 2 f \frac{i_b}{e} \frac{\sigma_z}{c} \gamma E_o \tag{1.12}$$

so thus the efficiency becomes

$$\eta_B = \frac{\mathscr{L}}{P_b} = \frac{2}{4\pi e c E_o} \frac{i_b^2}{\varepsilon i_b} F(\sigma_z/2\beta^*) = \frac{2}{4\pi e c E_o} \frac{i_b}{\varepsilon i_b} \bar{B} F(\sigma_z/2\beta^*) = \tag{1.13}$$

$$= (\frac{1}{8\pi e E_o c} \frac{i_b}{\varepsilon} \frac{\gamma}{\beta^*}) 2(\frac{2\beta^*}{\sigma_z}) F(\sigma_z/2\beta^*) \quad .$$

The first term in parentheses in (1.13) corresponds to (1.6) which is valid for $\sigma_z \ll \beta^*$. In Table I, I tabulate some values for $F(x)$ and $2F(x)/x$ as functions of x.

Table I: $F(x)$ and $2/x\, F(x)$ as functions of α.

x	$F(x)$	$\frac{2}{x} F(x)$
0.01	0.005	1
0.03	0.015	1
0.1	0.050	0.998
0.3	0.148	0.986
1	0.439	0.878
3	0.865	0.577
10	1.240	0.249
30	1.424	0.095

It may be seen that there is little point in demanding $\sigma_z \ll \beta^*$ as far as efficiency is concerned, $\sigma_z = 2\beta^*$ is certainly acceptable in that respect, as is $\sigma_z = 6\beta^*$, in all likelihood. However, the bunch length is limited by considerations of energy spread to a fairly small fraction of λ: for $|\Delta E/E| < 0.05$ $\sigma_z/\lambda = 1/\overline{B} < 0.14$, thus $\overline{B} > 7.2$, where E stands for energy. This then sets an upper limit to β^*:

$$\beta^* < 0.1 \lambda \ , \qquad \sigma_z < 0.14 \lambda$$

for good efficiency.

1.3. MULTIBUNCH OPERATION (Fig. 1)

The coupling between β^* and λ can be broken by switching to multibunch operation. The basic idea is to subdivide the long bunch discussed above into N_B small ones, each with a charge N/N_B particles and with a center to center distance, thus a new wavelength Λ, of

$$\Lambda = L/(N_B-1) \ . \tag{1.20}$$

Here is L the length the interaction region, which is equal to the length σ_z of the original long bunch. There are now N_B collision points, spaced at distances $1/2 \ \Lambda$ along the system axis, in which a total of N_B^2 bunch--bunch collisions occur per pulse. In calculating the overall luminosity and efficiency one has to calculate those quantities separately for each interaction point because the β's are different in the various points, and because the number of interactions per pulse in a particular point decreases linearly with its distance to the central one. Computer simulation shows little difference between the exact result and the analytical one for the long bunch, particularly if N_B is large. In this way I find for the efficiency of this multibunch mode

$$\eta_B = (\mathscr{L}/P_b)_{mb} = \frac{1}{2\pi e E_o c} \frac{i_b}{\varepsilon} F(L/2\beta^*) \frac{N_B}{N_B-1} \tag{1.21}$$

where I used (1.13) and (1.20) and where $N_B \geq 2$. This form is evidently independent of the choice of wavelength (although that variable may enter into the beam current i_b, as mentioned before), and nearly independent of that of N_B. It appears, therefore, that the efficiency in this mode is determined by the characteristics of the source, i.e., by i_b/ε and by the choice of $L/2\beta^*$, i.e., by the ratio of the length of the interaction region

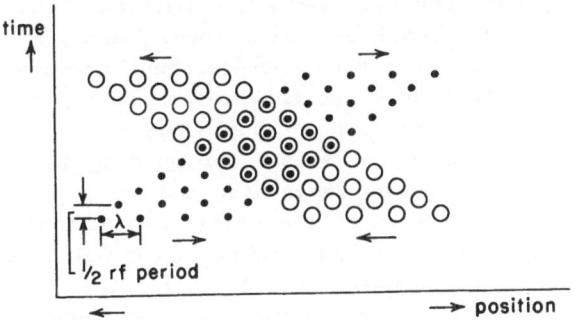

Fig. 1   Multibunch Interaction

and β^*. It is evident from Table I that there is not much point in choosing $L/2\beta^* \geq 10$.

Multiplication of any of the expressions for η_B, e.g., (1.21) with the beam power P_b yields the luminosity itself. Since

$$P_b = 2\ i_b \gamma E_o T f = 2\ i_b E_t \frac{L}{c} f \qquad (1.22)$$

where $E_t = \gamma E_o$ one finds that

$$\mathscr{L} \propto L\ i_b^2/\varepsilon\ .$$

1.4. COMMENTS

It has been assumed, so far, that single particle optics is valid. In practice the beam densities will be sufficiently high that this is not true. The computer model referred to before incorporates a linear approximation of this beam-beam or beam disruption effect and shows clearly that things change with increasing density. It describes each bunch/bunch interaction as the application of a focussing lens that transforms the emittance ellipse of each of the two participating bunches. The strength of that lens is adjusted according to the local beam radius and is proportional to the sum of the charges in the two bunches involved. The initial condition can be set up for each bunch individually so that Brian Montague's proposal for dynamic focusing[2] can be simulated, but this has not yet been tried. Another important factor that has been disregarded is the so-called beamstrahlung. Its effect is likely to be different in multibunch operation from what it is in short single bunch operation and will be studied.

2.1. ACCELERATION

We now consider some aspects of accelerators that might be useable for our purposes. The principal factors of interest are the energy spent per unit energy in the final beam and the average accelerating gradient. The first factor enters immediately into the operating cost of the facility and indirectly into its capital cost, the second one directly into the capital cost. The characteristics of those accelerators will depend on the choice of operating mode: single bunch or multibunch. Developments of the induction linac, wakefield accelerator or switched power linac might be suitable for single bunch operation while the multibunch mode needs something similar to the conventional electron linac. The first three accelerators are all single pulse, wide band, non-resonant devices; the electron linac is a high Q, narrow band, resonant device. This latter accelerator can also be used for the acceleration of single bunches but then its energy efficiency is low. A convenient parameter for guiding the choice of r.f. parameters is the ratio η_p of the luminosity and the r.f. power spent to produce it, or, equivalently, that of the integrated luminosity, integrated over a single beam-pulse and the r.f. energy invested in that pulse. We use the latter definition for convenience. An expression for the integrated luminosity was derived in the previous section, the present one addressed the determination of the r.f. energy. This requires a description of the accelerator installation. It appears that η_p, once an expression for it is available, can be maximized by proper choice of the r.f. frequency or wavelength and the coupling factor between the source of the r.f. power and the accelerator. This is a consequence of the fact that the system is pulsed to yield a beam-pulse length equalto the desired length of the inter action region. Each beam pulse is preceded by a filling period during which r.f. energy is spent in building up the fields in the accelerator to their nominal levels. This energy increases quickly with increasing wavelength and is lost after the last bunch of a pulse has passed if it is not recovered. The integrated luminosity due to a beam pulse of fixed duration increases also with wavelength because the tolerable beam current does. However, the energy expanded during the filing time increases faster than the integrated luminosity, so that an optimum wavelength, whose value depends on the length of the beam pulse, must exist.

Consider as a prototype an accelerator that consists of a string of independent cavities, each supplied by its power source via an ideal, i.e., lossless and reflection free, power transmission system. The length of that system is such that the time it takes a wavefront to travel from one of its ends to the other is more than half the length τ_s of the powerpulse. The beam pulse begins only a filling time τ_F after the arrival of the power pulse front at the cavity and terminates simultaneously with the power pulse. The restriction on the length of the transmission guide ensures that the power source is turned off before reflections due to mismatches between guide and cavity or any signals from the beam can reach it. It is therefore always loaded by the characteristic impedance of the transmission line during the pulse, regardless of the conditions at the cavity end while the source impedance seen by cavity and beam is always the characteristic impedance of the transmission line at its cavity side. It is convenient to represent the cardinal features of this prototype with the equivalent circuit diagram of Fig. 2: ℓ, γ, R_o represent the length, propagation constant and characteristic impedance of the transmission system and L, C, R the magnetic, electric and disspative parts of the cavity impedance, S is the power source, it produces a voltage $\bar{U}_s \cos \omega \tau$, U_g is the gap voltage and i_b is the beam current. The dissipative losses, which are primarily caused by eddy currents in the cavity walls, would be represented more accurately by a small resistor in series with the inductor than by the large parallel resistor shown, however, the difference is immaterial for the present purpose and the representation chosen is slightly more convenient. The beam consists of a stream of bunches whose lengths are short compared to the intra-bunch distances. The Fourier expansion of the beam current has therefore a dc component and many harmonics of the bunch repetition frequency, which we take to be identical to the r.f. frequency $\omega/2\pi$. We are only concerned about its fundamental component, which has an amplitude $\bar{i}_b$ that is

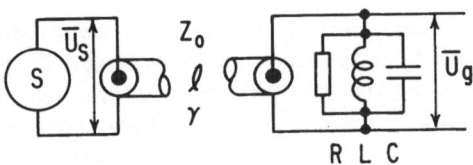

Fig. 2   Equivalent Circuit Diagram of Linac Cell

twice the dc component i_b. The beam current is represented as coming from a current source, this is an acceptable approximation if the particle energy γE_o is so high that fractionally small changes cause negligible fractional change in the velocity βc, i.e., if

$$\Delta\beta/\beta = \frac{1}{\gamma^2-1} \frac{\Delta\gamma}{\gamma} = 0 .$$

Via this model one finds for the behaviour of the gap voltage amplitude $\bar{U}_g$ as a function of time in the absence of beam:

$$\bar{U}_g(t) = \hat{U}_g (1 - e^{-t/\tau}) \tag{2.1}$$

t=0 represents the instant of arrival of the power pulse front at the cavity and the resonant frequency ω_r of the cavity is equal to the frequency of the power source:

$$\omega_r = \omega \tag{2.2}$$

$$\frac{2}{\tau} = \frac{1}{C} \left(\frac{1}{R_o} + \frac{1}{R} \right) \tag{2.3}$$

$$\omega_r^2 = 1/LC - 1/\tau^2 \tag{2.4}$$

ω_r is real if

$$\frac{LC}{\tau^2} = \frac{1}{2Q} \left(1 + \frac{R}{R_o} \right) < 1$$

thus if

$$R_o/R > 1/(2Q - 1) \tag{2.5}$$

where $Q = R\sqrt{\frac{C}{L}}$ represents the quality factor of the unloaded resonator.

The asymptotic value U_g of the gap voltage amplitude is related to the powersource voltage (as measured at the cavity) according to

$$U_g = 2 \bar{U}_s/(1 + R_o/R) . \tag{2.6}$$

It follows that the filling time τ_F necessary for a specific gap voltage amplitude $\bar{U}_g$ is

$$\tau_F = - \tau \ln(1 - \bar{U}_g/U_g) . \tag{2.7}$$

The effect of a beam current $\bar{i}_b$ may be described as a change in the amplitude of the effective power source voltage from $\bar{U}_s$ to $\bar{U}_s - 1/2 \, \bar{i}_b R_o$. This expression is valid if the source voltage and the beam current are in phase; the situation becomes algebraically much more complex if they are not: phase factors have to be added, the phase between cavity voltage and drive will change and switching transients will have to be taken into account.

52

Using the simple model one obtains for the new asymptotic gap voltage U_{gb}:

$$\hat{U}_{gb} = \frac{2\bar{U}_s - \bar{i}_b R_o}{1 + R_o/R} \, . \tag{2.8}$$

A beam current $\bar{i}_b$ is such that

$$\bar{U}_g = \hat{U}_g (1 - e^{-\tau_F/\tau}) = \hat{U}_{gb} \tag{2.9}$$

will therefore terminate the filling period and begin the accelerating period, during which the gap voltage amplitude remains $\bar{U}_g = U_{gb}$. This requires

$$\bar{i}_b = (\hat{U}_g - \bar{U}_g)(\frac{1}{R_o} + \frac{1}{R}) \tag{2.10}$$

$\bar{U}_g$ will continue to change until this condition is satisfied if $\bar{i}_b$ deviates from this value. Successive bunches that pass while $\bar{U}_g$ is changing will receive different gains in energy. Such differences increase the energy spread in the ultimate beam and are undesirable. It is therefore important that the power source, filling time, beam current and phases be matched properly.

The number N_c of cavities to reach a specified final energy γE_o is simply:

$$N_c = \frac{\gamma E_o}{\Delta E_p} = \frac{\gamma E_o}{\bar{U}_g F_{tr}} \tag{2.11}$$

where $\Delta E_p = \bar{U}_g F_{tr}$ represents the energy gain per particle per gap and F_{tr} the so-called transit time factor, introduced to account for the fact that a particle crosses a gap of non-zero length in a non-zero time, which may be as long as half an r.f. period. It corrects also for the non-uniformity of the gap field, due to end effects.

Before an expression can be written for the efficiency η_p a relation between beam current and emittance must be established. If we assume that the beam originates in a source of constant brightness, i.e., such that the density it produces is x, x', y, y' phase space is independent of the current, we may write

$$i_b = \tilde{B} \, \varepsilon^2 \tag{2.12}$$

where $\tilde{B}$ is the brightness of the particle source.

η_p is now calculated by performing the division:

$$\eta_p = \frac{\int_{\tau_B} \mathcal{L} \, dt}{2N_c \int_{\tau_s} P_s \, dt} \tag{2.13}$$

$\tau_B = L_B/c$ is the length in time of the beam pulse, L_B its physical length = length of interaction area, $\tau_s = \tau_F + \tau_B$ is the length in time of the power pulse; the factor 2 keeps account of the fact that there are two, presumably identical, beams. Rewriting (1.10) and (1.11) we have for the time integrated luminosity:

$$\int_{\tau_B} \mathscr{L}\,dt = \frac{i_b^2}{4\pi e^2 c^2} \frac{\gamma}{\varepsilon} \, L_B F(L_B/2\beta^*) \ . \tag{2.14}$$

The energy delivered by the power sources is:

$$N_c \int_{\tau_s} P_s\,dt = \frac{\gamma E_o}{\bar{U}_g F_{tr}} \frac{1}{2} \frac{\bar{U}_s^2}{R_o} (\tau_F + \tau_B) = \frac{\gamma E_o}{\bar{U}_g F_{tr}} \frac{1}{2} i_b^2 R_o \tau_B \ .$$

$$\cdot \left[1 + \frac{\bar{U}_g}{2i_b R_o} \left(1 + \frac{R_o}{R}\right) \right]^2 \left[1 + \frac{\tau}{\tau_B} \ln\left\{ 1 + \frac{\bar{U}_g}{2i_b R_o}\left(1 + \frac{R_o}{R}\right) \right\} \right] \tag{2.15}$$

where we used $\bar{i}_b = 2i_b$, (2.6), (2.7) and (2.10).

The expression for η_p may now be derived

$$\eta_p = \frac{\displaystyle\int_{\tau_B} \mathscr{L}\,dt}{\displaystyle 2N_c \int_{\tau_s} P_s\,dt} =$$

$$= \frac{\dfrac{1}{4\pi e^2 c^2} i_b^2 \sqrt{\dfrac{\bar{B}}{i_b}} \gamma\, \tau_B F(L_B/2\beta^*)}{2 \dfrac{\gamma E_o}{\bar{U}_g F_{tr}} \dfrac{1}{2} i_b^2 R_o \tau_B \left[1 + \dfrac{\bar{U}_g}{2i_b R_o}\left(1 + \dfrac{R_o}{R}\right)\right]^2 \left[1 + \dfrac{\tau}{\tau_B}\ln\left\{1 + \dfrac{\bar{U}_g}{2i_b R_o}\left(1 + \dfrac{R_o}{R}\right)\right\}\right]} =$$

$$= \frac{F(L_B/2\beta^*)F_{tr}\sqrt{\bar{B}}}{4\pi e^2 c E_o} \frac{\sqrt{i_b}\, \dfrac{U_g}{i_b R_o}}{\left[1 + \dfrac{\bar{U}_g}{2i_b R_o}\left(1 + \dfrac{R_o}{R}\right)\right]^2 \left[1 + \dfrac{\tau}{\tau_B}\ln\left\{1 + \dfrac{\bar{U}_g}{2i_b R_o}\left(1 + \dfrac{R_o}{R}\right)\right\}\right]} \tag{2.16}$$

(2.16) may be represented by

$$\eta_p = A \frac{x\sqrt{\lambda}}{(1+x)^2(1+a\lambda)} \tag{2.17}$$

Here is

$$x = \frac{\bar{U}_g}{2i_b R_o}\left(1+\frac{R_o}{R}\right) = \frac{2\pi F_{tr}}{\bar{a}}\frac{1}{R_o}\sqrt{\frac{L}{C}}\frac{\left(1+\frac{R_o}{R}\right)}{\sqrt{1-\left(\frac{1}{2Q}\right)^2\left(1+\frac{R}{R_o}\right)^2}} =$$

$$= \frac{2\pi F_{tr}}{\bar{a}}\frac{1}{R_o}\sqrt{\frac{L}{C}} \qquad (R \gg R_o,\ QR_o/R \gg 1)$$

$$a = \frac{2}{L_B}\frac{F_{tr}}{\bar{a}}\frac{\ln(1+x)}{x}$$

$$A = \frac{F(L_B/2\beta^*)F_{tr}\tilde{B}}{4\pi e^2 cE_o}\sqrt{\frac{\bar{g}\,\bar{G}\,\bar{a}}{4\pi F_{tr}}}\sqrt{\frac{C}{L}\left[1-\left(\frac{1}{2Q}\right)^2\left(1+\frac{R}{R_o}\right)^2\right]}\frac{2}{1+R_o/R} =$$

$$= 2\frac{F(L_B/2\beta^*)}{4\pi e^2 cE_o}\sqrt{\frac{g\,\tilde{B}\,\bar{G}\,\bar{a}\,F_{tr}}{4\pi}}\sqrt{\frac{C}{L}}$$

$$\bar{a} = \frac{\text{energy gain per bunch}}{\text{energy stored}}$$

with $g\lambda$ the length of the acceleration gap and $\bar{G}$ the field in it. At this level of approximation x, a, A, g and $\bar{a}$ are all more or less independent of the choice of the wavelength , particularly if $R_o/R \ll 1$ and if $(1+R/R_o) \ll 2Q$, with Q the quality factor of the unloaded cavities. We note that $L/C \lesssim 200\ \Omega$, $g \lesssim 0.5$, $F_{tr} < 1$ are all geometrical form factors whose exact values are set by the geometry of the accelerating cavities and the choice of operating mode (π, $2\pi/3$ or other). The source brightness $\tilde{B}$ and the accelerating field G should be maximized. The design of the accelerator cannot affect $\tilde{B}$, presumably, while $G \overset{?}{=} G(\lambda,T_B)$ is restricted by surface effects on the cavity walls: field emission of electrons, sparking, physical damage due to dissipation, etc. L_B, the length of the beam pulse, but also the length of the interaction region, is restricted by the users, as is $\bar{a} \overset{?}{=} \bar{a}(\lambda) < 0.05$, since the momentum spread in the beam tends to increase with $\bar{a}$ while the beam becomes less stable with increasing $\bar{a}$.

Returning to (2.17) one finds that η_p is maximized by choosing $\lambda = 1/a$ and $x \approx 1.4$. For those values

$$\eta_p = 0.12\ A/\sqrt{a}$$

x can be manipulated via the product $\bar{a}R_o$, thus

$$R_o = \frac{1}{x}\frac{2\pi F_{tr}}{\bar{a}}\sqrt{\frac{L}{C}} = 0.71\frac{2\pi}{\bar{a}}\sqrt{\frac{L}{C}}\ . \tag{2.18}$$

Substitution in (2.16) yields

$$\eta_p = 0.217 \frac{\bar{a}\, F(L_B/2\beta^*)}{4\pi\, ec\, E_o} \frac{g\, \bar{G}\, \tilde{B}\, L_B}{4\pi} \sqrt{\frac{C}{L}} \propto \bar{a}\sqrt{\tilde{G}\tilde{B}L_B} \quad , \qquad (2.19)$$

$$\lambda_{opt} = 0.8 \frac{\bar{a}}{F_{tr}} L_B \quad , \qquad R_o = 0.71 \frac{2\pi F_{tr}}{\bar{a}} \sqrt{\frac{L}{C}} \quad .$$

It is clear that for maximum η_p, $\bar{a}$, L_B, $\tilde{B}$ and G should all be as large as possible and that the optimum wavelength and source impedance are directly tied to the choice of α and L_B.

At the optimum wavelength the filling time and beam pulse length (in time) are equal, thus the source pulse length is then twice the beam pulse length. Since the length of the interaction region L_B will not be more than a few cm, the length of the power transmission system has to exceed those few cm, in order to validate our initial assumption. In practice it will be difficult to violate this condition, short of integrating the power source directly with the accelerating cavity. It seems that for optimum efficiency η_p one has to operate as close to the permissible limits in beam current (via $\bar{a}$) and accelerating gradient as possible. The instantaneous luminosity can then only be controlled via the final focus, i.e., β^*. The average luminosity can always be changed via the pulse repetition rate.

The results assembled in Table II illustrate the behaviour of systems of this type. We took

$\tilde{B} = 8\times10^{10}$ (A/(rad-m^2)

$\bar{G} = 10^9$ V/m

π mode, i.e.

$g = 0.5$

$F_{tr} = 2/$

$\sqrt{L/C} = 200$

$R/\sqrt{\lambda} = 5.3\ \Omega/m^{1/2}$

$F(L_B/2\beta^*) = 0.44$

$\beta^* = 1/2\ L_B$.

Table II

L_B (cm)	α	η_p (10^{22})	λ_{opt} (mm)	R_o (kΩ)	R (kΩ)	i_b (A)	P_s (MW)	$\partial P_s/\partial l$ (GW/m)	YN_{33}	N_B
1	0.01	2.5	0.12	57	59	0.4	0.025	0.4	4.8×10^{14}	80
10		8	1.25		188	4	2.5	4.0	1.5×10^{12}	
1	0.02	5	0.25	28	84	1.6	0.201	16	6.0×10^{13}	41
10		16	2.51		265	16	20.1	8.0	1.9×10^{11}	
1	0.03	7.5	0.38	19	103	3.5	0.68	3.6	1.8×10^{13}	28
10		24	3.77		325	35	68.0	36	5.6×10^{10}	
1	0.04	10	0.50	14	119	6.3	1.60	6.4	7.4×10^{12}	21
10		31	5.02		375	63	160	64	2.4×10^{10}	
1	0.05	12	0.63	11	133	9.8	3.15	100	3.8×10^{12}	17
10		39	6.28		420	98	315	100	1.2×10^{10}	

We chose two interaction lengths:

$$L_B = 1 \text{ cm}, \quad 10 \text{ cm}$$

corresponding with $\tau_s = \tau_F + \tau_B = 2\tau_B = 64$ psec, resp 640 psec, and five $\bar{a}$ values in the interval $0.01 \leq a \leq 0.05$.

In Table II:

$$p = \frac{\langle \mathscr{L} \rangle}{\langle P \rangle} = \frac{\int \mathscr{L} dt}{\int P_s dt} \qquad (\text{cm}^{-2} \text{ sec}^{-1} \text{ w}^{-1})$$

N_B number of bunches per pulse

P_s r.f. power/cavity during the pulse

$\partial P_s / \partial \ell = P_s / g\lambda$ r.f. power per unit length during the pulse

N_{33} pulse repetition rate for $L = 10^{33}$ cm^{-2}sec^{-1} at final energy γE_o.

It is noted that for L_B and $\bar{a}$ small our approximation $R \ll R_o$ breaks down. The associated values should be recalculated, taking this effect into account. It appears that heavy loading ($\bar{a}$ large) and long interaction lengths are favourable for efficiency, but also that that leads to very high peak powers P_s and $\partial P_s / \partial \ell$, for which there is, presumably, an upper bound. The pulse repetition rates are very high as a consequence of our very modest estimate of the achievable source brightness $\tilde{B}$, they decrease as $(\tilde{B})^{-1/2}$.

An important restriction is imposed by the beamstrahlung, i.e., by the changes in the energies of individual particles due to synchrotron radiation caused by the beam self field. So far this complex subject seems to have been studied only for single bunch interactions, but not yet for the multibunch mode discussed here[3]. There is a critical energy

$$E_{cr} = 3\hbar c r_e \gamma^2 N / (r\sigma_z) = 3 \frac{\hbar}{e} r_e i_b^{3/4} \gamma^{5/2} B \tilde{B}^{1/4} / \sqrt{\beta} =$$

$$= 2 \times 10^{-11} i_b^{3/4} \gamma^{5/2} B \tilde{B}^{1/4} / \sqrt{\beta} \qquad (\text{eV})$$

where

$\hbar$ $= h/2\pi$ Planck' constant,

r_e $=$ classical radius of electron,

$\bar{B}$ $=$ bunching factor (peak/average).

One finds, by simple scaling, that, in the crudest approximation, the relative energy change δ due to beamstrahlung should behave as $\bar{B}\lambda^{1/2}$ for $\gamma < \gamma_{cr}$ and as $\bar{B}^{-1/3}\lambda^{-1/2}$ for $\gamma > \gamma_{cr}$. Since $\gamma_{cr} \propto \gamma^{5/2}$ one is forced to operate in the second regime if the energy is sufficiently high. Using the earlier numerical assumptions one expects the changeover to occur somewhere in the 1–10 TeV range.

3.1. ACCELERATING STRUCTURES

A number of structures that support accelerating modes have been described[4] and model studies, which demonstrated the existence of such modes, have been performed on some of them. However, the support of accelerating modes, though necessary, is not sufficient. A second condition is that there be no beam deflection since such deflection would be the cause of energy loss due to synchrotron radiation. Such loss is proportional to γ^4/ϱ^2 with $\gamma = E/E_o$ and ϱ the local instantaneous radius of orbit curvature. Any curvature leads always, ragardless of its sign, to a loss because of the quadratic relationship, and the loss increases sharply with increasing energy. The radiation is directed along the beam and will have practically the same velocity as the electrons so that each bunch is a mixture of electrons (or positrons) and photons. Some of it will hit the accelerator structure, and cause emission of electrons and γ rays. Minimization of the synchrotron radiation requires evidently minimization of orbit curvature. The net field in the accelerator, basically a standing wave, may be described as a superposition of many modes. Only a few of these, if any, are accelerating modes, but nearly all can contribute to beam deflection. Requiring virtual absence of deflection presents therefore a severe restriction, which, however, can be met by imposing certain symmetry conditions on the transverse geometry, as demonstrated by existing linacs. These machines have circular cylindrical symmetry about the machine axis. Though their structures still support certain deflecting modes, such modes occur only due to asymmetry in the excitation, e.g., in the connection to the power source, or, in the case of collective effects, to beam—axis misalignments; they are generally weak. Circular cylindrical symmetry becomes increasingly difficult to arrange if the wavelength becomes smaller, but structures with two—fold and higher symmetry offer similar characteristics relative to deflection: although deflecting modes are possible, the nominal net field has no transverse dipole component in the vicinity of the axis.

Lately the emphasis of our studies has been on two structures which might be acceptable. One, the foxhole structure, can be seen as extrapolation from the conventional linac structure, the other, the colonnade, invented by Palmer, may be seen as a development of the gratings with which this enterprise began[4]. Both seem feasible down to and including $\lambda = 10\ \mu$m on the basis of the experience with micro machining by, e.g., ion etching, we have gained so far. Model studies are in progress for each.

3.2. FOXHOLE STRUCTURE (2π MODE)

Let me consider the foxhole structure first because it is relatively simple and closest to present practice. A sketch for its simplest version is given in Fig. 3. It shows a base plate in which rectangular holes, the foxholes, have been formed. The cross section of each hole is of order $1/2\lambda \times 1/2\lambda$, depth an integer multiple of $1/4\Lambda$ and the distance between centers is λ. Here is Λ the local wavelength in the foxhole. They are interconnected by slots through which the beam passes, each slot has a width of order 0.1λ and reaches from top to bottom. Geometries of this type can be realized, presumably with sufficient accuracy, by means of

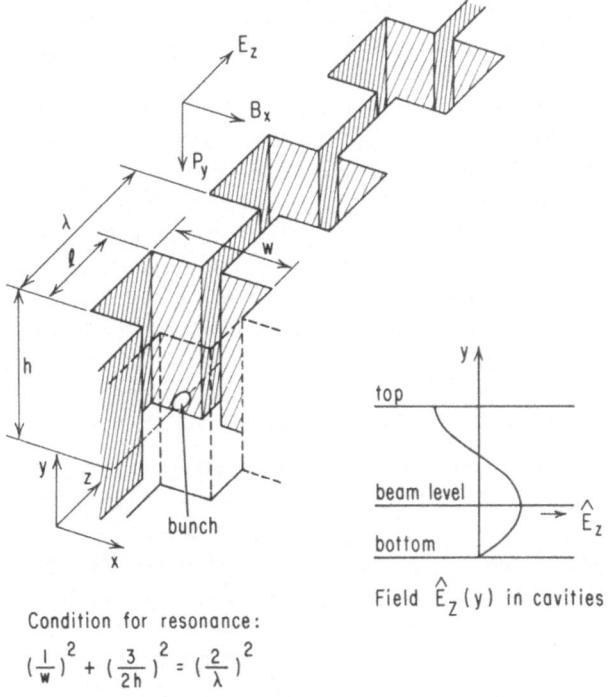

Condition for resonance:

$$(\frac{1}{w})^2 + (\frac{3}{2h})^2 = (\frac{2}{\lambda})^2$$

Field $\hat{E}_z(y)$ in cavities

Fig. 3   Foxhole Structure (2π mode)

techniques with which we have some experience at BNL. The structure is excited by a travelling wave of e.m. radiation that propagates perpendicularly to the base plate with its E vector directed along the slots. Each foxhole acts as a resonator, I shall neglect the coupling between them. The incident radiation generates a standing wave in each foxhole with a node for the magnetic field and a maximum for the electric one in the midplane at $1/4\Lambda$ from the bottom. The beam axis is located at that height. The fields in all resonators are in phase and the acceleration process is reminescent of that in an Alvarez proton linac, with the slots acting as drift tubes and the resonators as accelerating gaps. Regarding each resonator as a cavity with dimensions h x w x d, as indicated in Fig. 3, and assuming a transverse electric mode one finds for the resonant wavelength in free space

$$\lambda = \left[(\frac{1}{2w})^2 + (\frac{n}{4h})^2 \right]^{-1/2}$$

when n is the number of quarter wavelengths along the height. Choosing n=2 the electric field across the open end of the resonator will be zero if there is no energy loss in the cavity, for n=3 the magnetic field will be zer at that location. In both cases the incident radiation will reflect totally, the Poynting vector will be zero everywhere in this and surface and the amplitudes of the magnetic and electric fields in the cavity will be twice those in the incident wave. In actuality some energy is lost in the cavity to dissipation in the resistivity of the walls and to the beam. The lost energy is replenished by the incident wave, resulting in a small in phase component of the electric, resp. magnetic field in the open end plane of the cavity. There will be no reflection if the E/H ratio of the incident wave matches that at the mouth of the cavities and all its energy is absorbed. The first case represents a low impedance match with a small value for E/H, the second with its large electric field and low magnetic field represents a high impedance match. The value of the impedance $Z=E/H$ is easily calculable from the physical constants of the cavity, the matching conditions can be realized by proper arrangement of the source and of the optical system between it in the cavity orifice.

The power source would produce a beam which is focussed on the apertures of the resonators. Its cross section in that aperture plane would be a pulse length, i.e., $c\tau_s$, long in the direction of motion of the beam, and its width would cover the resonator apertures. The beam spot might be made to move synchronously with the train of bunches it is accelerating, or the accelerator could be built in sections, each with its own power source, as is standard practice for conventional accelerators. There would have to be many short sections for reasons of energy efficiency. The length of the radiation pulse would have to be longer than the beam pulse by $2N_s$ r.f. periods plus a filling time if a section is N_s wave lengths long: the last cavity is excited during N_s periods before the head of the beam reaches it while the first one is driven during N_s periods after the last bunch of a bunch train has left it.

I have disregarded so far the perturbation introduced by the slots. Slots in the boundary walls of wave guides and cavity resonators have been used for a long time and for various reasons. In this particular case their effect is thought to be small for the desired mode of operation, since there are no wall currents that have to cross them. The slots themselves act as wave guides that are driven in a higher mode, their characteristic impedance is low and they are close to a half wavelength long. Our model studies show the existence of the desired mode in a single cavity with the appropriate slots, which, however, are only a small fraction of a wave length long in the model used.

The presence of the slots, which divide the structure into two mirror symmetric halves suggests the possibility of constructing it in two halves. Doing so adds important flexibility to the design of the resonators and slots, they have no longer to be cylindrical and fabrication may be easier since the depresions to be generated are less deep by factors of 2 to 3.

3.3. FOXHOLE STRUCTURE (π AND $2\pi/3$ MODES)

The foxhole structure described above operates in the 2π mode, like most Alvarez linacs; its effective accelerating gradient is therefore only $F_{tr} \lessapprox 1/\pi = 0.31$ of the amplitude of the resonator field. Operation in the π, resp $2\pi/3$ modes would yield factors of $F_{tr} \lessapprox 0.62$ and $\lessapprox 0.86$, because they use the available space and field more efficiently than the 2π mode. Their realization requires the use of two or three resonators per wavelength and a large reduction in the lengths of the slots between successive resonators. the resonators would no longer run in phase but with phase differences of π rad (π mode) or $2\pi/3$ rad ($2\pi/3$ mode) between them. Although the latter is standard practice in conventional electron linacs it produces problems if the power source excites the resonators in parallel. These angles are too large to be obtained by simple detuning of the resonators relative to the frequency of the power source: the relative loss in field amplitude due to the detuning is larger than the gain derived from this mode of operation.

One solution for a π mode structure could be to dimension alternate resonators differently: both types would resonate with the source frequency but their heights would have the ratio $h_1/h_2 = 5/6$. As indicated in Fig. 4

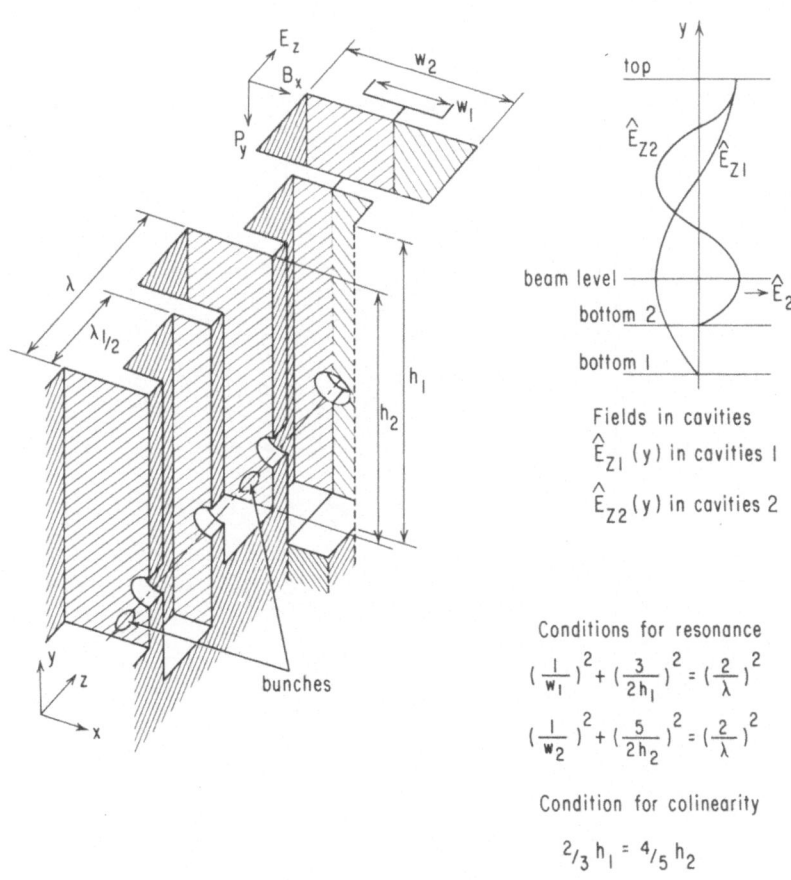

Fields in cavities
$\hat{E}_{Z1}(y)$ in cavities 1
$\hat{E}_{Z2}(y)$ in cavities 2

Conditions for resonance

$$\left(\frac{1}{w_1}\right)^2 + \left(\frac{3}{2h_1}\right)^2 = \left(\frac{2}{\lambda}\right)^2$$

$$\left(\frac{1}{w_2}\right)^2 + \left(\frac{5}{2h_2}\right)^2 = \left(\frac{2}{\lambda}\right)^2$$

Condition for colinearity

$$^2/_3\, h_1 = {}^4/_5\, h_2$$

Fig. 4   Foxhole Structure (π mode)

the shorter one would be 3/4 of a local wavelength long, the longer 5/4 of a different local wavelength. The field maximum in the longer one would occur at the same depth as the second field maximum in the shorter one, so that the beam sees a phase reversal. This trick cannot be used for the $2\pi/3$ mode since in essence the cavity fields are still in phase. Other solutions may be achieved by driving the resonators in groups of two (π mode) or three ($2\pi/3$ mode) while coupling the resonators within a group in a suitable manner. This is likely to require more complex geometries for the resonators than the simple cylindrical ones considered so far.

3.4. COLONNADE (2π MODE)

A second structure under study is Palmer's colonnate. This structure is a truly "open" one, in contrast with the foxholes, which are only semi-open at best. As shown in Fig. 5 it consists of a base plate on which two parallel rows of cylinders have been placed. The cylinders are not necessarily circular in cross section. The beam axis is located in the midplane between the cylinders at some distance from the base plate. In the simplest version, which operates in the 2π mode, the distance between successive cylinders along the axis is exactly λ, the distance between the rows about

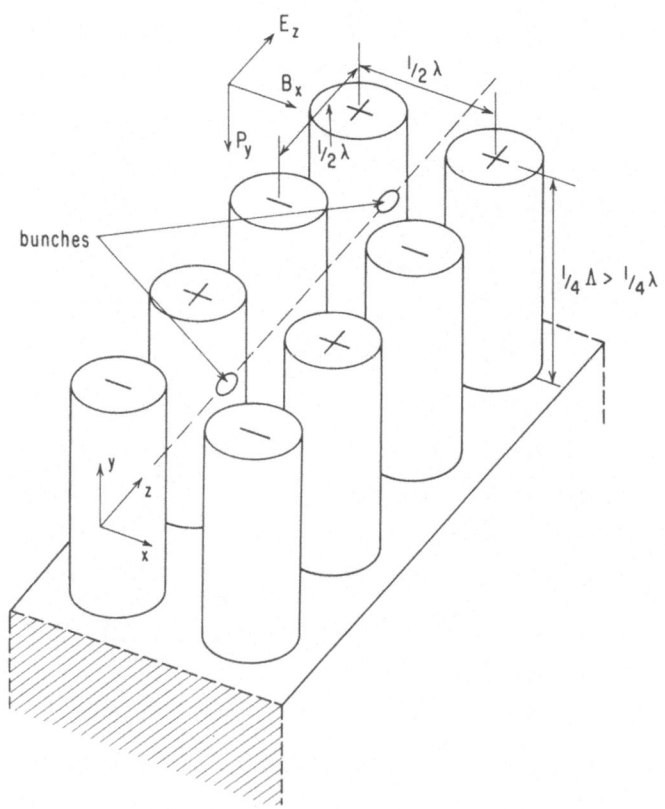

Fig. 5   Colonnade (π mode)

1/2 λ. Fig. 6 shows an experimental realization of a colonnade intended for 10 μm radiation. The structure is illuminated from above by a long source with its axis parallel to the beam path axis, the e.m. beam is focussed down to a narrow strip which covers the tops of the cylinders. The beam is polarized with its electric vector along the system axis. All cylinders oscillate in phase, and the distance between the rows is adjusted to prevent any net radiation perpendicular to the system axis. The system may be regarded as an antenna array and it is easy to see that it supports waves that travel in both directions along the axis[5]. The absence of radiation in the transverse plane depends upon the mutual cancellation of the elementary waves from each element and requires high dimensional accuracy. Mialignments from the design cause radiation which shows up as a reduction of the Q of the system. Such a depression has nothing to do with the dissipative losses in the cylinder.

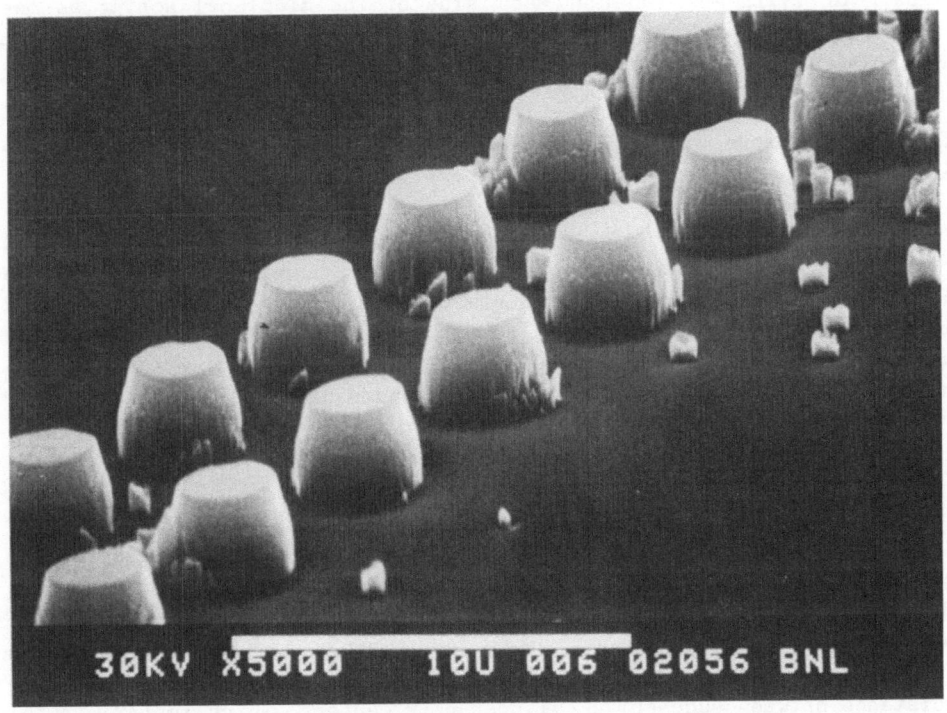

Fig. 6   Experimental Realization of Colonnade
for γ = 10 μm

(Courtesy J. Warren, BNL)

The cancellation of transverse waves and the presence of longitudinal waves suggest strong coupling between the cylinders. Local e.m. energy will be redistributed along the length of the structure with group velocity as in conventional linacs. This is wasteful since the beam is no more than a few mm long along the axis. Local energy concentrations become possible if the group velocity is sufficiently small. A convenient measure is that the power loss due to radiation from a cell (formed by 4 adjacent half cylinders) should be small compared to the dissipation in that cell.

Let me assume that a sufficiently long, if need be infinitely long, section of the colonnade is uniformly illuminated. The incident beam wave fronts will continue to travel towards the base plate after they have arrived at the tops of the cylinders. They will be reflected there and return forming a standing wave with planes of nodes and maxima that are parallel to the surface of the base plate and separated from it by multiple $1/4$ Λ's, with Λ the local wavelength, as modified from the free space λ by the presence of the cylinders. The system axis is defined by the intersection of the midplane with the first magnetic nodal plane.

The magnetic dipole field is zero in the vicinity of this axis while the amplitude of longitudinal electric field is maximum. The fields are periodic in time with the frequency of the incident radiation and periodic in space with the periodicity of the structure. The field distribution in the vicinity of the axis is similar to the one in the foxhole structure, each pair of cylinders (one on each side of the midplane) acting as one of the slot sections, longitudinal space between pairs as a foxhole. It is suitable for the acceleration of particles.

Neither the foxhole structure nor the colonnade depend for their operation on a resonance with the power source, but resonating them is essential for energy efficiency. Whenever there is a mismatch between the impedances of source and load, e.g., through lack of resonance, reflection occur and the energy in the rejected beam is wasted if it cannot be recovered. The colonnate can be tuned by adjustment of the length or height of the cylinders and the impedance it presents to the load can be made either low (cylinder height = $1/3\,\Lambda$) or high (cylinder height = $3/4\,\Lambda$). The actual value of the impedance is determined by the physical characteristics, among them the rate of energy loss to dissipation and radiation.

3.5. COLONNADE (π MODE)

A π mode colonnade is much more attractive than a 2π mode one for the same reasons that a π mode foxhole structure is to be preferred above a 2π mode one: a potential gain in effective accelerating gradient of close to two without serious loss of energy efficiency. It is thus well worth pursuing. It differs geometrically from the 2π colonnade in the longitudinal distance between successive cylinder pair centers, which is $1/2\,\lambda$ rather than λ. The electrical difference is that the accelerating field along the axis alternates between successive pairs. Even a single row colonnade will not radiate transversely when operated in the π mode, thus the transverse geometry of a two row colonnade is less critical in this respect for the π mode than it is for the 2π mode. The equivalent of an infinitely long mode single row colonnade with circular cylinders has been measured and the existence of an accelerating mode has been demonstrates. For this simulation a metallic circular cylinder was placed between two parallel metallic mirror sheets with its axis in the midplane between the mirrors. Field measurements were made as function of frequency, length and diameter of the cylinder. Lack of time has so far prevented similar measurements frombeing made for the 2π mode. These would require a more complex and less flexible model: identical half cylinders have to be attached to each mirror, properly located and oriented, facing each other.

The π mode structure requires that spacially alternating fields in the vicinity of the axis be obtained from a single power source. Palmer predicted that nearly any superperiodic perturbation with a super period of two periods, i.e., treating neighbours differently but next neighbours identically, will drive the π mode. Computer simulations of π mode structures of point dipoles have substantiated this[6]. However, the coupling factors must be sufficiently large to produce, from the same power source, and in a real structure, a mean accelerating gradient that is substantially larger than what can be achieved with the simple 2π mode. This is still a challenge.

References

1. B.Richter, Requirements for the very high energy accelerators, in "Laser Acceleration of Particles", Malibu, CA, 1985, ed. by C.Joshi et al., AIP Conference Proceedings, No. 130.
2. B.W.Montague, Multibunch non-disruptive focusing in linear colliders, CERN, Clic Note 11 (1986).
3. T.Himel and J.Siegrist, Quantum effects in linear collider scaling laws, in "Laser Acceleration of Particles", Malibu, CA, 1985, Ed. by C.Joshi et al., AIP Conference Proceedings, No. 130.
   K.Yokoya, Quantum correction to beam-strahlung due to the finite number of photons, KEK Preprint 85-53 (1985).
4. R.B.Palmer, Report of near field group;
   R.B.Palmer and S.Giordano, Preliminary results on open accelerating structures, in "Laser Acceleration of Particles", Malibu, CA, 1985, ed. by C.Joshi et al., AIP Conference Proceedings, No. 130.
5. J.A.Stratton, Electromagnetic Theory, Chapter 8.7-8.10, McGraw-Hill Book Company, Inc., New York, London, 1941.
6. R.C.Fernow, Brookhaven National Laboratory, private communication.

A TWO-STAGE RF LINEAR COLLIDER USING A SUPERCONDUCTING DRIVE LINAC

W. Schnell
CERN
Geneva, Switzerland

1. Introduction

The classical RF-driven, room-temperature, travelling-wave linac is capable of generating accelerating gradients approaching or exceeding 100 MV/m. If such a structure is to be used in a linear collider of the order of 2 × 1 TeV an economic source of peak power has to be found and the average power efficiency from mains input to beam power must be made acceptable. Clearly, to drive a structure with shunt impedance per unit length R' to an average accelerating gradient E_0 the peak power $\hat{P}_L$ per section length L must exceed E_0^2/R'. As this power is enormous the electron linac is always pulsed with a very small duty cycle. On the other hand, as the quality factor Q is much too small to conserve an appreciable fraction of stored energy over a realistic repetition period it is often assumed that the average RF input to one linac is roughly given by

$$\frac{\langle P_b \rangle}{\eta} \tag{1}$$

where $\langle P_b \rangle$ is the average beam power and η the fraction of stored energy extracted by a beam pulse. This fraction cannot be very large, certainly not above 10% if the beam's energy spread - about $\eta/2$ - should remain acceptable or at least correctible.

Dissipation during the fill time will make the efficiency from RF input to beam power worse than η. However, by making the structure fill time very short, the additional efficiency factor can be kept close to unity at the price of a further increase of peak power. This will be described in Section 2. Moreover, it might be possible to push the RF to beam efficiency substantially above 10% at acceptable energy spread by dividing the beam pulse into a train of closely spaced bunches and restoring the energy extracted by each bunch during the bunch interval[1]. This is discussed in Section 3.

A serious problem remaining is the economic generation of peak RF power of the order of two terawatts for a 1+1 TeV collider. The solution proposed and discussed in this paper is to employ a tightly bunched high-energy drive beam running alongside the entire main linac and directly powering it via short, side-coupled transfer cavities. The drive beam receives its energy from superconducting CW linac sections occupying only a fraction of the total length. This drive linac is operated at UHF frequency with an accelerating gradient much below E_0. The drive linac in turn is powered by a limited number of CW klystrons of proven design and the entire scheme may be viewed as one of pulse compression from CW to nanoseconds and harmonic conversion from UHF to microwaves. The power balance of this will be analysed and basic design equations derived in Sections 4 to 10.

The aim of this paper is an overview of basic constraints and effects estimated to be below 10% are generally neglected. The analysis is limited to the fundamental frequencies of the two stages of linear accelerator involved. Higher order wakefields will certainly be serious but are outside the scope of this paper. A 1+1 TeV collider with luminosity of 10^{33} cm^{-2}s^{-1} or more is taken for numerical examples.

2. The main linac

The main accelerating structure is assumed to be composed of classical travelling wave linac sections of length L, group velocity v and shunt impedance per unit length R' operated at frequency f = $\omega/2\pi$ and average accelerating gradient E_0. Constant group velocity is assumed unless stated otherwise. Other structure constants are the quality factor Q, the "R over Q per unit length" r' defined as

$$r' = \frac{R'}{Q} = \frac{E^2(z)}{\omega W'(z)} \tag{2}$$

where W'(z) is the stored energy per unit length z. A related parameter is the attenuation constant α for energy given by

$$\alpha = \frac{\omega L}{Qv} = \frac{L}{z_0} = \frac{\tau}{\tau_D} \tag{3}$$

where z_0 is the attenuation length, τ_D the decay constant for stored energy in an isolated cell and

$$\tau \;=\; \frac{L}{v} \tag{4}$$

the group delay or "fill time". Note that if a given structure geometry is scaled to different wavelengths

$$Q \propto \omega^{-\frac{1}{2}}, \quad R' \propto \omega^{\frac{1}{2}}, \quad r' \propto \omega \text{ and } \tau \propto \omega^{-\frac{3}{2}}.$$

Each section is powered by a square power pulse of peak power $\hat{P}_L$ and duration τ so that the wave front progressing in the section just reaches its end when the power is switched off. At this moment the beam pulse is made to pass. In this simple model the wave front is assumed to be sharp and the passage of the beam instantaneous, implying $v \ll c$. In reality the group velocity will amount to a fairly large fraction of c. In a forward wave structure this would actually permit reducing the duration of the power pulse by this fraction, in the more likley case of a backward wave the opposite is true and for the present analysis it has been ignored.

With all these assumptions and simplifications one finds the peak power per section length as

$$\boxed{\frac{\hat{P}_L}{L} \;=\; \frac{E_0^2}{g^2 \alpha R'}} \tag{5}$$

where

$$g \;=\; \frac{1 - e^{-\alpha/2}}{\alpha/2} \tag{6}$$

Clearly the total average RF power per linac $\langle P_{RF} \rangle$ equals $\hat{P}_L \tau$ times the total length and repetition rate. With the fraction of stored energy extracted by a charge bNe given by

$$\eta \;=\; \frac{bNe\omega r'}{E_0} \tag{7}$$

and the total beam power by

$$\langle P_b \rangle \;=\; bNeUf_{rev} \tag{8}$$

(eU being the final particle energy and f_{rev} the repetition rate), the average RF power can be written as

$$\langle P_{RF} \rangle = \frac{\langle P_b \rangle}{g^2 \eta} \qquad\qquad (9)$$

The total charge bNe may be contained in a single bunch per pulse or in a (small) number b of closely spaced bunches.

The classical choice is for the minimum of peak power occurring at α = 2.5 and $[g^{-2}\alpha^{-1}]_{min}$ = 1.23. But this implies g^{-2} = 3.1 and, hence, an intolerable wastage of average power. Clearly, since average power is of basic importance here a smaller value of α must be chosen, in spite of the concomitant increase of peak power. For all numerical examples in this paper α = 0.5 (hence g^{-2} = 1.28) will be chosen, implying that the peak power per section length is 2.56 times E_0^2/R'.

In Table 1 three examples of basic parameters are given for 6, 20 and 29 GHz, called cases A, B and C respectively. In all cases the beam power is 5 MW, the top energy (per linac) 1 TeV and the bunch population $N \sim 5.5 \times 10^9$ giving about 10^{33} $sm^{-2}s^{-1}$ luminosity with $\sigma_r^* \sim$ 80 nm beam radius and H $\sim$ 4.5 enhancement at collision. As the energy extraction

TABLE 1

Main linac parameters for three frequencies.   Parameters for one linac.

Case	A	B	C	
Final energy eU	1	1	1	TeV
Frequency f	6	20	29	GHz
Average accelerating gradient E_0	40	40	80	MV/m
Total active length L_{tot}	25	25	12.5	km
Shunt impedance per unit length R'	80	141	170	MΩ/m
Quality factor Q	9500	5030	4150	
R'/Q = r'	8.4	28	41	kΩ/m
Attenuation constant for power α	0.5	0.5	0.5	
Fill time τ	126	20	11.4	ns
Peak power per section length $\hat{P}_L/L$	51.1	29	96	MW/m
Bunch population N	6.3	5.68	5.35 $\times 10^9$	
Energy extraction per pulse η	0.08	0.08	0.08	
Number of bunches per pulse	10	1	1	
Repetition rate f_{rev}	0.496	5.5	5.8	kHz
Average RF power $\langle P_{RF} \rangle$	80	80	80	MW
Beam power $\langle P_b \rangle$	5	5	5	MW
Beam radius at collision σ_r^*	90	78.5	77	nm
Disruption D	2.2	1.3	1.3	
Pinch enhancement H	5.5	4.5	4.5	
Beam-beam radiation loss δ	0.13	0.23	0.21	
Bunch length σ_z	2	1	1	mm
Luminosity	1.06	1.03	1.01$\times 10^{33} cm^{-2}s^{-1}$	

per beam pulse is taken as 8% the average RF input per linac equals 80 MW in all three cases, i.e. the RF to beam efficiency is 6.25% The main question here is whether 4% energy spread is, in fact, correctible before the final focus is reached. It will be noted that at 6 GHz the common effort of ten successive bunches of typical population is required to achieve 8% extraction in spite of the very modest assumed gradient of 40 MV/m.

Final focus and beam emittance are not among the subjects of this analysis and the last six lines of Table 1 are added for illustration only. The beam-beam radiation is still essentially in the classical regime and the classical formula has been used. It is true that σ_z = 1 mm would be a little too long for good energy spread at 29 GHz. Whether a higher luminosity per unit beam power may be achievable by radical reductions of σ_r^* and σ_z (quantum regime) is outside the scope of this paper.

The structure constants assumed could be achieved with disc-loaded structures. However, the low group velocity of such structures (about 0.02 c) would lead to inconveniently short section lengths (especially for the higher frequencies) and to an unacceptable smear out of the wave front. Thus, suitable structures with stronger cell-to-cell coupling should be developed, taking into account mechanical and thermal problems, tolerances and possible manufacturing methods as well as longitudinal and transverse wakefields. For illustration a pure Jungle Gym structure for case C may be considered as a starting point. It would have about 5 mm diameter. At 3 kW/m average dissipation pairs of $\pi/2$ mode loading bars of 0.5 mm diameter (Cu) would reach about 30°C temperature rise.

For the small values of attenuation constant proposed here the use of a constant gradient (graded group velocity) structure changes very little. In equation (5) the factor $g^2\alpha$ in the denominator is replaced by α_0 (the attenuation constant at the input) alone. In equation (9) the factor g^{-2} is replaced by a factor

$$\alpha_0^{-1}\ln(1-\alpha_0)^{-1}\quad.$$

For α_0 = 0.4 the average power factor is 1.28 as above and the peak power factor is 2.50 as compared with 2.56. Constant gradient would be advantageous, however close to the breakdown limit, and would assure that the beam loading η does not change the spatial distribution of field (for as long as r' is constant along the graded structure).

In either case (and with $\eta \sim 0.1$) roughly half of the input energy reappears at the output. Instead of dissipating this electrical energy in a load resistor an attempt should be made to recover it. This may be done by reconverting it to d.c. in a rectifying load. Conversion from microwave power to d.c. has in fact been achieved[2], albeit with a continuous wave, at the level of tens of kilowatts and with about 80% efficiency. A superconducting drive linac offers the possibility of much easier energy recovery, as discussed in Section 10.

3. High efficiency by compensated multibunch operation

In principle, at least, the efficiency of energy transfer to the beam at acceptable energy spread can be increased substantially beyond equation (9) by employing a train of bunches occupying a certain fraction of the fill time τ. The charge of each bunch is limited by the maximum energy extraction η permitted by the concomitant energy spread but the bunch interval τ_b is adjusted so that the fresh influx of RF energy restores the average accelerating field from bunch to bunch[1]. It would appear that this "compensated multibunch" operation is a very promising scheme deserving detailed study. But since one of the main problems, the influence of higher order wakefields, is outside the scope of this paper anyhow the following analysis has been limited to loss-free wave propagation for simplicity. Dissipation is accounted for by applying the factor g^{-2} of equation (6) at the end. The simple model then is the following.

The first of a train of b bunches passes the structure at time $\chi\tau$ when the propagating wave front is at $\chi L < L$. If eN is the charge of this bunch it induces a decelerating field component ΔE given by

$$\frac{\Delta E}{E_0} = \frac{Ne\omega r'}{2E_0} = \frac{\eta}{2}$$ (11)

all over the full section length L. Here E_0 is the RF driven field within χL and η the energy extraction given by equation (7) for b = 1. The particles of this bunch gain an average voltage equal to

$$E_0 L\left(\chi - \frac{\eta}{4}\right)$$ (12)

To give the second bunch the same voltage the bunch interval τ_b must be chosen so that $\tau_b v E_0 = L\Delta E$ and hence

$$\tau_b = \frac{\eta}{2}\tau$$ (13)

The last one of b equidistant and uniformly populated bunches gains

$$E_0\left[\chi L + (b-1)\tau_b v\right] - \Delta E\left[(b - \tfrac{1}{2})L - \sum_{i=1}^{b-1} i\tau_b v\right]$$ (14)

The last term represents the fractions of beam-induced waves that have already left the structure at that time. If τ_b is adjusted according to equation (13), the last bunch experiences an absolute energy error with respect to the first one of

$$LE_0 \; \frac{\eta^2 b(b-1)}{8} \qquad (15)$$

It should be noted that, although η is the actual fraction of energy extracted by one bunch, the concomitant energy spread is $\eta/2\chi$ (for small η) since the RF driven field stretches over χL only while the charge-induced field stretches over the full length L. For this reason an equivalent energy extraction

$$\eta_\Delta \; = \; \frac{\eta}{\chi} \qquad (16)$$

[with η given by equation (7) for b = 1] is introduced as the parameter whose choice, 10% say, is governed by energy spread. Using this and equation (12) the fractional energy error of the last bunch with respect to the first one equals

$$\chi \; \frac{\eta_\Delta^2 b(b-1)}{8\left(1 \, - \, \dfrac{\eta_\Delta}{4}\right)} \qquad (17)$$

While the restriction to equidistant uniformly populated bunches may be unnecessary it appears prudent for the moment to restrict expression (17) to a few percent.

A good choice of filling factor χ may be 0.8 implying that 20% final energy of a given linac is sacrificed in order to gain luminosity for given RF power. Since the last bunch passage should coincide with the end of the fill time τ

$$b \, - \, 1 \; = \; 2 \, \frac{1-\chi}{\chi \eta_\Delta} \qquad (18)$$

With η_Δ = 0.1, χ = 0.8,

$$b = 6$$

and the fractional energy error of (17) becomes 3% which is likely to represent a limit. The same limit can be reached with smaller η (hence smaller energy spread) and a correspondingly larger number of bunches b. The approximate general criterion for this, derived by making expression

(17) equal to 3% for large b, small η_Δ and $\chi = 0.8$ is

$$\eta_\Delta b \sim 0.5 \qquad\qquad (19)$$

This means, however, that a given linac, operated in compensated multi-bunch mode without any other modification, can yield an approximate five-fold increase of beam power and luminosity at the price of a 20% reduction in energy compared with single-bunch operation. The missing energy can be restored by an increase of total active length and total average RF power $\langle P_{RF} \rangle$ by $\chi^{-1} = 1.25$. Thus it may be said that compensated multibunch operation holds the promise of RF to beam efficiencies at or above 25%.

Whenever the accelerating gradient E_0 can be increased at given bunch population the resulting decrease of η_Δ and concomitant energy spread $\eta_\Delta/2$ is now welcome, since efficiency can be maintained by increasing b and decreasing τ_b, so long as the final focus system permits this and $\tau_b f$ can be made an integer.

Table 2 gives two examples of 1 TeV compensated multibunch operation at 29 GHz. The first column represents case C of Table 1 lengthened by

TABLE 2

Case C of Table 1 modified for compensated multibunch operation.
Parameters for one linac

Case	C'	C''	
Final energy eU	1	1	TeV
Frequency f	29	29	GHz
Accelerating field E_0	80	160	MV/m
Filling factor for first bunch χ	0.8	0.8	
Average accelerating gradient χE_0	64	128	MV/m
Total active length L_{tot}	15.6	7.8	km
Peak power per section length $\hat{P}_L/L$	96	386	MW/m
Bunch population N	5.35×10^9	5.35×10^9	
Energy extraction η	0.08	0.04	
Energy spread within bunch $\eta_\Delta/2$	5%	2.5%	
Number of bunches per pulse b	6	11	
Repetition rate f_{rev}	5.8	3.2	kHz
Average RF power $\langle P_{RF} \rangle$	100	100	MW
Beam power $\langle P_b \rangle$	30	30	MW
Structure fill time τ	11.4	11.4	ns
Bunch interval τ_b (not adjusted for integer $\tau_b f$)	0.456	0.228	ns
RF cycles between bunches $\tau_b f$ approx.	13	7	
Beam pulse duration $(b-1)\tau_b$	2.28	2.28	ns
Luminosity	0.6×10^{34}	0.6×10^{34}	$cm^{-1}s^{-1}$

25%, the second one is the same linac at twice E_0. In both cases a luminosity of 0.6×10^{34} cm^{-2}s^{-1} is obtained with 100 MW average RF power per linac at 30% RF to beam efficiency.

Two classes of fundamental problem must, however, be solved before compensated multibunching can be attempted. Firstly, a final focus scheme must be found that can cope with multiple bunch crossings starting at a few centimetres' (3.5 cm in Case C") distance from the main collision point. Secondly, higher-order longitudinal wakefields must be either minimized or their time dependence tuned in such a way as to make the bunch-to-bunch variation of effective accelerating field tolerable. Whether this can be achieved by structure design alone is not certain. Modulations of input power, bunch population or bunch interval τ_b (modulo f^{-1}) might be additional elements of freedom. Contrary to the energy spread within the bunch the residual bunch-to-bunch energy variation is unlikely to be a monotonic function of time and, hence, difficult to correct.

In the following discussion of a two-beam scheme examples B and C of Table 1, in single bunch mode, will be used for numerical examples. However, if compensated multibunching became possible it could certainly be applied to this scheme.

4. A two-beam scheme

A very serious problem with linac parameters such as those shown in Table 1 is the generation of peak power. At least for the lower part of the frequency range considered, it is likely to become technically feasible[3] to power each linac section by an individual d.c. to RF converter, each one containing its own high-voltage input, cathode, gun, RF structure and collector. Above 3 GHz no suitable design is available yet but promising development work is going on[4]. However, in addition to yielding the required performance, such converters must also be producible at extremely low cost, as more than 10,000 units will be typically required for a 1+1 TeV collider.

Alternatively, a continuous drive beam, running all along the main linac and delivering energy to it at regular intervals may be considered. A specific scheme, in which RF energy is extracted from the drive beam by means of free electron lasers and restored by induction linacs has been proposed[5]. Other proposals combine free electron lasers with superconducting RF drive linacs[6] and RF structures for energy transfer with induction linacs[7].

What is analysed below is the combination of a tightly bunched high-energy drive beam directly furnishing energy to RF cavities and a superconducting RF drive linac.

Efficiency is of paramount importance for the drive chain. Therefore, and since the drive beam pulse cannot extract more than a fraction
of the drive linac's stored energy at tolerable energy spread the stored
energy must be conserved from pulse to pulse. It follows at once that
the drive linac has to be superconducting. This, however, is quite
acceptable provided the drive linac's gradient and operating frequency
can be made sufficiently low. Thus, the following concept presents itself for analysis.

The mains input power is converted to RF power at UHF frequency by
means of large CW klystrons and distributed via low-cost sheet metal
waveguides at atmospheric pressure. Klystrons of well over 1 MW CW output at efficiencies approaching 70% are available to-day[8]. A further
extension of power per klystron seems possible, so that the total number
of converters would hardly exceed 100 for a 1+1 TeV collider. The klystrons deliver power to a series of superconducting cavities very similar
to those developed for circular e^+e^- colliders at CERN[9] and elsewhere. Drive beam pulses of a duration equal to the main linac fill
time τ have their energy periodically restored by passing through this
superconducting drive linac. The drive beam is bunched at the UHF drive
frequency f_1 but the bunches are made so short as to interact directly
with travelling-wave transfer structures of main linac frequency f, each
transfer structure being coupled to the input of a main linac section.

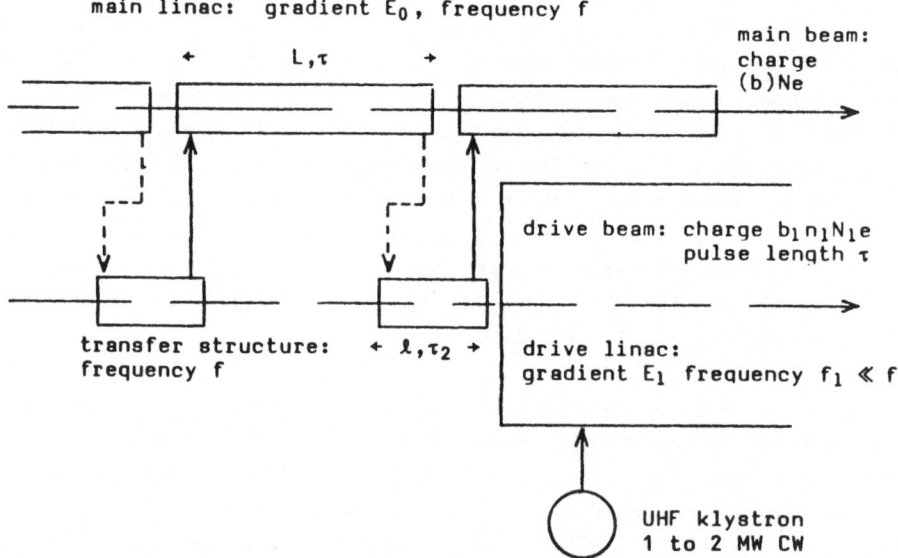

Fig. 1. Two-stage linear accelerator composed of a superconducting CW
drive linac at UHF frequency and a microwave main linac. Typi-
cal parameters might be:
Main linac: 1 cm wavelength, 80 MV/m, 1 TeV final energy
Drive linac: 1 m wavelength, 6 MV/m, 15 GV voltage gain
Drive beam: 3 GeV, 4×10^{11} per bunch, 1 mm bunch length

Figure 1 shows the arrangement.   Since the main linac section length will be under 1 m and the UHF drive linac cannot be sliced in such short sections, transfer structures and drive cavities cannot be inter-laced.   Thus, the drive linac length must be added to the total length unless two drive beams, each equipped with half the drive linac sections, are made to run along either side of the main linac powering it alter-nately from the left and from the right.

5.   The drive linac

Energy conservation along the drive beam demands that

$$\hat{P}_L \tau \;\; = \;\; \eta_2 n_1 b_1 N_1 ep E_1 mL \;\; .$$ (20)

Controlling the drive beam's energy spread and the mismatch at the drive linac's RF input means that

$$\eta_1 \;\; = \;\; \frac{p n_1 b_1 N_1 e \omega_1 r_1'}{E_1}$$ (21)

(the fraction of energy extracted from the drive linac) cannot be too large.   Here η_2 is the efficiency of energy transfer by the drive beam, $n_1 b_1 N_1 e$ is the total charge of the drive beam pulse (distributed over n_1 drive linac cycles and b_1 bunches per cycle), m is the fraction of active main linac length occupied by drive linac sections and ω_1, r_1' and E_1 are the angular frequency, R over Q per unit length and accelerating gradient respectively of the drive linac.   The factor p ($\leqslant$ 1) expresses the fact that the drive bunches may not ride on the crest of the drive-linac wave.   Together with equations (3) and (5) the two equations mean that

$$\left(\frac{E_1}{E_0} \right)^2 = \frac{1}{\eta_2 \eta_1 m q^2} \frac{\omega_1 r_1'}{\omega \; r'}$$ (22)

Since, obviously, E_1 should be much smaller than E_0 although η_2, η_1, m, g^2 are all smaller than unity it follows at once that ω/ω_1 must be made very large, as large as is technically possible.   At least the right-hand side of equation (22) is proportional to the square of the frequency ratio since $r_1' \propto \omega_1$ and $r' \propto \omega$ for given cell geometry.   In-troducing the superconducting cavities' quality factor Q_1 one finds

$$\langle P_1 \rangle \;\; = \;\; \frac{E_0}{\omega r' g^2} \frac{\omega_1}{Q_1 \eta_1 \eta_2} U$$ (23)

for the total drive linac dissipation, eU being the total main linac energy.   Note that the dissipation is independent of the drive linac's length so long as the limit for deterioration of Q_1 with increasing field E_1 is not reached. Note also that the dissipation does not depend on r_1'. Dividing by the overall cryogenic efficiency η_{cr} one obtains the cryogenic input power P_{cr} per drive linac.

The practical lower limit and approximate economic optimum of f_1 is likely to lie in the lower UHF range, just as for circular collider RF structures. An obviously interesting choice at CERN is 350 MHz. Inserting typical parameters one finds at once that the main linac frequency has to be near the upper end of the range considered so far - i.e. case B or C of Table 1 - to make the scheme viable. Table 3 shows two sets of drive linac parameters for 20 and 29 GHz main linac frequency, $\eta_{cr}^{-1} = 500$ and 1 TeV total energy per main linac. The actual drive linac is the same in both cases; $L_1 = 2.5$ km is its total length and $U_1 = 15$ GV its total voltage gain.

TABLE 3
Drive linac parameters for two main linac frequencies
Parameters for one linac

Case	B	C	
Main linac energy eU	1	1	TeV
Main linac frequency f	20	29	GHz
Main linac accelerating gradient E_0	40	80	MV/m
Main linac R over Q parameter r'	28	41	kΩ/m
Main linac fill time τ	20	11.4	ns
Fraction of main linac active length occupied by drive linac m	0.1	0.2	
Drive linac active length L_1	2.5	2.5	km
Drive linac frequency f_1	350	350	MHz
Number of drive bunch trains n_1	7	4	
Transfer efficiency assumed η_2	0.9	0.9	
Drive linac energy extraction η_1	0.1	0.1	
Drive linac R_1 over Q_1 parameter r_1'	270	270	Ω/m
Drive linac accelerating field E_1	6.2	6.0	MV/m
Drive linac total voltage gain U_1	15.5	15.0	GV
Drive linac quality factor Q_1	5×10^9	5×10^9	
Cryogenic efficiency assumed η_{cr}	2×10^{-3}	2×10^{-3}	
Total cryogenic input power $\langle P_1 \rangle / \eta_{cr}$	35.6	33.5	MW

The optimum choice of m (with twin drive beams as mentioned at the end of Section 4) will be that which equalizes the costs of the drive linac and of the main linac including its tunnel. The shortest overall length would result if both linacs were given equal lengths. With the values for structure constants and E_1 of Table 3, Case C, this would result in 5.6 km active length at $E_0 = 178$ MV/m. It seems unlikely that this extreme solution will coincide with the economic optimum. Nevertheless, main linac gradients E_0 well above 100 MV/m may become possible due to progress in superconducting RF technology permitting higher values of E_1 and Q_0.

6. The drive beam

The total required charge per drive beam pulse is given by equations (20) and (5) as

$$n_1 b_1 N_1 e \ = \ \frac{E_0^2}{\omega r' g^2 \eta_2 m p E_1} \ = \ \frac{E_0}{\omega r' g^2} \ \frac{U}{\eta_2 p U_1} \qquad (24)$$

Even with the drive bunches coinciding with the crest of the drive wave ($p = 1$) the total charge amounts to over $6 \times 10^{12} e$ for both cases of Table 3. It can, at least, be divided into

$$n_1 \ = \ \tau f_1 \qquad (25)$$

bunches but, as n_1 is small (7 and 4 in cases B and C respectively), the population of such bunches would still be very large whilst their lengths σ_{1z} must be of the order of 1 mm or less in order to interact with the transfer structure at frequency f. A further subdivision into trains of b_1 bunches per drive cycle will be discussed in Section 8 together with the necessity of $p < 1$.

Assuming that the drive linac sections are distributed along the main linac the mean energy of the linac (its injection and dump energy) can be chosen rather freely. A minimum condition is that energy spread and transverse emittance do not make the drive bunches drift apart. The first is prevented if

$$\langle \gamma_1^2 \rangle \ \geqslant \ \frac{L_{tot}}{2\sigma_{1z}} \qquad , \qquad (26)$$

the second if

$$\langle \gamma_1 \rangle \ \geqslant \ \frac{\varepsilon_{1n} L_{tot}}{4\sigma_{1z}} \left(\beta_{max}^{-1} + \beta_{min}^{-1} \right) \qquad (27)$$

obtained by integrating over the mean square of angular deviation for a beam with invariant emittance ε_{1n} in a FODO structure with β_{max} and β_{min} amplitude function. For $\varepsilon_{1n} = 10^{-3}$ m, $L_{tot} = 25$ km, $\beta_{max} = 7.5$ m, $\beta_{min} = 2.5$ m and $\sigma_{1z} = 1$ mm both conditions are satisfied above 1.8 GeV. If the corresponding maximum beam radius $\sigma_r = 1.5$ mm should be too large for the aperture of the transfer structure a higher injection energy combined with lower values of β must be chosen. In principle a large fraction of the injection energy can be regained by deceleration in the opposite injection linac (at the price of transporting the beam past the opposite drive and transfer sections). If, instead, the drive beam is dumped the overall RF drive power is increased by a factor $m_0 c^2 \langle \gamma_1 \rangle / e U_1$ (~ 20% for 3 GeV dump energy).

7.   The transfer structure

The transfer structures through which the drive beam is threaded and to which it delivers energy are assumed to be short sections of travelling wave structure - of length ℓ, group velocity v_2, quality factor Q_2 and R over Q per unit length r_2' - coupled to the inputs of the main linac section by short lengths of waveguide (Fig. 1). The group delay τ_2 is chosen so that

$$\tau_2 \;=\; \frac{\ell}{v_2} \;=\; \frac{\tau}{n_1} \;=\; \frac{1}{f_1} \tag{28}$$

The n_1 drive bunches (or bunch trains) of $b_1 N_1$ charge each succeed each other at that interval. (The subdivision into a train of b_1 smaller bunches per drive cycle will only be considered in the next section.) Therefore the transfer structure empties itself during the bunch intervals, each bunch finds an (essentially) empty structure and induces a wave train of duration τ_2. The power flow is made equal to $\hat{P}_L$ by proper choice of r_2'. The condition for this is

$$(b_1 N_1 e)^2 \omega f_1 \ell r_2' \;=\; (b_1 N_1 e)^2 \omega v_2 r_2' \;=\; 4\hat{P}_L \tag{29}$$

which is equivalent to

$$(b_1 N_1 e \omega)^2 \; r_2' \frac{v_2}{v} \;=\; 4 \frac{E_0^2}{q^2 r'} \tag{30}$$

This follows from the fact that each bunch (or bunch train) leaves behind an induced field

$$E_i \;=\; \frac{b_1 N_1 e \omega r_2'}{2} \tag{31}$$

and experiences half that field itself.

If $r_2' = r'$ and $v_2 = v$, i.e. if the transfer structure had the same characteristics as the main linac, equation (29) or (30) would imply

$$\left(b_1 N_1 \right)_{min} \;=\; \frac{2}{\eta g} N \tag{32}$$

80

where Ne is the charge that extracts a fraction η of the main linac energy. In practice the limitation of the drive linac voltage imposes a larger value of b_1N_1 via equation (24) so that $v_2r_2' \ll vr'$ by a large margin. This is fortunate here since it permits designing the transfer structure for very low r_2' and hence with a large aperture, relative to λ, for a fat drive beam. Another condition for the transfer structure is that the beam loading enhancement factor should be close to unity, i.e. that only a small fraction of the drive beam's energy loss should be due to higher-mode wakefields. Improvements in superconducting RF technology may permit an increase in E_1 and hence a decrease in drive beam charge but it seems unlikely that the transfer structure will become a limitation.

The attenuation constant of the transfer structure is given by $\omega\tau_2/Q_2$ and, hence, equals

$$\alpha_2 = \alpha \frac{Q}{Q_2 n_1} \tag{33}$$

Thus, provided $Q_2 \sim Q$, losses in the transfer structure will be quite small. They are assumed to be globally included in the transfer efficiency η_2.

8. <u>Drive beam energy spread and bunch charge</u>

Using a single drive bunch per drive RF cycle has two serious disadvantages. Firstly the charge per bunch becomes very large. Secondly, as the head of the bunch sees zero field and the tail twice the average, the energy spread becomes very large.

Both problems can be alleviated by distributing a number b_1 of drive bunches (not necessarily all of the same charge) over the rising slope of the drive wave so as to match the build-up of decelerating voltage in the transfer structure to the sinusoidal rise of drive linac field (Fig. 2a). The former equals

$$\frac{e\omega r_2' \ell}{2} \left[\frac{N_k}{2} + \sum_{i=1}^{k-1} N_i \left(1 - (k-i) \frac{f_1}{f} \right) \right] \tag{34}$$

for the kth out of b_1 bunches spaced in time by f^{-1}. The charge of the ith bunch is N_i and losses in the transfer structure are neglected. The second term under the sum accounts for the fact that a fraction f_1/f of the wave induced by each bunch leaves the structure during the bunch interval f^{-1}. By a suitable modulation of the bunch charges N_i (namely a gradual reduction towards the end of the bunch train)

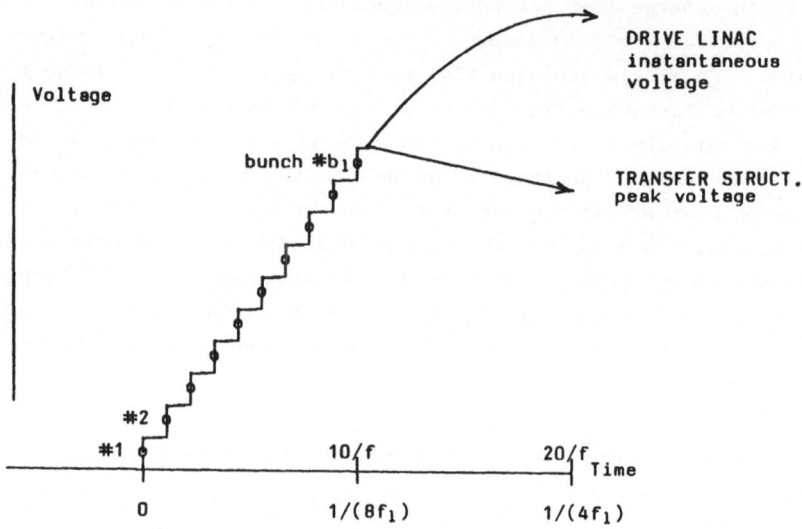

Fig. 2a.　Matching of transfer voltage to drive voltage, Case C.

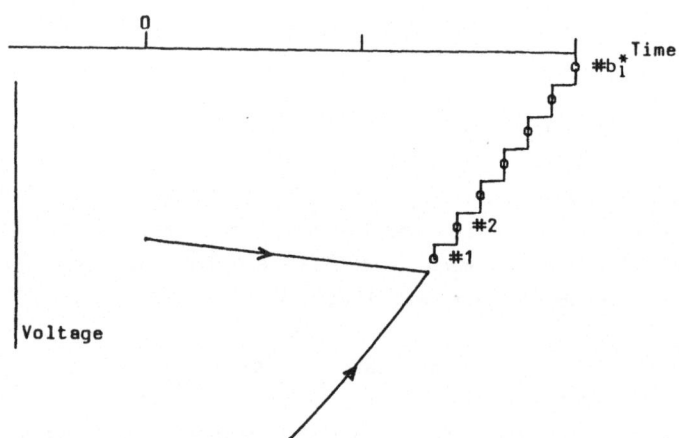

Fig. 2b.　Energy recovery via the transfer structure.

expression (34) can be matched to a quarter sine wave from bottom to top.

The match is, however, not exceedingly critical.　If ε is the maximum fractional mismatch the drive linac energy has to be increased by that fraction and the overall efficiency (including cryopower) decreases by the same fraction.　If, for simplicity, bunches of equal charge N_1 are spread over one eighth of the drive cycle $b_1 = f/8f_1$ and $\varepsilon \sim 4\%$ since more than half of the deviation of sin 45° from linearity is compensated by the $\sum i$ term of expression (34).

Since the drive bunches now ride on the slope of the drive wave the charge eb_1N_1 in equations (20), (21) and (24) (taken with $p = 1$) must be replaced by

$$\sum_{i=1}^{b_1} N_i \sin\left[(2i-1)\,\frac{\pi\omega_1}{\omega}\right] \tag{35}$$

For uniform bunch population N_1 and $b_1 = f/8f_1$ as proposed above this amounts to approximately

$$N_1 b_1 \frac{\int_0^{45°} \sin\phi\,d\phi}{\pi/4} = 0.373\ N_1 b_1 \tag{36}$$

Thus $p = 0.373$ and equation (24) gives the total drive charge as

$$n_1 b_1 N_1 e = 2.68\,\frac{E_0^2}{\omega r'g^2 \eta_2 mE_1} = 2.68\,\frac{E_0^2}{\omega r'g^2}\,\frac{U}{\eta_2 U_1} \tag{$\cdot$37}$$

Making use of $b_1 = f/8f_1$ and $n_1 = \tau f_1$ one finds

$$\boxed{eN_1 = 21.5\,\frac{E_0}{\tau f \omega r'g^2}\,\frac{U}{\eta_2 U_1}} \tag{38}$$

for the bunch charge. A small reduction of the numerical factor can be expected from a modulation of bunch charges as mentioned above. Note that $N_1 \propto E_0\,\omega^{-\frac{3}{2}}$ and that the only parameter of the drive linac entering is its total voltage gain U_1.

The drive bunch populations resulting from equation (38), namely,

3.5×10^{11} in case B
4.1×10^{11} in case C

(for $\eta_2 = 0.9$), are very high but do not seem obviously prohibitive for a beam whose transverse emittance can be rather large.

It should be noted that the peak field and dissipation of the drive linac - given by equations (22) and (23) - remain unaffected by the compensation of energy spread.

In reality the heads of all but the first drive bunch trains find a residual voltage in the transfer structure. Even if the travelling waves had perfectly sharp ends the residual voltage would amount to 1/16 with the scheme adopted above and dispersion will make it worse. This can be compensated by suppressing the first one or two of the b_1 bunches in all but the first of the n_1 bunch trains. An additional adjustment can be obtained by shortening τ_2 a little below f_1, at the price of a small modulation of power flow to the main linac.

What remains to be taken into account is the 10% drop of drive linac stored energy (with $\eta_1 = 0.1$) and the phase shift of the drive wave caused by the off-peak passage of the drive bunches. The former can be matched by a 5% decrease of charge from the first to the last drive bunch train (accepting a 5% droop of main linac field during τ). The latter amounts to $1.34\eta_1\cos(\pi/8)$ radian or $7.1°$ per drive pulse. The klystron drive should be stepped by that amount at every pulse to avoid reflection.

9. Intermediate energy storage

Some form of pulse compression might be envisaged in order to increase the peak power and hence the accelerating gradient. However, the only hope of achieving the necessary intermediate energy storage at tolerable losses rests with storage devices in TE-mode configuration or with making them superconducting. For instance the following scheme of two-fold compression might be considered.

In addition to the drive linac cavities and transfer structures the drive beam is threaded through passive, self-contained storage structures tuned to 2f. The drive pulse is preceded by a storage pulse of duration τ. The storage bunches are arranged on the rising slope of the drive cycle just as the drive bunches but have half the charge and 1/2f spacing so as to charge the storage structure but not the transfer structure and main linac. The subsequent drive bunches (with spacing 1/f) are shifted by c/4f with respect to the storage pulses. They receive additional energy from the storage structure which they have to empty for good efficiency.

In order to avoid the build-up of a large energy spread the storage device too consists of a short transfer structure emptying itself into the main storage structure in a time f_1^{-1}. The storage structure reflects the wave deposited therein so as to make it reappear in the transfer structure when the drive bunches pass. A biperiodic π-mode transfer structure is required for two-way operation at high group velocity. Compared with other schemes for pulse compression this one has the advantage of not interfering with the main linac "plumbing". The gain is the equivalent of a twofold increase in E_1. It may well turn out, however, that the losses of even an over-moded TE configuration at room temperature are excessive and that the aperture of a structure at 2f cannot be made large enough.

10. Energy recuperation

In a main linac structure designed for small loss during the fill time a large fraction of the input energy reappears at the output. In order to recuperate this energy the output of each linac section may be connected to an input created at the subsequent transfer section as shown in dotted lines in Fig. 1. A recuperation pulse follows the drive pulse and transfers the energy back into the drive linac.

Like the drive pulse the recuperation pulse consists of $n_1 = \tau f_1$ bunch trains but the bunches are situated on the decreasing decelerating slope of the drive cycle and the phasing with respect to the transfer structures is for acceleration (Fig. 2b).

In order to match the reduced field appearing in the transfer structure to the essentially unchanged drive linac field the charge of the recuperation bunches is left unchanged but their number b_1^* per drive cycle is reduced in proportion to the field attenuation (e.g. from $b_1 = 10$ to $b_1^* = 7$ in case C for 50% leftover power).

The forward direction of the drive beam determines the direction of phase and group velocities in the transfer structure. This should limit any leakage of drive energy back into the preceding linac section to a small fraction. In addition directional couplers may be used.

For single bunch operation with $\alpha = 0.5$, $\eta \leqslant 0.1$ this recuperation scheme permits, in principle, a factor two increase in repetition rate, beam power and luminosity per given average klystron power, at very little extra expense and complication.

11. Summary and conclusions

At the price of increased peak power, the RF to beam efficiency of a normal conducting linac can be raised above 5% by making the fill time very short. Compensated multibunch operation holds the promise of up to 30% efficiency but higher-order wakefield problems have to be solved and a suitable final focus must be found.

The worst remaining problem seems to be the economic and efficient generation of peak RF power. The scheme proposed here consists of a limited number of CW UHF klystrons, a superconducting UHF drive linac (occupying only a fraction of the total length) and a tightly bunched drive beam of several GeV average energy, transferring energy from the superconducting linac to the main linac via short sections of transfer structures. This scheme appears to have the following advantages:

- Power is converted from d.c. to RF by a limited number of CW klystrons of proven design and high efficiency.

- The subsequent transport of energy and its conversion to nanosecond
  pulses at microwave frequency can be carried out at an efficiency
  that might approach 90% except for the cryogenic power of the superconducting linac which - at the present state of the art - amounts
  to about 35 MW for a 1 TeV main linac.

- The repetition rate - the only parameter left free to adjust beam
  power in a given design - is determined by the drive beam and main
  beam injectors alone.

- Proper phasing of several tens of thousands of main linac sections
  is automatically assured by the highly relativistic drive beam.
  Precise adjustments of timing and populations of the drive bunches
  are required for optimum efficiency but these adjustments are all
  carried out at the drive beam gun.

- Most of the electromagnetic stored energy unavoidably left in the
  main linac after the passage of the beam can be recuperated and
  stored in the superconducting drive linac.

The main problem with this scheme is the generation and acceleration
to a few GeV of the very dense and intense drive bunches required (a few
times 10^{11} electrons in about 1 mm bunch length). Finding a transfer
structure with large aperture to wavelength ratio, good efficiency of
energy extraction at the fundamental frequency and high group velocity
may be another problem as well as the coupling of this structure to the
main linac. General problems affecting main linac structures at 1 cm
wavelength and longitudinal are transverse wakefields and tolerances.

Present-day performance of superconducting UHF cavities (about
6 MeV/m accelerating field and a few times 10^9 Q-factor at 350 MHz) is
already sufficient to make the scheme viable but limits the main linac
gradient to about 100 MV/m and makes cryo-power an appreciable contribution to overall dissipation. Any progress in the development of superconducting cavities will, therefore, permit a further improvement of main
linac performance.

References

1. P.B.Wilson, Linear Accelerator for TeV Colliders, SLAC-PUB-3674
   (1985).

2. W.C.Brown, Electronic and Mechanical Improvement for the Receiving Terminal of a Free-Space Microwave Power Transmission System, NASA CR-135194 (1977).

3. D.Prosnitz, Millimeter High Power Sources for High Gradient Accelerators, IEEE Trans. Nucl. Sci. NS-30:2754 (1983).

4. V.L.Granatstein et al., Design of Gyrotron Amplifiers for Driving 1 TeV e^-e^+ Linear Colliders, IEEE Trans. Nucl. Sci. NS-32:2957 (1985).

5. D.B.Hopkins and R.W.Kuenning, The Two-Beam Accelerators: Structure Studies and 35 GHz Experiments, IEEE Trans. Nucl. Sci. NS-32:3476 (1985).

6. U.Amaldi and C.Pellegrini, private communication (1986).

7. R.Marks, A relativistic Klystron, LBL-20918/UC-34A (1985).

8. H.Frischholz, Generation and Distribution of Radio-Frequency Power in LEP, IEEE Trans. Nucl. Sci. NS-32:2791 (1985).

9. G.Arnolds-Mayer et al., A Superconducting 352 MHz Prototype Cavity for LEP, IEEE Trans. Nucl. Sci. NS-32:3587 (1985).

THE MICRO LASERTRON[*]

An Efficient Switched-Power Source of mm Wavelength Radiation

R. B. Palmer[†]

Stanford Linear Accelerator Center
Stanford University, Stanford, California 94305

ABSTRACT

An extension of W. Willis' "Switched Power Linac" is studied. Pulsed laser light falls on a photocathode wire, or wires, within a simple resonant structure. The resulting pulsed electron current between the wire and the structure wall drives the resonant field, and rf energy is extracted in the mm to cm wavelength range. Various geometries are presented, including one consisting of a simple array of parallel wires over a plane conductor. Results from a one dimensional simulation are presented.

1. INTRODUCTION

The genesis of the idea reported here is the proposal by Bill Willis[1] for a "Switched Power Linac" (see Fig. 1). In this idea a single burst of electron current is switched by a single pulse of laser light. That burst as it crosses a circumferential gap generates a pulse of electromagnetic radiation which is focussed by the cylindrical geometry and used to accelerate particles on the axis.

In the present proposal, the same concept is used, except that a train of light pulses is employed, and these are used to excite fields in a resonant cavity (see Fig. 2). The resulting fields can again be used to accelerate a particle beam, but a conventional accelerating cavity would be used. The switched power device has become an rf power source, and shares the basic concept of a "Lasertron"[2] (see Fig. 3). However, the idea differs significantly from the Lasertron, and since it could be made to generate shorter wavelengths, I have chosen to refer to it as a "micro-lasertron".

---

[*] Work supported by the Department of Energy, contract DE-AC03-76SF00515.
[†] On leave from Brookhaven National Laboratory.

2. DESCRIPTION

The basic concept is represented (Fig. 2) by a long vacuum-filled rectangular cavity with a wire passing down the center. The wire is held at a high potential relative to the enclosure. By illuminating a photocathode on one side of the wire, electron bursts are allowed to pass from the wire to one wall of the box. Repeated pulsing of the photocathode at the cavity resonant frequency results in the excitation of a transverse electric mode within the cavity. Energy can then be extracted by coupling the cavity to a waveguide.

This simple concept may be compared to that of the lasertron (Fig. 3) in which the pulsed photocathode is used to generate a bunched accelerated beam. The energy is then extracted as the beam is decelerated passing one or more ring cavities. Since the beam's kinetic energy will be lost when it hits the anode, high efficiency is only obtained if the beam is decelerated to as near rest as possible.

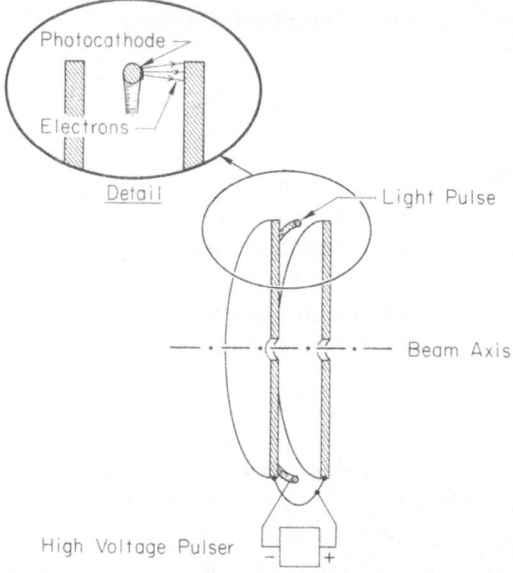

Fig. 1. Switched power accelerator concept as proposed by W. Willis (Ref. 1).

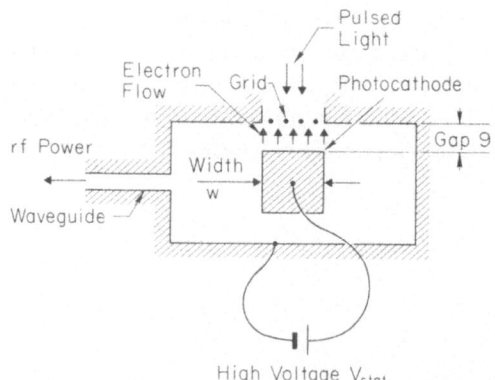

Fig. 2. Micro-lasertron concept as discussed here.

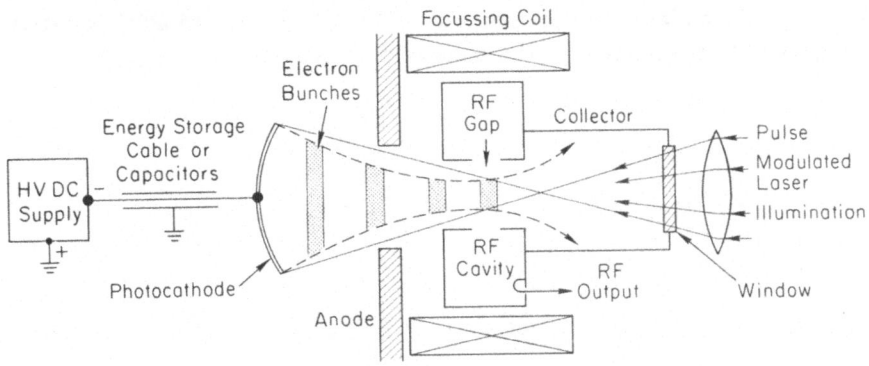

Fig. 3. The conventional Lasertron concept.

In the micro-lasertron, the acceleration and deceleration occur within the same gap. High efficiency is again obtained when the electrons are brought nearly to rest as they arrive at the anode. But the processes of acceleration and subsequent deceleration are controlled by the phase of the field rather than the position along the beam.

The grid indicated in Fig. 2 may not be practical in a short wavelength realization of the idea. One finds, however, that the grid can be omitted so long as a relatively deep slot is enclosed between conducting walls (see Fig. 4a). An extension of this idea has an array of parallel wires under a single slotted cover (see Fig. 4b). In this case we find that the slots do not need to be deep. A drive frequency is chosen so that the distance between wires is approximately one half wavelength and the light pulses entering alternate slots we arranged to be 180° out of phase. At all finite angles radiation from the slots cancel. Even in the direction parallel with the plane the fields do not sum because the loading from the wires causes the phase velocity in the cavity to be less than the speed of light. Fields leaking through the slots will also have a phase velocity less than the speed of light and will thus be evanescent, falling exponentially from the slotted wall.

The above observation leads to a further extension of the idea (see Fig. 4c). Now the upper cover of the cavity has been removed, the photocathodes placed on the remaining surface and the pulsed light brought down at an angle between the wires. Again, because of the presence of the wires, the phase velocity along the surface is less than the speed of light, and the field excited will be evanescent, falling exponentially from the surface. Figure 5 shows the approximate form of such fields.

As in the single wire case, electromagnetic energy could be taken out of the multiwire cavities by coupling them to waveguides. However, it is possible, and perhaps more convenient, to arrange that they couple to plane parallel waves emanating from the surface. In both the cases of Figs. 4b and 4c, this can be achieved by disturbing the periodicity of the wires and corresponding slots (see Fig. 4d).

The attraction of the micro-lasertron is its simplicity, and the possibility of scaling it down to produce very short wavelengths. It also appears from the following analysis that very high powers and efficiencies may be obtainable.

In the following discussion we will consider an equivalent circuit which will allow us to make approximate calculations of performance. The analysis is strictly applicable only to the plane parallel examples of Figs. 2, 4c and 4d although similar results are probably obtainable for the cases of Figs. 4a and 6.

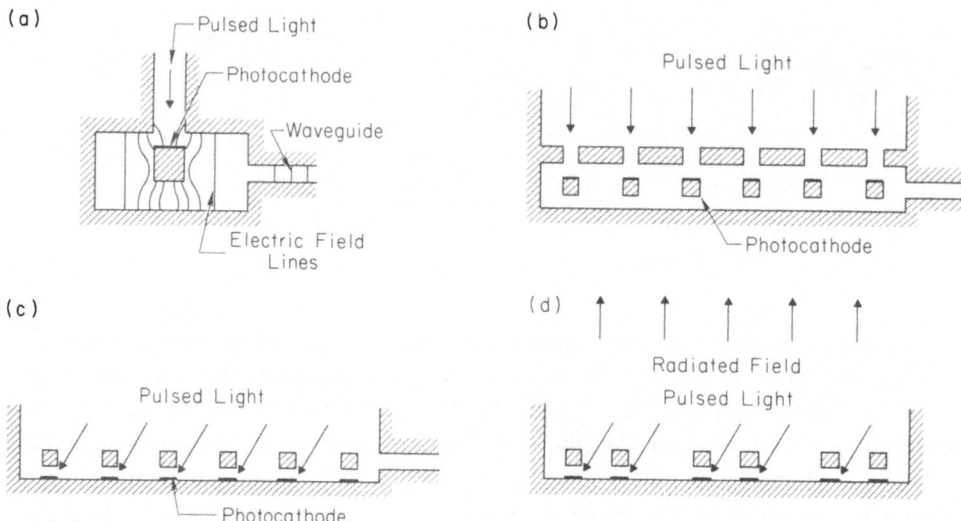

Fig. 4. Realizations of the micro-lasertron: (a) single cavity; (b) multi cavity; (c) grating cavity; (d) modification to grating cavity to couple to an outgoing wave.

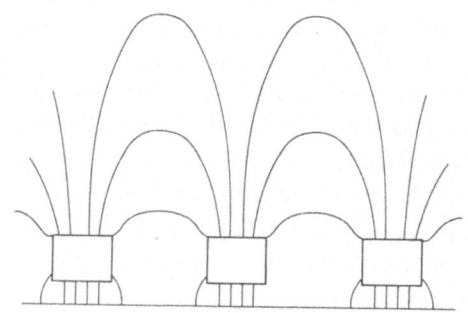

Fig. 5. Approximate field pattern about grating cavity.

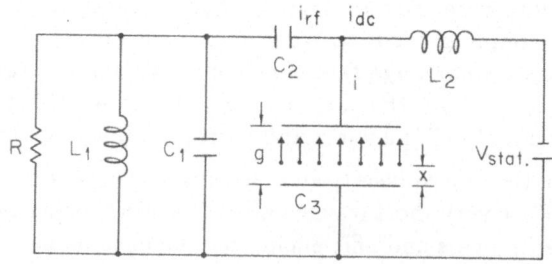

Fig. 6. Equivalent circuit of micro-lasertron.

3. ANALYSIS

3.1 Equivalent Circuit

Under the conditions where

$$g \text{ (gap)} \ll w \text{ (gap width)}$$
$$v \text{ (electron velocity)} \ll c \text{ (velocity of light)}$$
$$w \text{ (gap width)} \ll \lambda \text{ (wave length)}$$

then the device of Fig. 2 may be thought of as separate diode gap coupled to a resonant circuit (see Fig. 6).

For our purposes we can think of L_2 as an infinite inductance that allows a D.C. current i_{DC} to flow from the high voltage supply V_{stat}. C_2 may be thought of as an infinite capacitor that isolates the D.C. from the resonant circuit.

The current i may be divided:

$$i = i_{res} + i_v$$

where i_{res} is a resonant oscillatory current associated with the resonant circuit of L_1 and $(C_1 + C_3)$; and i_v which is associated with motion of charges in the gap.

$$i_v = q \, v/g$$

where q is the charge in the gap and v the velocity of that charge across the gap.

With these definitions we have

$$i_{rf} = I_{rf} \sin \omega t + q \, v/g \tag{2}$$

The voltage V across the gap will again have two components:

$$V = V_{rf} \cos \omega t + V_{stat} \tag{3}$$

The energy transferred to the resonant system per cycle will be

$$\Delta E_{res} = \int_o^{\lambda/c} i_v(t) V_{rf} \cos \omega t \, dt \tag{4}$$

The phase θ of the resonance may also be perturbed. In a steady state, the change in phase per cycle will be:

$$\Delta \theta = \frac{2\pi}{Q} \, \tan \chi \tag{5}$$

where

$$\tan \chi = \frac{\int\limits_{0}^{\lambda/c} i_v(t)\sin(\omega t)dt}{\int\limits_{0}^{\lambda/c} i_v(t)\cos(\omega t)dt} \qquad (6)$$

If driven at the resonant frequency ω_0, then clearly χ must be equal to zero. But if the driven frequency is off the resonance, then we have

$$\omega - \omega_0 = \Delta\theta \cdot \frac{\omega}{2\pi} = \frac{\omega}{Q}\tan\chi \qquad (7)$$

3.2 Efficiency

The energy used from the D.C. source per cycle will be

$$\Delta E_{dc} = \int\limits_{0}^{\lambda/c} i_v V_{stat}\, dt = q\, V_{stat} \qquad (8)$$

the difference between this and ΔE_{res}, the energy going into the rf field, is lost as heat in the anode as electrons give with finite kinetic energy. Thus:

$$\Delta E_{dc} - \Delta E_{res} = \frac{q}{2}\frac{m}{e}v_f^2 \qquad (9)$$

where v_f is the velocity with which the electrons hit the anode, m is the electron mass and e its charge. Thus the efficiency ϵ of transferring energy from the D.C. source to the resonant circuit is

$$1 - \epsilon = \left[\left(\frac{mc^2}{e}\right)\cdot\frac{v_f^2}{2\,c^2\,V_{stat}}\right]$$

$$1 - \epsilon = \left[.51\times10^6\,\frac{v_f^2}{2\,c^2\,V_{stat}}\right]\,(\text{mks}) \qquad (10)$$

To get v_f we obtain first the acceleration of a charge in the gap which will be

$$\frac{dv}{dt} = \frac{e}{m}\left[\mathcal{E}_s + \mathcal{E}_{rf}\cos(\omega t + \phi)\right] \qquad (11a)$$

where $\mathcal{E}_s = V_{stat}/g$, $\mathcal{E}_{rf} = V_{rf}/g$ and ϕ is introduced as an arbitrary phase so that we can define $t = 0$ as the time when the charge is initially released from the cathode. Integrating Eq. (11a) we obtain the velocity of the electrons at time t:

$$v(t) = \frac{e}{m}\left[\mathcal{E}_s\,t - \frac{\mathcal{E}_{rf}}{\omega}\sin(\omega t + \phi) + \frac{\mathcal{E}_{rf}}{\omega}\sin\phi\right] \qquad (11b)$$

Integrating again we obtain the distance travelled:

$$x = \int\limits_{0}^{\tau} v(t)dt = \frac{e}{m}\left|\frac{\mathcal{E}_s t^2}{2} - \frac{\mathcal{E}_{rf}}{\omega^2}\cos(\omega t + \phi) + \frac{\mathcal{E}_{rf}}{\omega}\sin\phi\right|_0^\tau \qquad (11c)$$

Setting $x = g$ in Eq. (11c) will give the transit time $\tau = t$ which when substituted into Eq. (11b) gives the final velocity v_f and thus the energy loss and efficiency using Eq. (10).

3.3 Low rf Field Case

If $\mathcal{E}_{rf} \ll \mathcal{E}_{stat}$, then

$$g \approx \frac{e}{2m} \, \mathcal{E}_{stat} \, \tau_0^2$$

and

$$\tau_0 \approx \left(\frac{2m}{e}\right)^{1/2} \left(\frac{g}{\mathcal{E}_{stat}}\right)^{1/2}$$

This may be compared with the cycle time λ/c and we define this ratio as:

$$F_\tau = \frac{\tau_0}{\lambda} \, c = \left(\frac{2mc^2}{e}\right)^{1/2} \left(\frac{1}{V_{stat}}\right)^{1/2} \left(\frac{g}{\lambda}\right) \tag{12a}$$

$$F_\tau = 1.01 \times 10^3 \left(\frac{1}{V_{stat}}\right)^{1/2} \frac{g}{\lambda} \; (\text{mks}) \tag{12b}$$

We will continue to use this definition of F_τ even when $\mathcal{E}_{rf}$ is not less than $\mathcal{E}_{stat}$. F_τ then becomes a useful scaling fact or rather than an actual ratio of transit time to cycle.

3.4 Short Transit Time Approximation

If we assume

$$F_\tau \ll 1 \tag{13}$$

then $\omega\tau \ll 1$ and Eq. (11c) reduces to

$$g \approx \frac{e\tau^2}{2m} \left(\mathcal{E}_s - \mathcal{E}_{rf} \cos\phi\right) \tag{14a}$$

and

$$v_f \approx \frac{e}{m} \, \tau \left(\mathcal{E}_s - \mathcal{E}_{rf} \cos\phi\right) \tag{14b}$$

Maximum efficiency is realized if v_f is minimum, i.e. the least kinetic energy is dumped on the anode. This will occur for $\phi = 0$. Referring to Eq. (6) and noting that throughout the transit $\omega t \approx \phi$, we see that this maximum efficiency condition also implies $\chi = 0$, i.e., that the structure is driven on resonance: $\omega = \omega_0$.

So for $\phi = 0$, and remembering that $\tau = F_\tau \lambda/c$ we obtain:

$$1 - \epsilon \approx \frac{mc^2}{e} \left(\frac{g}{\lambda}\right)^2 \frac{1}{F_\tau^2 \, V_{stat}} \tag{15a}$$

$$1 - \epsilon \approx 1.02 \times 10^6 \left(\frac{g}{\lambda}\right)^2 \frac{1}{F_\tau^2 \, V_{stat}} \; (\text{mks}) \tag{15b}$$

If for example we chose $\lambda = 6$ mm, $g = .5$ mm, $\epsilon = 50\%$ and $F_r = .38$, then we require

$$V_{stat} \approx 50,000 \text{ Volts}$$

and

$$\mathcal{E}_s \approx 100 \text{ M Volts/m.}$$

Now from the references quoted by Willis,[1] we note that[3] for times less than 5 nsec, voltages as high as 160,000 volts have been held over 1 mm - 2 mm gaps ($\mathcal{E} = 160$ MV/m) and for times less than 1 nsec,[4] 80,000 volts could be held across a 27 micron gap (3000 MV/m). Thus, although the values of our example are high, they are by no means unreasonable for times less than 5 nsec.

However, the above example assumed an efficiency of only 50%. Higher efficiencies would seem to imply even higher fields. It also assumed a transit time rather long (.38) compared with the cycle time. This hardly satisfies condition (13). Clearly we should calculate the effect of finite transit times.

4. ONE DIMENSIONAL SIMULATION

In order to study this problem for finite transit times a small computer program was written that would emit and track the electrons as a function of time in the varying fields. Initially, we consider short pulses and ignore space change effects.

4.1 *With Short Pulses* $\chi = 0$ *(On Resonance)*

Keeping the driving phase $\chi = 0$, i.e., driving at a frequency equal to the cavity resonant frequency, we calculate the efficiency as a function of the ratio of $\mathcal{E}_{rf}/\mathcal{E}_{stat}$. The results are shown in Fig. 7a for the case where F_r (defined by Eq. (12)) is 0.38 (e.g., for V = 50,000 v, g = .5 mm). As expected the efficiency rises as the rf field rises. As expected, the efficiency is somewhat less than that expected for $F_r \to 0$ (dotted line). (Due to the finite transit time, the electrons do not feel the maximum deceleration field over their full transit). What is surprising however is that the efficiency remains finite even when $\mathcal{E}_{rf}$ is greater than $\mathcal{E}_{stat}$.

If we examine the variation of electron velocity (and thus current) as a function of the rf phase for a case of low rf field (e.g., $\mathcal{E}_{rf}/\mathcal{E}_s = .5$, see Fig. 8a), then we see the velocity rising more or less linearly and hitting the anode at its maximum. This then is much as predicted by Eq. (14b).

For higher values of the rf field the situation becomes more complicated (Fig. 8b for $\mathcal{E}_{rf}/\mathcal{E}_{stat} = 1.5$). Now the particles are started at a phase when the rf field is helping the static field. The electrons are thus initially accelerated, then decelerated for awhile, and finally accelerated again to arrive at the anode at a phase which again corresponds to the $\mathcal{E}_{rf}$ helping the $\mathcal{E}_{stat}$.

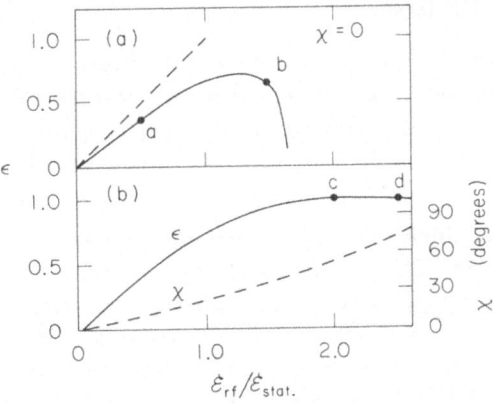

Fig. 7. Efficiency vs. strength of rf field (a) for phase advance $\chi = 0$ and (b) phase advance adjusted to give maximum efficiency.

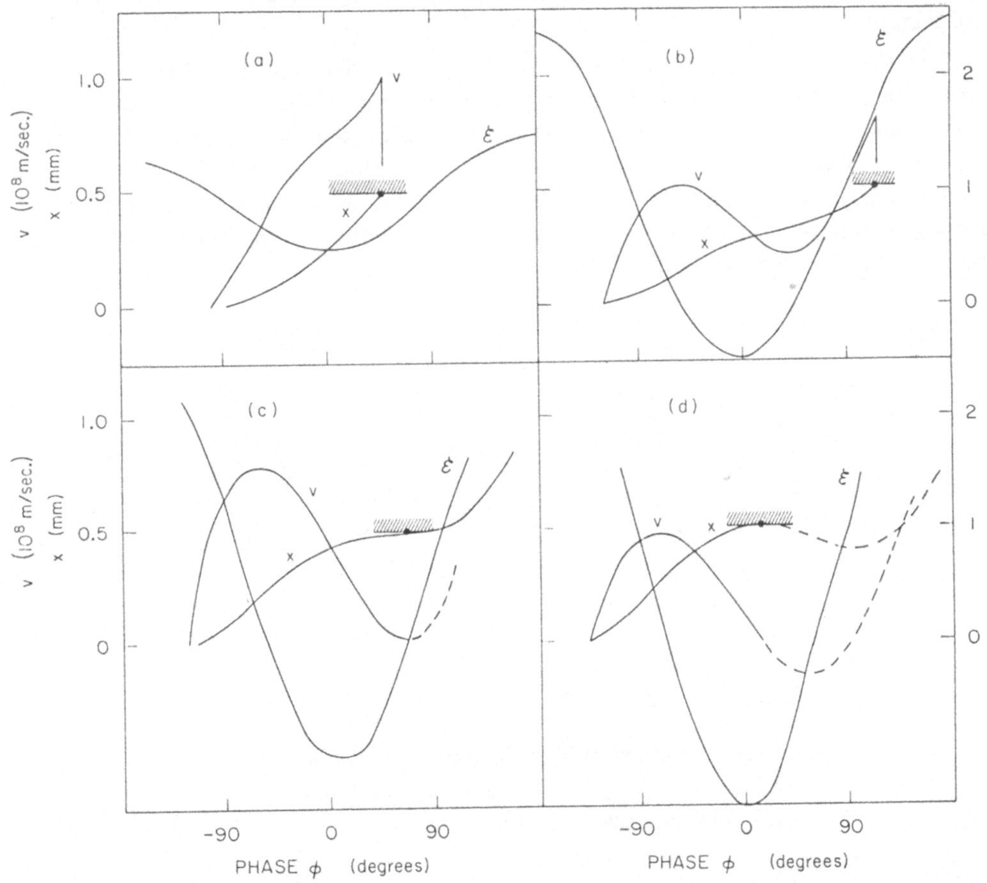

(a)

(b)

(c)

(d)

PHASE ϕ (degrees)

PHASE ϕ (degrees)

Fig. 8. Motion of electrons in the gap as function of rf phase, with electric field and electron velocity also shown: (a) for $\mathcal{E}_{rf}/\mathcal{E}_{stat} = .5$ and $\chi = 0$; (b) for $\mathcal{E}_{rf}/\mathcal{E}_{stat} = 1.5$ and $\chi = 0$; (c) for $\mathcal{E}_{rf}/\mathcal{E}_{stat} = 2$ and χ adjusted for maximum efficiency. This represents the 'magic' condition; (d) $\mathcal{E}_{rf}/\mathcal{E}_{stat} = 2.5$ and χ adjusted for maximum efficiency.

4.2 Short Pulse Driven Off Resonance

As we noted above (Eq. (7)), it is not necessary to operate at $\chi = 0$. If we drive the photocathode at a frequency ω different from the resonant frequency ω_0, then the stable phase χ is finite. If we adjust ω and thus χ to give maximum efficiency (i.e., minimum arrival electron velocity), then we obtain efficiencies as plotted on Fig. 7b.

Now we observe that efficiencies of a 100% are achieved as $\mathcal{E}_{rf} \approx 2\mathcal{E}_s$. We examine this case in Fig. 8c. The particles are again initially accelerated and then decelerated. The parameters, however, are such that the electrons come to rest just as they arrive at the anode. We note further that for this "magic" case the electrons are not only at rest as they touch the anode but that their acceleration is also zero. This situation should be contrasted with that obtained at an even higher rf field (e.g., $\mathcal{E}_{rf}/\mathcal{E}_s = 2.5$, Fig. 8d). In this case the electrons are also at rest as they approach the anode, but they are at that point being decelerated. An electron that just missed the anode would move away from it and only arrive some time later and at a finite velocity. This is a less desirable operating point than the "magic" one of Fig. 8c.

All of the above analysis was done for $F_\tau = .38$. We find a different magic point for each value of F_τ and these are plotted on Fig. 9. Remember that F_τ is the fractional cycle time taken for transit in the absence of an rf field. $F_\tau \to 0$ corresponds to very high fields and a small gap. In that case 100% efficiency is obtained at $\mathcal{E}_{rf}/\mathcal{E}_s = 1$ and no phase lag χ is needed. For larger transit times a phase lag is required and larger rf fields are needed to obtain the magic condition. In our following studies we will assume

$$F_\tau = .38$$

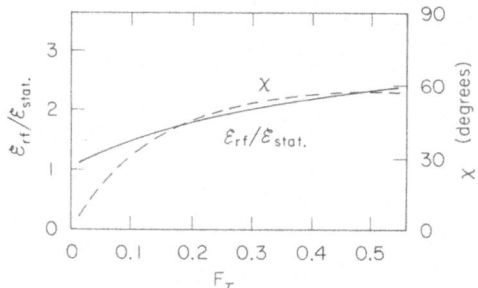

Fig. 9. Values of $\mathcal{E}_{rf}/\mathcal{E}_{stat}$ and the phase advance χ to give the 'magic' condition for different transit time factors F_τ.

4.3 Finite Pulse Lengths

The above calculations have all been performed for electrons emitted at one phase, i.e., for light pulses on the photocathode of arbitrarily short duration. If pulses of finite extent are used then it is no longer possible to bring all the electrons to rest at the anode, and the efficiency varies as a function of the initial phase of each part of the bunch (see Fig. 10). We see that for the magic condition (curve a) the variation is very rapid with the efficiency dropping to 50% when the phase is wrong by only $+4°$ or $-10°$ from the magic value. At lower values of the rf field, although the maximum efficiency is less, the sensitivity to phase is weaker (curve b and c). As a result a plot of efficiency vs $\mathcal{E}_{rf}$ for finite pulse lengths (Fig. 11a) shows maximum efficiencies being obtained at progressively lower values of $\mathcal{E}_{rf}$. Figure 12a shows the maximum efficiencies as a function of pulse duration ($\Delta\phi$) in degrees.

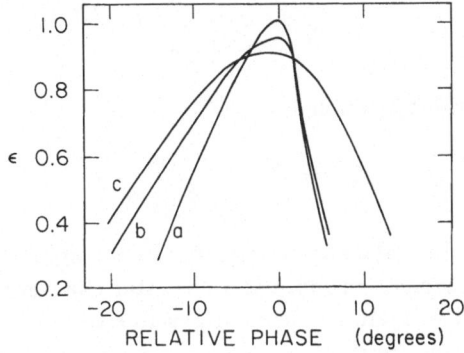

Fig. 10. Efficiency vs. relative phase for (a) the magic condition with $\mathcal{E}_{rf}/\mathcal{E}_{stat} = 2.0$; (b) $\mathcal{E}_{rf}/\mathcal{E}_{stat} = 1.5$; (c) $\mathcal{E}_{rf}/\mathcal{E}_{stat} = 1.0$.

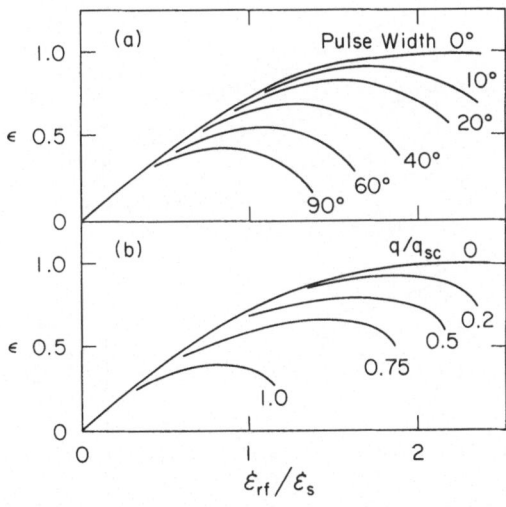

Fig. 11. Efficiency vs. $\mathcal{E}_{rf}/\mathcal{E}_{stat}$ for (a) different pulse lengths and (b) different ratios F_{sc} of charge q to the space charge limit q_{sc}.

4.4 Space Charge Effects

If a charge density q in coulombs/m^2 exists just above the cathode surface then the induced electric field behind this charge $\mathcal{E}_{sc}$ is given by

$$\mathcal{E}_{sc} = \frac{q}{\epsilon_0}$$

where ϵ_0 is the dielectric constant of free space in mks units ($1/36\,\pi\,10^{-9}$). It is convenient to define a threshold space charge limit, q_{sc}, by:

$$q_{sc} = \mathcal{E}_{stat}\,\epsilon_0$$

and define a scale invariant fraction F_{sc}:

$$F_{sc} = \frac{q}{q_{sc}} = \frac{q}{\mathcal{E}_{stat}\,\epsilon_0}$$

Clearly if $F_{sc} \ll 1$ the effect of space charge will be negligible. As F_{sc} increases the earlier charges will be accelerated more and the later charges less than in the small charge case. As a result it again becomes impossible to bring all charges to rest at the anode and the efficiency suffers.

Figure 11b shows the efficiency for different amounts of charge F_{sc} and different $\mathcal{E}_{rf}$. As with the long pulse case, maximum efficiency occurs at fields less than the magic value. Figure 12b shows maximum efficiencies as a function of the charge.

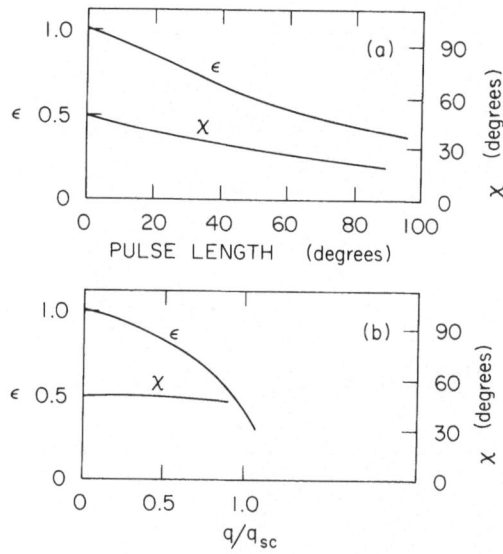

Fig. 12. Maximum efficiencies and corresponding phase advances χ as a function of (a) pulse length and (b) charge.

At first it would seem natural to pick a charge of the order of .4 times the space charge limit, for which the efficiency would still remain above 70% even with a pulse length of 18°. Such a large charge however would have two difficulties:

(1) The current density in our example ($V = 50$ KV, $g = .5$ mm, $\lambda = 6$ mm) would be 44 KA/cm² which is probably excessive.

(2) The instantaneous voltage drop of the wire would be large. In a closed device as illustrated in Fig. 4a or 4b, assuming equal gaps above and below the wire, then the fractional voltage drop would be $q/(2q_{sc})$ or 20%. In the open structure case of Figs. 4c or 4d the capacity is only half and the voltage drop would be 40%.

We would thus probably chose a somewhat smaller value of q/q_{sc}, such as 0.1, for which the resulting inefficiency is very small.

5. A POSSIBLE PARAMETER LIST

5.1 The Cavity

I will consider an open structure of the type illustrated in Fig. 4d, with 1 mm square wires and a total area of 10 cm by 10 cm. For a wavelength of ~ 6 mm the wires would be placed about 2.5 mm apart. Thus there would be ~ 40 wires. The total photo cathode area is about 40 cm². I will consider a pulse length of 18° and $q/q_{sc} = 0.1$. I then obtain:

$$V_{stat} = 50 \text{ KV}$$
$$g = .5 \text{ mm}$$
$$\mathcal{E}_{stat} = 100 \text{ MV/m}$$
$$\lambda = 6 \text{ mm}$$
$$\lambda/c = 18 \text{ p sec}$$
$$t_{pulse} = .9 \text{ p sec } (18°)$$
$$q_{sc} = 8.8 \times 10^{-4} \text{ c/m}^2$$
$$F_{sc} = .1$$
$$q = 8.8 \times 10^{-5} \text{ c/m}^2$$
$$Q_{pulse} = 3.5 \times 10^{-7} \text{ coulombs}$$
$$\mathcal{E}_{rf} = 150 \text{ MV/m}$$
$$\epsilon = 80\%$$
$$\phi = -130°$$
$$\chi = -38°$$
$$i_{photo\ cathode} = 10 \text{ KA/cm}^2$$
$$W_{output} = .78 \times 10^9 \text{ watts .}$$

The time for the micro lasertron to fill requires knowledge of both the stored energy in the structure and the average efficiency during the fill, neither of which we have. Rough estimates suggest:

$$\tau_{las\ fill} \approx .8 \text{ nsec}$$

Now if is this power source is to fill a semi conventional accelerating cavity then the pulse duration must be less than the natural "fill time" of that cavity. If I scale from SLAC's $\tau_{fill} = .8$ μsec at $\lambda = 10$ cm then I obtain

$$\tau_{acc\ fill} = 12 \text{ nsec} \quad (\text{at } \lambda = 6 \text{ mm})$$

Reference 3, however, suggests that breakdown could occur after about 5 nsec and thus the pulse length and energy per pulse would have to be reduced to:

$$\tau \approx 5 \text{ nsec}$$
$$n_{cycles} \approx 250$$
$$J_{output} \approx 3.9 \text{ Joules}$$

This still represents a large total energy output for such an apparently simple device.

5.2 Laser and Photocathode Requirements

The laser would be required to deliver 1 p sec pulses approximately 18 p sec apart for trains lasting of the order of 5 n seconds. The optic frequency and power required would depend on the photo cathodes. If a conventional S 20 type of cathode could be employed and if $\epsilon_q = 10\%$ quantum efficiency is assumed at a wavelength of the order of 500 n meters ($V_{wf} = 2.5$ volts) then the power required would be

$$J_{opt} = \frac{J_{rf}}{\epsilon} \frac{V_{wf}}{V_{stat}} \cdot \frac{1}{\epsilon_q}$$

$$= 2.4 \text{ mJ/train of pulses} \quad (\text{c.f. 3.9 Joules output})$$

$$= .06 \ \% \ J_{rf}$$

Alternatively if a more rugged photocathode such as C_sI were used which might be expected to have $\epsilon_q = .01$ and use 200 nm light ($V_{wf} = 6$ eV) then

$$J_{opt} = 58 \text{ mJ/train of pulses} \quad (\text{c.f. 3.9 Joules output})$$

$$= 1.5 \ \% \ J_{rf}$$

In this latter case the laser efficiency must be significantly better than 1.5% to avoid the total energy going to generate the light being comparable to the total rf power output.

6. CONCLUSION

The above analysis has shown that a micro-lasertron might be a source of powerful mm radiation but there are many assumptions that need to be demonstrated:

(1) Firstly we have assumed that a gradient of the order of 100 MV/m can be maintained over a .5 mm gap for 5 nsec. This is consistent with extrapolations from experimental results using metal electrodes but may not be possible if one of the electrodes has a low work function surface (photo cathode). Experimental work is required.

(2) We have assumed current densities of the order of 10 K amps/cm^2 for 1 p sec pulses, and average currents of the order of 500 A/cm^2 for the order of 5 nsec. Both values are higher than observed values of a conventional photocathode. Values of the required order have been observed from metal photocathodes, but the quantum efficiency in this case would probably be unacceptable.

(3) The analysis ignores the finite width of the diode gaps and also ignores the direct radiation from the electrons into the cavity. Full two dimensional numerical calculation is required.

(4) We have not considered the effect (pointed out by J. Claus) of the magnetic field generated by the feed current in the wires. Rough calculations suggest that as long as the wires are only a few cm long the effects are small, but they need calculation.

(5) There has been no discussion of how to end the cavities described. It seems reasonable to believe that open ends will suffice but modelling is needed.

(6) Details of switching the primary current have not been discussed. Laser activated solid state switches may be suitable, but much work remains to be done.

(7) Details of rf windows, heat removal and a thousand other questions have not yet been addressed.

Despite these questions the potential of the proposed device seemed to justify its presentation now. Work on many of these problems is being pursued at various labs and future publication will hopefully answer some of the questions.

I would like to thank W. Willis whose original idea started this work and whose continued interest and suggestions have nurtured it. I also wish to thank J. Clause, U. Stumer, V. Radeka, T. Rao, and many others for their contributions.

REFERENCES

1. W. Willis, "Switched Power Linac," Laser Acceleration of Particles (Malibu, 1985), AIP Conf. Proc. 130, p. 242.

2. E. L. Garwin et al., "An Experimental Program to Build a Multimegawatt Lasertron for Super Linear Colliders," 1985 Particle Accelerator Conf. (to be pulished in IEEE Trans. Nucl. Sci. NS-32); also SLAC-PUB-3650. Y. Fukushima et al., " Lasertron, a Photocathode Microwave Device Switched by Laser," 1985 Particle Accelerator Conf. (to be published in IEEE Trans. Nucl. Sci. NS-32).

3. F. T. Warren et al., "Current Evolution in a Pulsed Overstressed Radial Vacuum Gap," IEEE Trans. Elec. Insulation. EI-18, No. 3, p. 226 (1983).

4. B. Juttner et al., "Zerstörung und Erzeugung von Feldemittern auf ausgedehnten Metalloberflächen," Beitrage für Plasmaphysick 10, p. 383 (1970).

DISCUSSION

JOHO:

What is the current that must be switched on to the anode wires?

PALMER:

Per wire the current would be 30000 amps for 5 ns or 150 microcoulombs.

SCHEMPP:

I believe that the realization of your generator will present same technical problems, particularly about the matching of a 5 ns, 50 kV, 30 kA pulse to a sensitive discharge chamber, the spark erosion, the reproducibility and the lifetime of the discharge resonator.

COLLIDER SCALING AND COST ESTIMATION[*]

R. B. Palmer[†]

Stanford Linear Accelerator Center
Stanford University, Stanford, California 94305

1. INTRODUCTION

The primary motivation for high energy physicists to study new acceleration mechanisms is to find a way to build colliders at energies above the SSC at costs less than the SSC. Cost considerations are, unfortunately, crucial. It is simply not useful to know how to build an accelerator that would cost 100 billion dollars. Although it is difficult to make cost estimates without knowing the technology, I believe the attempt is useful.

In comparing a linear collider, assumed to be electron positron, with the SSC, a circular proton-proton machine, we need to know the parameters of each that will attain "equivalent" physics. Strictly there is no such equivalence. The two machines have different strengths and weaknesses and are in many ways complimentary. Nevertheless we can establish equivalent parameters for the production of particular final states and make some kind of average over different such states.[1] For the purposes of this lecture I will assume the following parameters to be equivalent:

	SSC	e^+e^- Collider
Beam energy	20 + 20 TeV	1.5 + 1.5 TeV
Luminosity	10^{33} cm^{-2} sec^{-1}	10^{33} cm^{-2} sec^{-1}

In discussing the cost scaling I will often refer to cost estimates for this "SSC equivalent" e^+e^- collider. These costs should be compared with a value of about 2 billion dollars for the SSC. This is the SSC cost without detectors, site, contingency or escalation. As in the SSC case the real cost would be about a factor of two higher than the values given.

---

[*] Work supported by the Department of Energy, contract DE–AC03–76SF00515.
[†] On leave from Brookhaven National Laboratory.

2. SCALING LAWS AND COST ESTIMATION

It would be technically feasible to construct two SLAC-like linear accelerators, producing beams of electrons and positrons, respectively, up to 1.5 TeV. The problem is that if one bases cost estimates on a simple scaling of the existing SLAC linear accelerator parameters, then costs are excessive. Optimization of parameters for a linear collider is very likely to lead to numbers different from those pertaining to SLAC. Specifically, gradient, wavelength, mechanical tolerances, structural parameters, focusing systems, to name but a few would have to be quite different. Whether research and development based on such an optimization of parameters would lead to a practical machine whose cost is lower than that of the SSC is far from certain, but is not excluded. I try in this lecture to perform such an optimization despite the relative lack of detailed costs.

Costs for existing linear accelerators are associated with physical length, average power, peak power, and energy storage per pulse, and the scaling laws associated with each of these parameters as a function of wavelength and accelerating gradient can be determined. If we rather arbitrarily assume linear relations between costs and each of these parameters then one can obtain cost estimates as a function of the parameters and look for those values that would minimize the cost. It must be remembered that the exercise leads us to parameters and technology very far from existing linear colliders and cannot, therefore, be treated as a realistic estimate of actual cost. It is nevertheless an interesting exercise and may indicate where efforts should be directed.

We are considering only technologies that employ radio frequency power sources, driving near field accelerating structures. I will examine, in turn, the requirements on accelerating gradients, total stored RF energy, peak RF power, and finally average power consumption.

2.1 Accelerating Gradient Requirements

There is a rather obvious relationship between length and the cost of a linear collider and most early efforts at developing new technology were aimed at increasing the accelerating gradients in order to reduce these costs. The length proportional costs might lie somewhere in the range between $10,000 and $100,000 per meter (civil construction alone would be near the lower figure; the cost of SLAC in current dollars is somewhere in the middle of the range). For the purposes of cost optimization I will assume $30,000 per meter.

A gradient of 20 MeV per meter (as at the SLC), would imply, for our SSC equivalent, a total length of the order of 150 kilometers and a total linear cost of the order of 5 billion dollars. Gradients as high as 150 MeV per meter have been achieved in a SLAC structure. With this the length would be reduced to 20 kilometers, but the cost remains still relatively high: of the order of .6 billion. If we are aiming for costs substantially less than the SSC, it would seem prudent to aim for an accelerating gradient somewhat, but not greatly, larger than this.

The limit set by breakdown is believed to rise as the inverse wavelength to the 7/8 power (see Fig. 1). Another limit is set when heating in the accelerating structure would cause momentary melting of its surface. This limit has been discussed by Norman Kroll[2] and Perry Wilson[3] and occurs somewhere between 300 and 1000

MeV per meter at 10 cm and scales as the inverse wavelength to the 1/8th power (Fig. 1 is taken from Perry Wilson's paper[3]). This scaling law assumes that the cavity is filled for a time scaled from that used in SLAC; shorter fill times raise this limit. When the wavelength falls below 100 microns, the scaling law changes and rises as the inverse 1/4 power; the change arising because the temperature becomes limited by the specific heat of the surface material instead of its conductivity. At a wavelength of 10 microns, fields over a plane mirror as high as 4 GeV per meter have been recorded without damage to the surface.

As an example, if we assume a wavelength of less than 1 cm, we can hope for gradients of the order of 500 MeV per meter, and in that case, the linear costs of our SSC equivalent would be of the order of 180 million dollars; substantially less than the SSC.

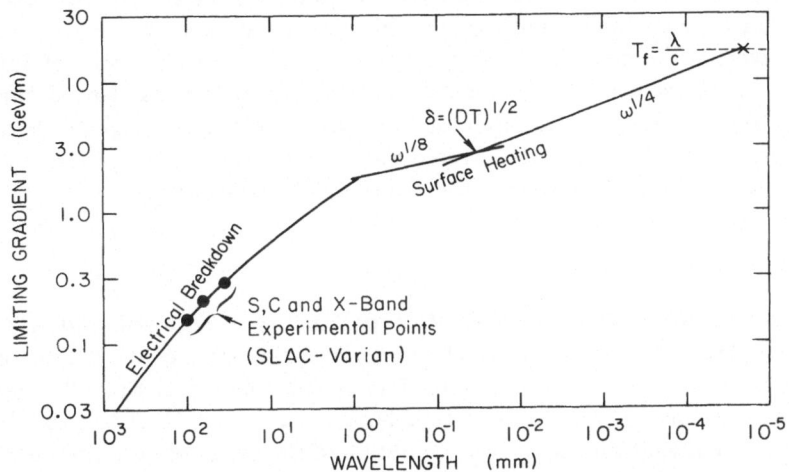

Fig. 1. Limitations on gradient as a function of wavelength due to electric field breakdown and surface heating in a SLAC-type disk-loaded structure.

2.2 Total Stored RF Energy Considerations

If, as in the SLC, the wavelength is 10 cm, but the accelerating gradient 500 MeV per meter, then the total RF energy required would be approximately 40 million Joules. If the costs associated with this stored energy are similar to those at SLAC, i.e., about \$1000 per Joule,[4] then this would cost 40 billion dollars, clearly unreasonable.

The stored energy J is reduced by lowering the gradient E_a but such reduction will increase the linear costs. The stored energy is also reduced by reducing the wavelength (by the square):

$$J \text{ (both beams)} = \sqrt{S}\, \frac{E_a \lambda^2}{4c_1} \quad \left(\text{mks if}\sqrt{S} \text{ and } E_a \text{ are in volts}\right) . \qquad (2.1)$$

The c_1 depends on the structure geometry, for SLAC:

$$c_1 = k_1 \lambda^2 \approx .2 \times 10^{12} \text{ Vm/coulomb} . \qquad (2.2)$$

The only limit to how small the wavelength can be, is set by wake field effects. These, as we shall see below, set a limit on the number of particles that can be accelerated without having their emittance excessively increased. This number of particles per bunch, determines both the average power requirement and the beamstrahlung. All these will be discussed later and will require the wavelength to be of the order of a millimeter. If I arbitrarily take $\lambda = 1.4$ mm as an example then for $E_a = .5$ GeV/m and a SLAC-like structure

$$J = 4000 \text{ Joules}$$

If I assume 50% filling efficiency then the RF source must supply a total of 8000 Joules. This to be compared with 40 million Joules for $\lambda = 10$ cm.

In estimating the costs of providing this RF energy at 1 mm, we cannot use the price associated with the 10 cm wavelengths. Estimates of the cost of, for instance, a Free Electron Laser, which could provide this wavelength are harder to come by. The recent Livermore experiments[5] suggest that it will be of the order of $10,000 per Joule ($\sim$ 10 times that of a klystron), and the resulting cost would then be around 80 million dollars, far less than for the conventional wavelength even at the higher cost per Joule. Despite the extreme uncertainty in such cost estimates, the above illustrates the need to research the use of short wavelengths.

2.3 Peak Power Consideration

If an acceleration gradient of 500 MeV per meter is required and a 10 cm wavelength employed, then the stored RF energy is of the order of 40 million Joules. If the power is supplied over a fill time of .8 microseconds (as at SLAC), then the peak power requirement of the RF source is 50 terawatts. The cost to provide this peak power with conventional klystrons would be of the order of 40 billion dollars.[4] The peak power, as well as the stored energy cost is excessive.

If the wavelength is reduced, then the stored energy goes down as the square of the wavelength, but the fill time also goes down as the 3/2th power, resulting in the peak power going down as the root:

$$\tau \propto \lambda^{2/3} ; \qquad P(\text{peak}) = \frac{J}{\tau} \propto \frac{\sqrt{S} E_a \lambda^{1/2}}{c_2} \qquad (2.3)$$

for a SLAC-type structure made of copper

$$c_2 = .35 \times 10^6 \text{ V}^2 \text{ m}^{-1/2} \text{ W}^{-1} . \qquad (2.4)$$

In the case described above, the wavelength is 1.4 mm, the fill time scales to 1.3 nanoseconds and the peak power requirement is 6 terawatts. The cost per peak power for klystrons and an induction linac driven free electron laser (FEL) are similar[5] ($\sim 7 \times 10^{-4}$ $/watt), and thus, the cost would still be of the order of 4 billion dollars. Unless the cost per unit of peak power can be significantly reduced from this value, the total cost, even after optimization at a lower gradient, remains considerably higher than that of the current SSC. As a result, there is great interest in power sources which have much lower cost per unit of peak power.

Possible new technologies that would work at short wavelengths include lasers, bunch compression prior to a free electron laser, laser-driven photo-diode amplifiers, laser-driven solid state amplifiers; and finally, almost any power source followed by pulse compression schemes.

The cost per unit of peak power for a short pulse laser is already far lower (of the order of 1/1000)[6] than that for conventional RF power sources, and it is this low cost per peak power that has made lasers attractive sources for the generation of high accelerating gradients. Unfortunately, their very short wavelength (of the order of 10 microns) makes it difficult to accelerate bunches of the required magnitude in conventional cavities, although the use of plasma beatwaves and other unconventional structures might overcome this problem. The second difficulty, as noted in Section 2.2, is that the cost[6] per unit of average power for such lasers is very much higher than for more conventional RF sources.

Perhaps the most promising technique, as of this review, is the use of an induction linac driven free electron laser, together with binary pulse compression[7] by a factor of say 32 (five stages). If the capital cost per unit of peak power could be reduced by a factor of twenty, then the peak power cost in our example would be reduced to of the order of 200 million dollars.

Another promising solution is to use a laser driven photo-diode amplifier. The "Lasertron" is such a device operating at normal microwave frequencies. It is similar to a conventional klystron but employs a gun employing a photo-cathode illuminated by RF modulated laser light. Scaling the lasertron, as now conceived, down to millimeter wavelengths seems impractical, but work on a microlasertron that could be so scaled has been inspired by a proposal of W. Willis and is being worked on at BNL and SLAC.[8]

2.4 Average Power Consumption

Simple extrapolation of current technology would need thousands of terawatts of average beam power. Besides the inevitable operating cost of such energy, there are clearly related capital costs in the conversion of electrical power first to RF fields and then to the beam. Somehow the average beam power must be reduced.

The beam power $P = NfE$ is given by the number of particles per bunch (N), the frequency (f) and the energy (E). The requirements for N and f are set by the need for a high luminosity (10^{33} cm^{-2} sec^{-1} for an SSC equivalent). The luminosity for round beams is given by $L = N^2 f/A$, where A is the cross-sectional area of the two beams at the final focus. Combining these two relationships, and noting that the energy times the cross-sectional area of the beams is given by the invariant beam emittance (ϵ_n) times the focus parameter (β), we obtain for the power per beam

$$P_B \approx 1 \times 10^{-12} \frac{L\epsilon_n \beta}{N} \quad \text{(mks)} . \tag{2.5}$$

This relationship implies that a large number of particles per bunch (N) is desirable. But a limit to this is set by two considerations: (1) beamstrahlung and (2) wake fields. I will take the approach of leaving discussion of beamstrahlung for the moment, assuming that it is not a problem, and consider only the wake field constraints.

If N is too large then for a given structure and wavelength, wake fields will cause (1) excessive energy spread of the beam (longitudinal wake) and (2) blow up of the transverse beam size if the initial beam is slightly off axis. I consider them in turn:

Longitudinal Wake

The longitudinal wake fields generate an energy spread in the accelerated beam given[9] by

$$\frac{\Delta E}{E} = e \, \frac{B_0(\sigma_z/\lambda) \cdot C_0 N}{E_a \, \lambda^2} \tag{2.6}$$

where B_0 depends on the geometry of the structure and on the bunch length σ_z. As σ_z/λ goes to zero, B_0 goes to a constant of the order of 4. The C_0 is the wavelength independent loss parameter ($k_0 = C_0/\lambda^2$), which is dependent only on the structure geometry.

It is instructive to substitute for N, an expression using the fraction η of energy that is extracted from the cavity by the bunch, divided by the total stored energy in the cavity: ($\eta = 4eNC_0/\lambda^2 E_a$):

$$N = \frac{\eta \lambda^2 E_a}{e4C_0} \tag{2.7}$$

then

$$\frac{\Delta E}{E} = \frac{B(\sigma_z/\lambda)}{4} \, \eta \tag{2.8}$$

which for $\sigma_z/\lambda \to 0$

$$\frac{\Delta E}{E} \approx \eta \; . \tag{2.9}$$

Thus we see that in order to control the longitudinal wake fields we have to limit the fraction η of energy extracted from the cavity.

Transverse Wake

The transverse wake scaling seems at first to be more complicated. If the bunch is perfectly on the structure axis, no transverse wake exists, but if the bunch entering a structure is displaced by ϵ then transverse fields are excited that can cause a sideways displacement of the tail of the bunch by a distance δ given by[9]:

$$A = \frac{\delta}{\epsilon} \propto \frac{N\beta z \, u_0(\sigma_z/\lambda)}{4E\lambda^3} \tag{2.10}$$

where β is the average focusing strength in the structure, z the distance along the structure and u_0 a variable dependent on the structure and the bunch length. For small values of the variable σ_z/λ, u_0 will be proportional to σ_z/λ.

The above equation is true providing the β is the same for the head and tail of the bunch. Luckily this is unlikely to be the case. The longitudinal wake effect will cause the tail to have a lower energy than the head and this will result in the β's for head and tail being different. In this case the amplitude A does not rise linearly with z but oscillates with a maximum amplitude given by:

$$A = e \; \frac{N\beta^2 \; u_0(\sigma_z/\lambda)}{4E\lambda^3 \; \Delta p/p} \; . \tag{2.11}$$

It is now instructive to substitute for N as above and also substitute for β assuming that the focusing is provided by RFQ fields and that the magnitude of these fields is some fixed fraction of the accelerating fields E_a. In this case

$$\beta = B_0(\lambda E/E_a)^{1/2} \tag{2.12}$$

where B_0 is a constant for the cavity geometry. Substituting into the equation for the transverse wake gives

$$A = \left[\frac{1}{8} \; \frac{u_0(\sigma_z/\lambda) \cdot B_0}{c_0} \right] \frac{\eta}{dp/p} \tag{2.13}$$

and for small σ_z/λ

$$A \approx \text{const} \cdot \frac{\sigma_z}{\lambda} \cdot \frac{\eta}{dp/p} \; . \tag{2.14}$$

As in the case of the longitudinal wake we find that the transverse wake, expressed in this way, is independent of λ (except for the fraction σ_z/λ) and scales only with the fraction of energy taken from the cavity η.

Consequences for Average Power

We have seen that wake field considerations will set a limit on the number of particles per bunch which is related to the structure, the fraction η of energy extracted, the wavelength and the accelerating gradient by Eq. (2.7). Combining this with Eq. (2.1) we obtain

$$P_B \propto \mathcal{L} \; \frac{\epsilon_n \beta C_0}{E_a \lambda^2 \eta} \; . \tag{2.15}$$

If we insert a value for C_0 equal to that for a SLAC-type structure:

$$C_0 = k_0 \lambda^2 = 20 \times 10^{12} \cdot (.1)^2$$
$$= 2 \times 10^{11} \; \text{Volt meters/coulomb}$$

then

$$P_B = 1 \times 10^{-19} \mathcal{L} \; \frac{\epsilon_n \beta}{E_a \lambda^2 \eta} \qquad \text{(mks)} \; . \tag{2.16}$$

If as an example I choose $\eta = 5\%, \lambda = 1.4$ mm and $E_a = .5$ GeV/m (we will see why I choose such a small λ later) then for $\epsilon_n = 1.2 \times 10^{-8}$ meters, $\beta = 1$ mm and $\mathcal{L} = 10^{33}$ cm^{-2} sec$^{-1}(10^{37}$ m^{-2} sec^{-1}), I obtain

$$P_B = .3 \; \text{megawatt} \qquad \text{(per beam)}$$
$$\text{and} \quad N = 4 \times 10^8 \qquad \text{(per bunch)} \; .$$

If we take 10% efficiency RF-to-beam, then, for this example, we need 6 megawatts average RF power. The capital cost associated with such power will depend on the technology. For SLAC klystrons it can be as little as $2 per watt, but for a new short wavelength source such as an FEL, it is probably a factor of ten higher: say $20 per watt. Thus for 6 MW the capital costs might be in the range of 120 million dollars, quite reasonable.

For a laser power source, the cost is higher ($\sim$ $100 per average watt) yielding an estimated capital cost of 600 million dollars. Whether such a high cost for average power can be offset by the laser's peak power advantage remains to be seen.

3. COST OPTIMIZATION

In the above section we have discussed the possible costs for a collider of $\sqrt{S} = 3$ TeV center of mass energy, luminosity of 10^{33} cm^{-2} sec^{-1}, accelerating gradient of 500 MeV/m and wavelength of 1.4 mm. The parameters and costs of the example are summarized in Table I. The choice of gradient and wavelength was justified only on qualitative grounds. In this section I wish to examine the dependency of overall cost on their choice in a more quantitative manner.

Making all the same assumptions as above but allowing the energy $\sqrt{S}$, the accelerating gradient E_a, and the wavelength λ to be variables, then one obtains the following contributions to the total machine cost. I have assumed that the luminosity must rise linearly with S.

$$\$(\text{length}) \approx 30\text{M}\$ \cdot \frac{(S(\text{TeV}^2))^{1/2}}{E_a(\text{GeV/m})} \propto \frac{1}{E_a} \tag{3.1}$$

$$\$(\text{Ave. Power}) \approx 13\text{M}\$ \cdot \frac{S(\text{TeV}^2)}{E_a(\text{GeV/m}) \cdot (\lambda(\text{mm}))^2} \propto \frac{1}{E_a \lambda^2} \tag{3.2}$$

$$\$(\text{RF energy}) \approx 30\text{M}\$(S(\text{TeV}^2))^{1/2} E_a(\text{GeV/m})(\lambda(\text{mm}))^2 \propto E_a \lambda^2 \tag{3.3}$$

$$\$(\text{Peak Power}) \approx 100\text{M}\$(S(\text{TeV}^2))^{1/2} E_a(\text{GeV/m})(\lambda(\text{mm}))^{1/2} \propto E_a \lambda^{1/2} \tag{3.4}$$

Note that the constants were always obtained on assumptions that may turn out to be unreliable and that they should be considered to have errors of about a factor of three both up and down. Note also that the above does not include costs for the cooling and injector systems (that may be very large), nor for experimental areas, labs, contingency etc.

The dependence of cost on E_a and λ can be illustrated graphically as shown in Fig. 2(a). Lines of constant component cost are shown as a function of the variables E_a and λ. The minimum cost for the sum of the four components will be at the geometrical center of the four-sided inner figure. We see that this cost minimum ($\lambda = 1.6$ mm and $E_a = .4$ GeV/m) is very close to the example of Table I (not, of course, a coincidence).

We can also examine the sensitivity of the minimum to our cost assumptions. If the cost of average power were ten times higher, or that of RF energy ten times lower, then the wavelength for minimum cost only rises by a factor of 1.8 to 2.9 mm. If the linear cost rises by a factor of three, or the cost of peak power falls by the same amount, then the accelerating gradient rises by 1.7 to .7 GeV/m. Clearly it will require quite radical changes in our assumptions to change the general conclusion that the accelerating for minimum cost should be of order .5 GeV/m, and that the wavelength for minimum cost is of the order of a few millimeter.

As the energy increases all components rise linearly except that for average power, which rises as the energy squared. As a result the cost minimum moves to somewhat higher wavelengths. Figure 2(b) shows the situation for $\sqrt{S} = 10$ TeV where λ for cost minimum is 2.5 mm.

Table I

Example parameters of a linear collider with energy 1.5 plus 1.5 TeV and luminosity 10^{33} cm^{-2} sec^{-1}.

	Conventional (SLC)	Example
Gradient, E_a	20 MeV/m	500 MeV/m
Length, l		6 km
Cost per Meter	~ 30 K$	~ 30 K$
Total Length Cost		~ 180 M$
Invariant Emittance, ϵ_n	3×10^{-5}	1.2×10^{-8} m
Focus Parameter, β	5 mm	1 mm
Spot radius, $\sigma_x = \sigma_y$		2 nm
Bunch Length, σ_z	1 mm	10 μm
Beamstrahlung, $\Delta E/E = \delta$		.16
Ave. Beam Power (both), P_b		.6 MW
Wall Plug to Beam Eff., η_w	~ .1%	1%
Wall Power		60 MW
RF Source to Beam Eff., η_{RF}	2%	10%
RF Ave. Power		6 MW
Capital Cost per Ave. Watt	~2$	~ 20$
Capital Cost per Ave. Power		~ 120$
Particle per Bunch, N	5×10^{10}	4×10^8
Bunch Energy (Both), E_b		200 Joules
Stored RF to Bunch Eff., η	~ 5%	5%
RF Stored Energy, E_{RF}		4000 Joules
RF Source - RF Stored Eff., η_t	~ 50%	50%
RF Energy/Fill, E_f		8000 Joules
Source Frequency, f_s	120 Hz	750 Hz
RF Source Cost/Joule	~ 1 K$/J	~ 10 K$/J
RF Source Stored Energy Cost		~ 80 M$
Wavelength	10 cm	1.4 mm
Fill time	.8 μsec	1.3 nsec
RF Peak Power		6 TW
Cost per Peak Watt	~ 700×10^{-6}	~ 35×10^{-6} $/watt
Peak Power Cost		~ 200 M$

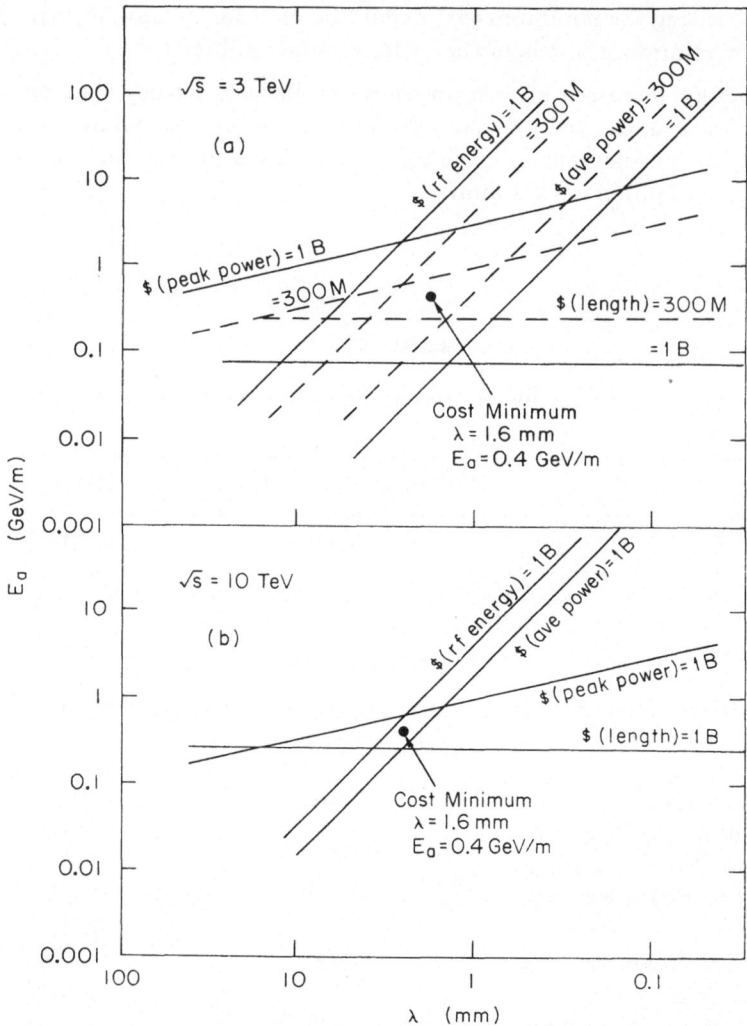

Fig. 2. Plot of lines of constant component costs as a function of accelerating gradient (E_a) and the microwave wavelength (λ). (a) Is for a 1.5 on 1.5 TeV collider and (b) is for a 5 on 5 TeV collider.

4. BEAMSTRAHLUNG CONSIDERATIONS

So far we have ignored consideration of beamstrahlung at the final collision. Beamstrahlung is the name given to synchrotron radiation emitted by the particles of one bunch as they are deflected by the fields (both electric and magnetic) within the opposite bunch.

If the bunches are sufficiently long and the energy sufficiently low then the normal classical synchrotron radiation formulae apply and the average fractional energy loss δ_c of one bunch passing through the other is given[10] by:

$$\delta_c \approx 5 \times 10^{-45} \frac{N^2\gamma^2}{\epsilon_n \beta \sigma_z} \qquad \text{(mks)} \tag{4.1}$$

where σ_z is the rms bunch length.

The spectrum of radiation emitted has the familiar synchrotron form with a critical photon energy of

$$E_{\text{crit}} \approx (\sqrt{3}\hbar c \ r_e) \cdot \frac{\gamma^2 N(r)}{r\sigma_z} \tag{4.2}$$

where $N(r)$ is the number of particles inside the radius r.

As σ_z gets less or γ increases eventually E_{crit} becomes larger than the initial electron energy. At this point the classical formulation breaks down and one enters a quantum beamstrahlung region where the fractional energy loss is given[11] by:

$$\delta_{qm} \approx 7 \times 10^{-9} \left(\frac{N^2 \sigma_z}{\epsilon_n \beta} \right)^{1/3} . \tag{4.3}$$

In the example listed in Table I we had $N = 4 \times 10^8$, $\epsilon_n = 1.2 \times 10^{-8}$, $\beta = 1$ mm and $\gamma = 3 \times 10^6$ and for these parameters we get δ as a function of σ_z as shown in Fig. 3. At very large values of σ_z the classical formula holds and δ decreases with increasing σ_z; at small σ_z the quantum formula is applicable and the reverse is true. The dotted transition between the regions was added by eye.

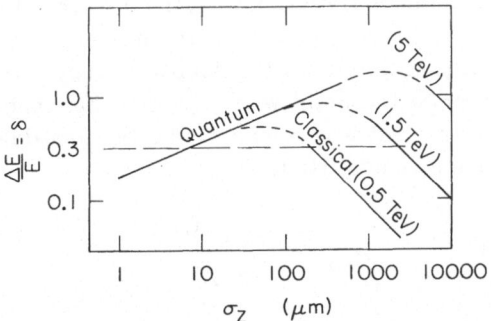

Fig. 3. Fractional energy loss due to beamstrahlung $(\Delta E/E = \delta)$ for the example given in Table I, plotted versus the rms bunch length σ_z.

We see that the fractional energy loss δ is reasonable, (i.e. less than 30%), for bench lengths less than 10 μm, or longer than about 1 mm. Unfortunately the long bunch solution has too great a disruption parameter and is not practical. In addition, of course, it is not possible to accelerate a 1 mm long bunch in a structure employing a wavelength of the same order. Thus we find ourselves compelled to use a short bunch and operate in the quantum beamstrahlung region. We note also that at energies above the SSC equivalent, the clasical beamstrahlung even at $\sigma_z = 1$ mm is excessive and the use of short bunches and the quantum regime is imperative.

We have in this analysis chosen the wavelength to minimize cost and obtained from those considerations a number of particles per bunch. With this we have calculated beamstrahlung and chosen a bunch length to give a reasonable energy spread. We could have gone the other way.

If we had fixed the acceptable energy spread δ then we could have expressed the number of particles per bunch:

$$N \approx 1.7 \times 10^{12} \left(\frac{\epsilon_n \beta \delta^3}{\sigma_z} \right)^{1/2} \quad \text{(mks)}, \qquad (4.4)$$

and then combining this with Eq. (2.5) obtain:

$$P_{\text{beam}} \approx .5 \times 10^{-24} \frac{\mathcal{L}(\epsilon_n \beta \sigma_z)^{1/2}}{\delta^{3/2}} \quad \text{(mks)}. \qquad (4.5)$$

This is a particularly interesting relationship. As the machine energy rises, then the costs related to the average power rise faster than all other costs (assuming the need for luminosity to rise as the square of energy). Eventually these average power costs will dominate and Eq. (4.5) indicates the importance for the three-dimensional emittance $\epsilon_n \sigma_z$ and strong focusing, i.e. a low β.

5. FOCUSING OPTICS

5.1 Introduction

As has been discussed above there is a relation (Eq. (4.5)) between the average beam power and, among other things the beam emittance (ϵ_n) and final focusing strength (β). The parameter β defines the "depth of focus" or distance along the axis from the focal center to the point when the spot has increased in radius by a factor of two. It is a convenient way of expressing the focusing strength and gives the spot radius $\sigma_{x,y}$ for a given emittance:

$$\sigma_{x,y} = \left(\frac{\epsilon_n \beta}{\gamma} \right)^{1/2}. \qquad (5.1)$$

We see that reduction of either ϵ_n or β will reduce the σ and thus the cross-sectional area of the spot. This in turn increases, for a given beam current, the rate of interactions when two beams collider; (i.e. increases the "luminosity").

Focusing of particle beams is done by magnetic fields which bend the particles towards the axis. If the fields rise linearly with their distance from the axis, i.e. have a gradient, then the resulting bends will result in focusing the beam to a spot, just as a conventional lens focuses an optical beam. Unfortunately it is not possible in a vacuum to magnetically focus simultaneously in both horizontal and vertical directions. "Quadrupole" magnets are used whose fields focus in one direction but defocus by an equal amount in the other. By the "strong focusing principle", when two or more of such magnets are employed with finite spacing, focusing in both planes can be achieved.

5.2 Conventional

The current SLC focusing system employs quadrupoles with gradients of the order of 10 kG per cm and achieves a final beta of about 5 mm. A fairly sophisticated chromatic correction system cancels first and second order effects and achieves, with .5% momentum spread, a final spot little larger than that given by geometric considerations. If such a design is scaled in length by the root of gamma/gradient then the beta scales by the same factor:

$$\beta \propto \left(\frac{\gamma}{\text{gradient}}\right)^{1/2} \tag{5.2}$$

and the relative correction of chromatic effects remains unchanged. Unfortunately this implies that with the same gradients, the final beta for a 1.5 TeV collider will have risen to 3 cm. With exotic superconductors a gradient of the order of 40 kG/cm might be possible. This would reduce the beta to about 1 cm. More efficient packing and correction of higher order chromatic effects might be able to further lower this to 1 mm[12] (Erickson), but the required higher order correction remains to be demonstrated. Alternatively, as we will see below, one could use very small "conventional" quads and obtain about the same value.

Discussions in earlier sections have indicated that if the average beam power is to be limited in a high luminosity collider then the emittance, as well as beta will have to be very small. If we assume such a very small emittance (10^{-8} m) then the beam size, even at its largest point in the focus system, is less than a tenth of a millimeter. It is tempting then to fix the pole tip fields (B) and scale the quadrupole apertures to a fixed factor over this beam size. Doing this, we obtain for the same chromatic correction and the same quadrupole packing:

$$\beta \propto \left(\frac{\gamma\epsilon_n}{B^2}\right)^{1/3} . \tag{5.3}$$

For an emittance (epsilon) of 10^{-8}, and 10 kG pole tip field, then at 1.5 TeV one would again obtain $\beta = 1$ mm, but this is now obtained without third order chromatic correction. The required quads are indeed very small, but no smaller than the gaps in recording heads, and there do seem to be possible ways to build them to the required optical precision. Once again however it has yet to be demonstrated.

5.3   Laser Focusing

With the short wavelengths and high powers of lasers one hopes to be able to obtain accelerating gradients of perhaps as high as 5 GeV per meter. Whatever the economics of such fields for acceleration they might be useful for focusing. In a radio frequency quadrupole (RFQ) such fields would correspond to pole tip bending fields of 150 kG and, by the above scaling Eq. (5.3), reduce the final beta by another factor of six. This option has not however been fully studied.

5.4   Super Disruption

It has been known for some time that when two bunches of finite length collide the fields induced by one will focus or pinch the other. The result is an effective reduction of beta and a consequent enhancement of the luminosity by a factor of the order of four. Unfortunately the use of such long bunches is inconsistent with the need to limit the energy loss from beamstrahlung.

117

It has been suggested, however, that the enhancement could be recovered by the use of two or more short bunches[13] (Leith/Palmer). In this case the earlier bunches can be used to focus the oncoming final bunch and greatly enhance the luminosity achieved when these two final bunches collide. The luminosity enhancement (h) for bunches with uniform current distribution is given approximately by:

$$h \approx .5 \times 10^{30} \left(\frac{N}{\epsilon_n}\right)^2 \quad \text{mks} . \tag{5.4}$$

Note: new beta = old beta/$(4h)$.

In this equation N is the total number of particles, divided half and half in the two bunches and epsilon is the emittance in meters. For an N chosen to give a beamstahlung energy loss of .3, and an emittance of 10^{-8}, h would be approximately 25, and the beta reduced by 100. Even higher gains are calculated when more than two bunches are used. The enhancements are more limited when a less ideal current distribution is assumed, but a significant factor is still obtained even with Gaussian distributions.

A second advantage that comes with this "self focusing" is that the two beams do not need to be aligned to the same accuracy as would be required if the same final spot were obtained by external focusing. This relaxation in tolerance is by a factor of four for two bunches and nearly seven with three. There is of course no relaxation on the tolerance on the alignment between the first and second bunches.

One must note from Eq. (5.4) that neither the enhancement nor the reduction in tolerance can be obtained without first obtaining the very small emittance. It cannot be used as a substitute.

5.5 Conclusion

There seems at first sight to be several ways of obtaining a beta of 1 mm, or even less, provided a very small emittance is available. There is however at least one possible problem that has not been looked at. After the collision the bunches will be greatly refocused by their passage through one another and these "debris" may well not pass through the apertures of the opposite focusing quads. Will this produce unacceptable background? Or even damage the magnets? Clearly much more study is required.

6. CONCLUSION

The main conclusion of this discussion is that the best wavelength for an SSC equivalent electron positron collider is in the few millimeter region. If conventional wavelengths of a few centimeter are used then the likely cost of providing the microwave energy is excessive. If a laser is employed the likely cost of providing the average beam power is excessive. Efforts should be made to develop efficient sources in the millimeter range.

A second conclusion might be, providing the right choice of wavelength is made and providing an FEL and bunch compression scheme is practical, that the likely cost of an SSC equivalent electron positron collider would be in the region of 500 million dollars (see Table I). We might conclude therefore that the overall cost of an SSC-equivalent linear collider could, in principle, be lower than that of the SSC. The problem is much more complex than this indicates. The considerations of luminosity and beam power have led to a requirement for extremely high particle density at interactions (of the order of 10^{27} electrons per cc). This, in turn, requires more difficult mechanical tolerances in the accelerating structure beyond that attained in the past and establishes a need for novel methods of achieving final focusing. The formal optimization based on unit costs indicates the necessity to use wavelengths shorter than those at SLAC. This, in combination with the increased mechanical tolerances, brings us into a new region of technology whose cost implications have not been studied. Thus, although cost scaling laws applied to the different parameters give a useful guide as to how an overall cost minimum for a large linear collider might be attained, one should be loath to simply add the costs associated with those parameters. There are still so many cost elements associated with things we do not as yet know how to do at all, such that overall cost estimates would not be meaningful.

I am nevertheless encouraged by this study and, despite all the reservations, believe that we will, in time, learn how to make linear colliders at costs that will enable us to go well beyond the SSC. The emphasis should however be on the "in time". There is much work to be done.

I have freely used ideas from many sources. In particular I wish to acknowledge the contribution of W. Panofsky (some of whose prose has been used), and of P. Wilson.

REFERENCES AND NOTES

1. B. Richter, "Requirements for Very High Energy Accelerators," SLAC-PUB-3630 (1985).

2. N. Kroll, "Surface Heating by Short Bunches of Radiation," 'Laser Acceleration of Particles,' *AIP Conference Proceedings #130*, p. 296 (1985).

3. P. B. Wilson, 'Linear Accelerators for TeV Colliders'; IBID, p. 560, and also SLAC AAS-Note 2.

4. For the purposes of this analysis I have taken the cost of a SLAC klystron to be $100,000 and its modulator and feeders to cost $200,000. I have further taken the klystron cost to be associated with peak power need, and divided the modulator cost into $100,000 for average power and $100,000 for stored energy. I was guided in this by conversations with Greg Loew of SLAC. The performance assumed, including SLED, was: stored energy 120 Joules, pulse length .8 μsec, and peak power 150 M watts. This yields $800/Joule (rounded to 1000), 7×10^{-4} dollars per peak watt, and $4 /average watt. Without SLED the cost per average watts is $2.

5. For induction linac driven FEL costs I have had to rely on conversations with members of the Two Beam Accelerator group who have demonstrated such a source. I have assumed that a peak power of 1 GW can be obtained (as reported in the Wall Street Journal). I have assumed a pulse length of 20 nsec and thus a stored energy of 20 Joules. I have assumed that such a system could be made to cycle at 1 kHz and that it would cost 1.4 million dollars. If I divide this cost into .2 million dollars for stored energy, .4 millon dollars

for average power and .8 million dollars for peak power; then I obtain costs of \$10/joule, \$20/ave. watt, and $.8 \times 10^{-3}$ dollar/peak watt. Despite the Wall Street Journal publication the performance of the system is classified and thus no reference is available for the 1 GW operation. A reference to a lower performance is: T. J. Orzechowski et al., Phys. Rev. Lett. **54** p. 889 (1985). See also J. S. Wurtele, p. 305, (1985).

6. For average power costs I used some numbers quoted by D. D. Lowenthal of Spectra Technology Inc. from a study by that firm for another application. Values as low as \$50 per watt and as high as \$1,000 per watt were given depending on optimism and time scale. I have taken \$100/watt. For peak power I note that Corkum achieved, for 2 psec pulses, power densities of the order of 10^{12} watts/cm^2. Assuming one tenth of this density, and taking \$100,000 as the cost of a 1 cm^2 final amplifier, I obtain 10^{-6} dollar/watt for peak power. The stored energy cost of \$10,000 per Joule is obtained by considering the cost of lasers designed for high storage capacity. It must again be emphasized that these estimates have errors of a factor three or so either up or down. The Corkum reference is: P. B. Corkum "High Power, sub-psec ten μm Pulse Generation", Optics Letters **8**, 514 (1983). See also D. Lowenthal and J. Slater, 'Laser Acceleration of Particles,' *AIP Conference Proceedings #130*, p. 818 (1985).

7. Z. D. Farkas, "Binary Power Multiplier" SLAC-PUB-3694.

8. The 'switched power linac' was proposed first by W. Willis, 'Laser Acceleration of Particles' *AIP Conference Proceedings #130*, p. 421 (1985); see also *Proceedings of the ECFA/INFN Workshop*, CERN 85/07 (1985). F. Villar has a similar idea: SLAC-PUB-3804. In general, these papers concern a single pulse of radiation generated by a pulse of current from a wire photocathode to a high voltage anode. The idea can be extended to the excitation of a resonant standing wave in a small cavity by fast pulsing a photocathode wire within such a cavity. This idea, which I refer to as a "microlasertron" may be compared with a conventional lasertron in which a photocathode is again pulsed, but in which the electrons produced are focused into a bunched beam and energy extracted as in a klystron by ring cavities about the beam. See E. L. Garwin et al., "An Experimental Program to Build a Multimegawatt Lasertron for Super Linear Colliders", 1985 Particle Accelerator Conference (to be published), IEEE Trans. Nucl. Sci. NS-32; also SLAC-PUB-3650.

9. P. B. Wilson, "High Energy Electron Linacs" SLAC-PUB-2884 (1982); also Ref. 3.

10. M. Bassetti and M. Gygi-Hanney, LEP-Note-221, CERN, Geneva (1980).

11. T. Erber and G. B. Baumgartner Jr., *Proc. 12th International Conference on High Energy Accelerators* (Fermilab, August 1983), p. 372; T. Himel and J. Siegrist, "Quantum Effects in Linear Collider Scaling Laws," 'Laser Acceleration of Particles,' *AIP Conference Proceedings #130*, p. 602 (1985). See also Ref. 3.

12. Roger Erickson, "Final Focus", SLAC AAS Note #6 (1985).

13. R. Palmer "Super Disruption", SLAC-PUB-3688.

COOLING RINGS FOR TEV COLLIDERS*

R. B. Palmer[†]

Stanford Linear Accelerator Center
Stanford University, Stanford, California 94305

1. INTRODUCTION

We are now familiar[1] with the relation for the beam power for a quantum beamstrahlung limited collider:

$$N \approx 1.7 \times 10^{12} \left(\frac{\epsilon_n \beta^* \delta^3}{\sigma'_z} \right)^{1/2} \quad \text{(mks)} \tag{1a}$$

$$f \approx 3.6 \times 10^{-24} \frac{\mathcal{L}\sigma'_z}{\gamma \delta^3} \quad \text{(mks)} \tag{1b}$$

$$P_{beam} \approx .5 \times 10^{-24} \, \mathcal{L} \, (\epsilon_n \, \beta^* \, \sigma'_z)^{1/2} \, \delta^{-3/2} \quad \text{(mks)} \tag{1c}$$

Burt Richter and others have rather arbitrarily considered various desirable values for these constants. I will try:

$$\beta^* = 1 \text{ mm} \quad (10^{-3} \text{ m}) \quad \text{(final focus strength)}$$
$$\sigma'_2 = 1 \, \mu \quad (10^{-6} \text{ m}) \quad \text{(bunch length at collision)}$$
$$\delta = .16 \quad \text{(beamstrahlung fractional energy loss)}$$
$$\mathcal{L} = 10^{33} \text{ cm}^{-2} \text{ sec}^{-1} \quad (10^{37} \text{ m}^{-2} \text{ sec}^{-1}) \quad \text{(luminosity)}$$
$$\epsilon'_n = 1.35 \times 10^{-8} \text{ m} \quad \text{(normalized emittance)}$$
$$\gamma' = 3 \times 10^6 \quad (1.5 \text{ TeV}) \quad \text{(collision energy)}$$

With these values one obtains a

$$N \approx 4 \times 10^8 \quad \text{(particles per bunch)}$$
$$f \approx 3.0 \text{ kHz} \quad \text{(repetition frequency)}$$
$$P_{beam} \approx .3 \text{ M Watts} \quad \text{(power per beam)}$$

---

* Work supported by the Department of Energy, contract DE–AC03–76SF00515.
† On leave from Brookhaven National Laboratory.

A value which yields a total wall plug power for both beams at 1% efficiency (c.f. SLAC eff. is less than 10^{-3}) of:

$$P_{wall} \approx 60 \text{ M Watts}$$

which is reasonable.

The question I want to address here is: can one obtain $\sim \epsilon_n = 10^{-8}$ in any plausible cooling ring. In order to answer this one must consider not only quantum fluctuations but also intra beam scattering, cooling rates and ring acceptance.

2. COOLING RATE

Cooling arises in a ring because the synchrotron energy loss occurs not only longitudinally, but also, if the beam has a finite angular divergence, transversely. The rf cavities make up the longitudinal component but leave the loss of transverse component.

The rate of cooling of transverse momentum is proportional to the rate of loss of energy (mostly longitudinal and made up by the rf). Thus the time τ_q to lower the transverse momentum by "e" is given by

$$2.718 \approx \text{"e"} = \int \frac{\Delta E}{E} = \frac{2\ e^2}{3\ m_0 c} \frac{\beta^4 \gamma^3}{\rho^2} \tau_q\ F_m$$

$$\approx 9 \times 10^{-7} \frac{\gamma^3}{\rho^2} \tau_q\ F_m \quad \text{mks}$$

where F_m is the fraction of the ring filled with magnets. Thus

$$\tau_{qx,y} \approx \frac{3 \times 10^6}{J_{x,y}} \frac{\rho^2}{F_m \gamma^3} \qquad \text{mks} \tag{2a}$$

J_x is the partition function[2] in the bending plane which is equal to 1 for a separated function lattice. In any case:

$$J_x + J_y + J_z = 4 \tag{2b}$$

J_y is hard to shift from 1. J_L is 2 in a separate function lattice and can, at best be lowered to say .5, at which point $J_x \approx 2.5$.

Equation (2) assumes no mixing between horizontal and vertical emittance. Or alternatively it implies that both are being cooled simultaneously, as for instance is true initially. As equilibrium is approached, however, the horizontal emittance is being blown up by fluctuations and intrabeam scattering, while the vertical is not. Under these conditions equation (2) is only true in the absence of mixing. If we introduce a mixing parameter ς which is =0 for no mixing and =1 for full mixing then J_x can be substituted by $J_x + \varsigma J_y$. However, this is true only when the vertical emittance is cold. Initially, we must set $\varsigma = 0$ whether there is or is not mixing.

Adding this term and substituting the field B for ρ:

$$\rho \approx 1.7 \times 10^{-3} \ \gamma/B \tag{3a}$$

$$\tau_q \approx \frac{8.3}{J_x + \varsigma J_y} \ \frac{1}{B^2 \ \gamma \ F_m} \quad \text{mks} \tag{3b}$$

For instance the SLAC cooling ring has $B \approx 2$ tesla, $\gamma \approx 2.4 \times 10^3$, $F_m \approx .36$, $J_x \approx 1$ and since we are considering initial cooling, $\varsigma = 0$. The equation then gives $\tau \approx 2.4 \times 10^{-3}$ sec. This may be compared with the published[3] value of 3×10^{-3} sec, which is near enough for our purposes.

3. EQUILIBRIUM EMITTANCE FROM QUANTUM FLUCTUATIONS

The existence of an equilibrium emittance arises because of the existence, in a ring, of a dispersion η. Different momenta have different orbits and when a sudden charge of momentum occurs due to the radiation of a photon, the particle finds itself in a position away from its equilibrium. Before it can be re-accelerated by the cavity it starts oscillating about its new orbit and, as a result, gains transverse momentum. This effect, balanced against the cooling, yields[2] an equilibrium emittance ϵ_{qn} (the q is for quantum, the n is for normalized)

$$\epsilon_{qn} = \frac{C_q}{J_x + \varsigma J_y} \ \gamma^3 \ \left\langle \frac{H}{\rho} \right\rangle \tag{4a}$$

where

$$C_q = \frac{55}{32\sqrt{3}} \ \frac{\hbar}{mc} \approx 3.8 \times 10^{-13} \ \text{m} \tag{4b}$$

Since

$$\rho \approx 1.7 \times 10^{-3} \ \frac{\gamma}{B}$$

$$\tag{4c}$$

$$\epsilon_{qn} \approx 2.2 \times 10^{-10} \ \frac{1}{J_x + \varsigma J_y} \ \gamma^2 \ \langle HB \rangle$$

The function H depends on lattice parameters round the ring:

$$H = \frac{1 + \beta'^{\ 2}/4}{\beta} \ \eta^2 - \beta' \eta \eta' + \beta \eta'^{\ 2} \tag{4d}$$

β and η are the lattice parameters in the bending plane, the hyphen indicating the differential with respect to length. Note that where $B = 0$ it does not matter what H is.

Obviously the average over the lattice of a function like H is rather complicated and depends on the lattice. For a given number of bending magnets n it can be minimized and, assuming $J_x = 1$, $\xi = 0$ one obtains[4]

$$\epsilon_q \approx 8.3 \times 10^{-15} \ \gamma^3 \ \left(\frac{2\pi}{n} \right)^3 \quad \text{(m)} \tag{5}$$

Unfortunately the lattice required to achieve this minimum involves a relatively large amount of length devoted to manipulating β, β', η etc between each magnet.

As a result it would tend to have a low fraction of magnets F_m. This is not only bad for the cooling rate (see equation (2)) but will be bad for intrabeam scattering also. I will therefore choose to consider a more conventional ring with F_m as large as possible and with a sufficiently small phase advance per cell that I can take the approximation that β_x and η are constants around the ring. I will however, following Steffen, introduce one novelty[5]:

I will assume that each bending magnet is really a wiggler whose average bending field $\bar{B}$ is finite but less than average absolute field B. I define $\alpha_1 =$ average B in a magnet/local absolute B's

Remembering the definition of F_m, the average radius (R) of the ring is given by:

$$R = \rho/(\alpha_1 \; F_m) \tag{6}$$

Given these assumptions then

$$H \approx \eta^2/\beta_x \tag{7a}$$

since

$$\eta \approx R/Q^2 = \beta_x^2/R \tag{7b}$$

Thus[6]

$$H \approx \frac{\beta_x^3}{R^2} = \frac{\beta_x^3}{\rho^2} \; \alpha_1^2 \; F_m^2 \tag{7c}$$

And using

$$\rho \approx 1.7 \times 10^{-3} \; \frac{\gamma}{B}$$

$$H \approx 3.5 \times 10^5 \; \frac{\beta_x^3 \; B^2 \; \alpha_1^2 \; F_m^2}{\gamma^2} \tag{7d}$$

putting this into equation (4c):

$$\epsilon_{qn} \approx 7.7 \times 10^{-5} \; \frac{\beta_x^3 B^3 F_m^2 \alpha_1^2}{J_x + \varsigma J_y} \tag{8}$$

To see how good this approximation is I again consider the SLC cooling ring[3] for which $\bar{\beta}_x \approx .77$ m, $B = 2$ Tesla, $F_m = .36$, $\alpha_1 = 1$, $J_x = J_y = 1$ and $\varsigma = 1$ *which gives* $\epsilon_{qn} \approx 1.8 \times 10^{-5}$. *The published value is* 2×10^{-5}.

A slightly more familiar form of equation (8) may be obtained by noting again $Q = R/\beta_x$ *then for* $J_x = 1$ *and* $\varsigma = 0$:

$$\epsilon_{qn} \approx 3.8 \times 10^{-13} \; \frac{\gamma^3}{Q^3} \; \frac{1}{F_m \; \alpha_1} \tag{9}$$

Further, if I assume a 65° phase advance per half cell (SLC) then the bending angle θ per cell is

$$\theta = \frac{2\pi}{360} \frac{65}{Q}$$

and

$$\epsilon_{qn} \approx \frac{2 \times 10^{-13}}{J_x} \frac{\gamma^3 \theta^3}{F_m \, \alpha_1} \tag{10}$$

which may be compared with P. Wilson's equation[7]

$$\epsilon_{qn} \approx 4.8 \times 10^{-13} \frac{\gamma^3 \theta^3}{F_m}$$

Thus my number is more optimistic than his, but also agrees better with the SLC ring.

There is an obvious condition when using the wiggler. The local change in η within the wiggler must be kept small compared with the average η in the ring.

For small α_1 the orbits within the wiggler will consist of alternating arcs on either side of an essentially straight axis. The maximum orbit deviation from the axis, a, is given by

$$a = \ell_p^2 / 8\rho$$

where ℓ_p is the length of one arc, i.e. the length of an individual pole of the wiggler.

The change in dispersion, η', for unit dp/p, will be equal to a, and η' should be held to some small fraction f_ω of the average η. Thus

$$\eta' = a = \frac{\ell_p^2}{8\rho} = f_2 \eta = \frac{f_\omega \beta_x^2}{R} = \frac{f_\omega \beta_x^2 \alpha F_m}{\rho}$$

and thus

$$\ell_p = \beta_x (8\alpha F_m f_\omega)^{1/2} \tag{11}$$

ℓ_2, the length of one pole of the wiggler magnet, can be compared with the total length of wiggler ℓ_ω

$$\ell_\omega = F_m \, \phi \, \beta_x \tag{12}$$

where ϕ is the phase advance per 1/2 cell (i.e. per bending magnet).

The minimum number n_ω of wiggles per wiggler is thus

$$n_\omega \geq \sqrt{\frac{\phi^2 F_m}{8 \, \alpha_1 \, f_\omega}} \tag{13a}$$

or for $\phi \approx 1$ radian, $F_m = 1/2$, $f_\omega = 1/4$

$$n_\omega \geq \sqrt{\frac{1}{4\alpha_1}} \tag{13b}$$

4. INTRABEAM SCATTERING

In the above sections we have assumed that the beam current is small and scattering of particles within a bunch is negligible. If the current is raised then eventually this intrabeam scattering becomes significant and will eventually determine the equilibrium emittance independent of the quantum fluctuation limit of equation (8).

In principle, it may be argued, intrabeam scattering within a spherical phase space will not charge that phase space and should not lead to a blow up. In practice, however, in any plausible electron cooling ring the phase space is very far from spherical. For instance a longitudinal momentum spread of 10^{-3} at 3 GeV corresponds to a longitudinal Δp_ℓ of $.3 \times 10^{-6}$. This must be compared with the transverse momentum spread Δp_t which, even at $\epsilon_n = 10^{-8}$ and $\beta = 1$ m, is 1.7×10^{-6}. Thus $\Delta p_t \gg \Delta p_\ell$ and scattering transfers transverse phase space into the longitudinal. The resulting fluctuations in the momentum would perhaps be harmless but for the dispersion. As for the quantum effect the fluctuations in momentum in the presence of dispersion cause orbit jumps and result in a blow up of the transverse emittance.

The rate of growth due to these effects has been given[8] by

$$\frac{1}{\tau_c} = C_c \frac{I \gamma^2}{\varsigma \epsilon_{cn}^2} \left\langle \frac{H^{1/2}}{\sigma_p \gamma^3 \beta_y^{1/2}} \right\rangle \tag{14}$$

where $C_c \approx 10^{-10}$ m^2/(Amp sec) and $\epsilon_{vert} = \varsigma \epsilon_{horiz}$. The H here is the same as that above (equation (4d)) but the average is of course different.

Equilibrium is reached if this growth rate equals the quantum cooling rate $1/\tau_q$ thus

$$\epsilon_{cn} = \left\{ \frac{C_c I \tau_q}{\sigma_p \gamma \varsigma^2} \left\langle \frac{H^{1/2}}{\beta_y^{1/2}} \right\rangle \right\}^{1/2} \tag{15}$$

since

$$I = \frac{e N c}{\sqrt{2\pi} \sigma_z} \approx 1.9 \times 10^{-11} \frac{N}{\sigma_z}$$

$$\tau_q \approx 8.3/(B^2 \gamma F_m) \qquad \text{from (3)}$$

thus

$$\epsilon_{cn} \approx \frac{1.2 \times 10^{-10}}{\varsigma} \left(\frac{N}{\sigma_z B^2 \gamma^2 F_m \sigma_p} \left\langle \frac{H^{1/2}}{\beta_y^{1/2}} \right\rangle \right)^{1/2} \tag{16}$$

c.f. the SLAC cooling ring: $N = 5 \times 10^{10}$, $\sigma_z = 6 \times 10^{-3}, B = 2$, $\gamma = 2.4 \times 10^3$, $F_m = .36$, $\varsigma = 1$, $\sigma_p = 7.3 \times 10^{-4}$, $H = .017$, $\bar{\beta}_y = 1.7$, giving $\epsilon_{cn} \approx 1.3 \times 10^{-6}$ or 4% of ϵ_{qn}.

Defining the normalized longitudinal emittance:

$$\epsilon_{zn} = \gamma \sigma_p \sigma_z \tag{17}$$

$$\epsilon_{cn} \approx \frac{1.2 \times 10^{-10}}{\varsigma} \left(\frac{N}{\epsilon_{zn} B^2 \gamma F_m} \left\langle \frac{H^{1/2}}{\beta_y^{1/2}} \right\rangle \right)^{1/2} \tag{18}$$

Finally I can substitute for H from equation (7d):

$$\epsilon_{cn} \approx 2.9 \times 10^{-9} \frac{1}{\varsigma\gamma} \left(\frac{N\,\alpha_1}{\epsilon_{zn}\,B}\right)^{1/2} \left(\frac{\beta_z^3}{\beta_y}\right)^{1/4} \tag{19}$$

Equation (19) would be correct if the blow up due to quantum fluctuations were negligible. If the two are comparable one obtains:[8]

$$\epsilon_n \approx \frac{1}{2}\left[\epsilon_{qn} + \left(\epsilon_{qn}^2 + \epsilon_{cn}^2\right)^{1/2}\right] \tag{20}$$

In the following I will consider rings in which:

$$\epsilon_{qn} = \epsilon_{cn} \tag{21a}$$

and thus from equation (20):

$$\epsilon_n \approx 1.2\,\epsilon_{qn} \tag{21b}$$

5.   LONGITUDINAL EMITTANCE

Synchrotron radiation not only cools in the transverse directions but also in the longitudinal. High momentum particles radiate more than low momentum ones and thus the momentum spread tends to reduce. Balanced against this the quantum fluctuations of the process itself tends to increase the momentum spread. An equilibrium is reached given[2] by:

$$\frac{\Delta p}{p} = \sigma_p \approx \left(\frac{2}{J_z}\right) 1.1 \times 10^{-5} \left(\gamma\,B\right)^{1/2} \quad \text{mks} \tag{22}$$

J_z is the longitudinal partition function which for normal separate function machines has the value 2. As noted above (equation (2b))

$$J_2 + J_z + J_y = 4$$

as in general $J_y = 1$, thus

$$J_z \approx 3 - J_z \tag{23}$$

The total normalized longitudinal emittance ϵ_{zn} is

$$\epsilon_{zn} = \gamma\,\sigma_p\,\sigma_z \le \gamma'\,\sigma_p'\,\sigma_z' \tag{24}$$

and is related then to the minimum momentum spread σ_p' and bunch length σ_z' at the final collider energy γ'.

The bunch length σ_z is of course determined by the strength and frequency of the r.f.:

$$\sigma_z = \frac{R\,\alpha}{Q_s}\,\sigma_p \qquad (25a)$$

where Q_s is the synchrotron tune

$$Q_s = \left(\frac{e\,\alpha\,U\,h}{2\pi\,E}\right)^{1/2}$$

$$\alpha \approx 1/Q_z^2$$

thus

$$\sigma_z \approx \frac{R}{Q_z}\left(\frac{2\pi\,\gamma\,(mc^2)}{U\,h}\right)^{1/2}\sigma_p \qquad (25b)$$

where (mc^2) is the electron mass in electron volts, U is the voltage energy gain per revolution, h is the harmonic number (number of r.f. cycles per revolution), σ_p is beam momentum spread $\partial p/p$, and Q_z is the horizontal tune.

For our purposes however we can regard (Uh) as a free parameter and simply select ϵ_{zn} from bunch length and energy spread at full collider energy (equation (24)).

6. RINGS WITH QUANTUM AND INTRABEAM EMITTANCE MATCHED

Recall equation (4c)

$$\epsilon_{qn} \approx 2.2 \times 10^{-10}\,\frac{1}{J_z + \varsigma J_y}\,\gamma^2\,\langle HB\rangle$$

and equation (18)

$$\epsilon_{cn} \approx \frac{1.2 \times 10^{-10}}{\varsigma}\left(\frac{N}{\epsilon_{zn}\,B^2\gamma\,F_m}\langle\frac{H^{1/2}}{\beta_y^{1/2}}\rangle\right)^{1/2}$$

Setting $\epsilon_{qn} = \epsilon_{cn} = \frac{\epsilon_n}{1.2}$ and assuming that H is uniform about the ring, we can eliminate H and obtain:

$$\gamma = 3.57 \times 10^{-8}\,\frac{N^{1/2}(J_z + \varsigma J_y)^{1/4}}{\epsilon_{zn}^{1/2}\,\varsigma\,F_m^{1/2}\,\epsilon_n^{3/4}\,B^{5/4}\,\beta_y^{1/4}} \qquad (26)$$

For instance, if for a linear collider we chose the parameters listed in section 1:

$$\left.\begin{array}{l}\epsilon_n' = 1.35 \times 10^{-8} \\ \sigma_2' = 10^{-6} \\ \gamma' = 3 \times 10^6 \\ \sigma_p' = .33\% \\ \delta = .16 \\ N = 4 \times 10^8\end{array}\right\} \; \epsilon_{zn}' = \epsilon_{zn} = 10^{-2}\text{ m}$$

In order to operate at $\epsilon_n' = 1.3 \times 10^{-8}$ we need on equilibrium emittance ϵ_n

some what lower than this. I take

$$\epsilon_n = 3/4 \ \epsilon' \approx 1 \times 10^{-8} \text{ m rad}$$

choosing

$$\varsigma = 1 \qquad i.e. \text{ full } x, y \text{ mixing}$$

$$J_\gamma = J_y = 1 \qquad \text{normal partition functions}$$

$$F_m = .5 \qquad 50\% \text{ full of bending magnets}$$

$$B = 2 \qquad (20 \text{ Kg})$$

and

$$B_y = 1.4 \text{ m}$$

then

$$\gamma = 4.8 \times 10^3$$

$$E_e = 2.4 \text{ GeV}$$

The β_x we can now obtain from equation (8) turned around, and with equation (21)

$$\alpha_1^{2/3} \ \beta_x \approx 22 \ (J_x + \varsigma J_y)^{1/3} \ \frac{\epsilon^{1/3}}{B \ F_m^{2/3}} \qquad (27)$$

In our example $\epsilon_n = 10^{-8}$, $B = 2$, $F_m = .5$, $J_x = J_y = \varsigma = 1$:

$$\alpha_1^{2/3} \ \beta_x = 4.8 \times 10^{-2}$$

Now if α_1 the wiggler parameter were equal to 1 this implies a β_x of 5 cm which is not very reasonable at $E = 2.4$ GeV. So what is a reasonable β_x? At SLC $\bar{\beta}_x \approx .77$ m, at a γ of 2.4×10^3 and quadrupole apertures, a, of 2.5 cm. Scaling gives

$$\beta_x \propto (a \ \gamma)^{1/2}$$

Or normalizing to the SLC cooling ring

$$\beta_x \geq .1 \ (a \ \gamma)^{1/2} \qquad (\text{mks}) \qquad (28)$$

If we take the aperture, a, to be 2.5 mm (note that the beam will be only tens of microns in diameter) then for 2.4 GeV

$$\beta_x(\text{reasonable}) \approx .34 \text{ m}$$

and thus $\alpha_1 \approx .06$.

The ring radius R is given by

$$R \approx 1.7 \times 10^{-3} \frac{\gamma}{B} \frac{1}{F_m \alpha_1} \qquad (29)$$

For our example $\gamma = 4.8 \times 10^3$, $B = 2$, $F_m = .5$ and $\alpha = .06$ thus

$$R = 130 \text{ m}$$

Now

$$Q_x = \frac{R}{\beta_x} \approx 390$$

$$Q_y = \frac{R}{\beta_y} \approx 100$$

$$\frac{Q_x}{Q_y} \approx 4$$

Now we can look at what σ_p and σ_z are. For σ_p I will use equation (22), which is for quantum fluctuations only. It will at least give the right order of magnitude

$$\sigma_p \approx \left(\frac{2}{3 - J_x} \right) 1.1 \times 10^{-5} \, (\gamma \, B)^{1/2}$$

which for $J_x = 1$, $\gamma = 4.8 \times 10^3$ and $B = 2$ gives

$$\sigma_p \approx 1.0 \times 10^{-3}$$

σ_z is then given by

$$\sigma_z = \epsilon_{zn}/(\gamma \, \sigma_p)$$

which for $\epsilon_{zn} = 10^{-2}$, $\gamma = 4.8 \times 10^3$ gives

$$\sigma_z = 2 \times 10^{-3} \text{ m}$$

Finally we calculate the cooling rate given by equation (2)

$$\tau \approx \frac{8.3}{J_x} \frac{1}{B^2 \, \gamma \, F_m}$$

which for $\gamma = 4.8 \times 10^3$, $B = 2$, $F_m = .3$, $J_x = 1$:

$$\tau = .9 \times 10^{-3} \text{ sec}$$

We note that the diameter is not so unreasonable, it is less than PEP. The cooling rate is relatively fast and most parameters are not unreasonable. But the tunes are very high. Will such a ring have any acceptance?

7. ACCEPTANCE

I know of no generally accepted scaling law or equation for the acceptance of a lattice. What follows is therefore not to be taken too seriously. I will assume that the acceptance is limited by non-linear effects coming from sextupoles inserted to correct chromaticity (i.e. changes of Q with momentum). As before I will assume a lattice with essentially constant values of β, ς etc, i.e. a lattice with a sufficiently small phase advance per cell that I can think of the focussing as being continuous.

I define k to be a focussing strength, ℓ_q the quadrupole lengths and G the quadrupole field gradients:

$$k = \ell_q\, G \tag{30}$$

$$\beta \propto 1/k^{1/2} \tag{31}$$

and note

$$2\,\frac{d\beta}{\beta} = -\frac{dk}{k} = \frac{dp}{p} \tag{32}$$

In order to correct this variation of β with momentum we insert sextupoles around the ring. Again we assume that the phase advance is so small that the sextupole effect is essentially continuous and corresponds to a variation Δk of the focussing strength k with the average radial position ΔR of the beam:

$$\Delta k = S\; dr = S\,\eta\,\frac{dp}{p} \tag{33}$$

where S is the sextupole strength. Adding this term to equation (32) we obtain

$$2\,\frac{d\beta}{\beta} = -\frac{dk}{k} + \frac{\Delta k}{k}$$

$$= \frac{dp}{p} - \frac{S}{k}\eta\,\frac{dp}{p} \tag{34}$$

So, for no charge in tune β $(d\beta = 0)$ we require

$$S = \frac{k}{\eta} \tag{35}$$

If the sextupole strength is provided by sextupoles of length ℓ_s at every quad length ℓ_q then we note that

$$S = \ell_s\,\frac{dG_x}{dr} \tag{36}$$

and since in a sextupole

$$B = \frac{B_p}{(a/2)^2}\cdot r^2$$

$$G_s = \frac{B_p}{(a/2)^2}\cdot 2r$$

$$\frac{dG_s}{dr} = 2\cdot\frac{B_p}{(a/2)^2}$$

so

$$S = \ell_s \, 2 \, \frac{B_p}{(a/2)^2} \tag{37}$$

and since

$$k = \ell_q \, \frac{B_p}{a/2} \tag{38}$$

and

$$S = \frac{k}{\eta}$$

we find

$$\frac{\ell_s}{\ell_q} = \frac{a}{4\eta} \tag{39}$$

For our example $a = 2.5$ mm, $\eta = 9 \times 10^{-4}$ so $\ell_s/\ell_q = .7$ which means a lot of sextupole!

Now for a small enough emittance the effect of the sextupole strength is only seen as a charge in quadrupole strength. As the emittance rises however the more extreme orbits will see the nonlinear effects of the sextupoles. The relative magnitude of these non linear effects can be assessed by looking at the charge of focussing strength $\Delta k'$ arising from the maximum amplitude of oscillation σ.

$$\Delta k' = -S \, \hat{\sigma} \tag{40}$$

My assumption will be that non linear effects will become serious when this shift in focussing strength becomes a significant fraction f_S of the normal focussing strength k

$$f_S = \frac{\Delta k'}{k} = \frac{S \hat{\sigma}}{k} \tag{41}$$

now

$$\hat{\sigma} = \sqrt{\frac{\hat{\epsilon}_n \hat{\beta}}{\gamma}} \qquad \text{and} \qquad S = k/\eta$$

so

$$\hat{\epsilon}_n = \gamma \, \frac{f_S^2 \eta^2}{\hat{\beta}} \tag{42}$$

Now in order to reduce intrabeam scattering it is desirable to have $\beta_y > \beta_x$ and for the same reason one likes strong mixing so that $\epsilon_y \approx \epsilon_x$. Under these circumstances $\hat{\sigma}$ will be in the vertical direction y:

$$\hat{\epsilon}_{ny} = \frac{\gamma \, f_S^2 \, \eta^2}{\beta_y} \approx \frac{\gamma \, f_S^2 \, \beta_x^4}{R^2 \, \beta_y} \tag{43a}$$

$$\hat{\epsilon}_{ny} = \gamma \, f_S^2 \, R \, \frac{Q_y}{Q_x^4} \tag{43b}$$

the fraction, or fudge factor, f_S we can obtain from the SLC example

$$\hat{\epsilon}_n(SLC) = 10^{-2} = 2.4 \times 10^3 \ f_S^2 \ 5.6 \ \frac{3.25}{7.25^4}$$

$$f_S = 2.5 \times 10^{-2}$$

(44)

i.e. our scaling law implies that when the non linear focussing is more than 2.5% of the linear focussing the orbits become unstable. A not unreasonable conclusion.

Our scaling law thus predicts:

$$\hat{\epsilon}_{ny} \approx 6 \times 10^{-4} \ \gamma \ R \ \frac{Q_y}{Q_x^4}$$

(45)

For our example $\hat{\epsilon}_{ny} = 1.6 \times 10^{-6}$ which is very small, but still 160 times the equilibrium emittance.

8. CONCLUSIONS

I have summarized the assumptions in our example in Table I, and the calculated parameters in Table II, together with those for the SLC ring. As noted before there seems nothing impossible about such a ring although the magnet apertures of 2.5 mm, the tune of 390, and acceptance of 20 microns are certainly daunting.

Table I

Assumed Parameters of Example (A)

Including Variations Assumed in Later Examples

Collider Energy	E'	$1.5 + 1.5$		TeV
Collider Luminosity	$\mathcal{L}$	10^{33}		cm^{-2} sec^{-1}
Final Focus	β^*	1.0		mm
Final Bunch Length	σ_z'	1.0		μm
Final Mom. Spread	σ_p'	3.0×10^{-3}	(D : 3.3×10^{-4})	
Beamstrahlung Mom. Loss	δ	.16	(B : .32)	
Horizontal-Vertical Mixing	ς	1		
Partition Function	J_x	1	(C : 2)	
Dipole Fraction of Circ.	F_m	.5		
Dipole Field	B	2	(E : 4)	Telsa
Tune Ratio	Q_x/Q_y	4	(G : 40)	
Magnet Apertures	a	2.5	(F : 10)	mm
Phase Advance/$\frac{1}{2}$ Cell	ϕ	65°		
$d\eta/\eta$ in Wiggle	f_w	.25		
sext./quad. Strength	f_s	2.5×10^{-2}		

Table II

Calculated Parameters for Example A
and Comparison with SLC Cooling Ring

		Ex − A	SLC	
ϵ_{qn}	Quantum ϵ_n Equilibrium	$.8 \times 10^{-8}$	2×10^{-5}	m
ϵ_{cn}	Coulomb ϵ_n Equilibrium	$.8 \times 10^{-8}$	1.3×10^{-6}	m
N	Particles/Bunch	4×10^8	5×10^{10}	
f	Pulse Repetition	3×10^3	120	Hz
P	Power/Beam	.3 MW	70 KW	
E	of Cooling Ring	2.4	1.2	GeV
R	Radius of Ring	130	5.6	m
α	Wiggler $B/\bar{B}$	.06	1	
ℓ	Wiggler	19	32	cm
ℓ	Pole	≤ 8	—	cm
η	Chromaticity	9×10^{-4}	1.7×10^{-2}	m
$\bar{\beta}_x$		.34	.77	m
$\bar{\beta}_y$		1.4	1.7	m
Q_x		390	7.25	
Q_y		100	3.25	
$\ell_{\text{sext}}/\ell_{\text{quad}}$		1.0		
$\hat{\epsilon}_n$	Acceptance	1.6×10^{-6}	1×10^{-2}	m
$\langle\hat{\sigma}\rangle$	Acceptance	20 μm	2.6 mm	
$\hat{\epsilon}_n/\epsilon_n$		160	500	
σ_p	dp/p in Ring	1×10^{-3}	$.73 \times 10^{-3}$	
σ_z	in Ring	2	5.9	mm
τ	Cooling Time Constant	.9	3	msec

Table III

Calculated Parameters of Various Cooling Rings

	Example		A	B $\delta_{beamstrahlung}$ = .32 (.16)	C Partition $J_z = 2$ (1)	D Mom. Spread at Final Focus $\sigma'_p = 3.3\times10^{-4}$ (3.3×10^{-3})	E Magnetic Field $B \equiv 4$ Telsa (2)	F Magnet Aperture $a = 10$ mm (2.5 mm)	G Ratio of Tunes $Q_x/Q_y = 40$ (4)	H Emittance at Final Focus $\epsilon'_n = 1.35\times10^{-9}$ (1.35×10^{-9})	I Emittance at Final Focus $\epsilon'_n = 1.35\times10^{-7}$ (1.35×10^{-8})
ϵ_n	Equilibrium	m	10^{-8}	10^{-8}	10^{-8}	10^{-8}	10^{-8}	10^{-8}	10^{-8}	10^{-9}	10^{-7}
N	Electrons/Bunch		4×10^8	1×10^9	4×10^8	4×10^8	4×10^8	4×10^8	4×10^8	1.3×10^8	13×10^8
f	Pulse Frequency	kHz	3	.37	3	3	3	3	3	3	3
P	Power/Beam	MW	.3	.1	.3	.3	.3	.3	.3	.09	.9
E	of Ring	GeV	2.4	3.8	2.6	6.6	1.1	2.1	1.4	6.7	.8
R	of Ring	m	130	300	130	800	50	290	54	2500	7
α_1	Wiggler B/B		.06	.04	.07	.03	.04	.02	.09	.009	.4
l_w	Length Wiggler	cm	19	24	20	32	13	36	15	32	12
l_p	Length of Pole	cm	8	9	9	9	4.5	10	8	5	13
η	Chromaticity	m	9×10^{-4}	6×10^{-4}	10×10^{-4}	4×10^{-4}	11×10^{-4}	14×10^{-4}	13×10^{-4}	1.3×10^{-4}	60×10^{-4}
β_x		m	.34	.43	.36	.57	.23	.6	.27	.57	.2
β_y		m	1.4	1.7	1.4	2.3	.9	2.5	10.6	2.3	.8
Q_x	Horizontal Tune		390	690	350	1400	210	450	205	4400	34
Q_y	Vertical Tune		100	173	90	350	50	110	5	1100	8
l_{ext}/l_{quad}			.7	1.0	.6	1.6	.6	1.8	.5	4.8	.1
ξ_n	Acceptance	m	1.6×10^{-6}	1×10^{-6}	2.2×10^{-6}	$.57\times10^{-6}$	1.7×10^{-6}	1.8×10^{-6}	$.3\times10^{-6}$	6×10^{-8}	4.4×10^{-5}
$\hat{a}_y$	Acceptance	μ	20	15	25	10	27	34	30	3	150
ϵ_n/ϵ_n			160	100	200	57	170	180	26	56	440
σ_p	dp/p in Ring		1×10^{-3}	1.3×10^{-3}	2.2×10^{-3}	1.7×10^{-3}	1×10^{-3}	1×10^{-3}	$.8\times10^{-3}$	1.8×10^{-3}	$.6\times10^{-3}$
σ_x	in Ring	mm	2	1.0	.8	.04	4.3	2.4	4.2	.4	9
τ	Cooling Time	msec	.9	.6	.8	.3	.5	1.0	1.5	.3	2.5

In order to see how the ring depends on the assumptions, I have calculated a number of rings changing each assumption in turn (see Table III). What do I conclude:

1. Only a small gain is obtained (example C) by messing with the partition functions.

2. A very significant gain is made by using higher (presumably superconducting) bending fields. Example E using 4 Tesla magnets has a radius reduced from 130 to only 40 meters and the Q has dropped from 390 to 210. The physical acceptance has gone up a bit (20 μ to 27 μ) and the cooling rate has gone up too. Whether such advantages would compensate for the great complication of superconducting magnets I do not know, but this should be studied.

3. A reduction in the ring diameter is obtained (example G) by allowing β_y to be much larger than β_x. For $\beta_y/\beta_x = 40$ the diameter has dropped from 130 to 54 meters. But the acceptance has dropped and is now only 26 times the equilibrium value. This is not in principle unacceptable, the ring could be fed from another pre cooling ring, but we must remember that the acceptance law is not reliable and only lattice tracing would tell us how bad this example is.

4. As would be expected the ring gets bigger if the magnet apertures are increased (example F).

5. Far more serious, however, is the ring diameter increase if the momentum spread of the beam is reduced (example D). This is a serious question. I had assumed .3% $\Delta p/p$ at 1.5 TeV and no dilution. This implies 3% $\Delta p/p$ at 150 GeV if the final bunching were performed at this energy. The short bunches (1 μ) are desirable to suppress wake field effects but some have suggested that small momentum spread may also be required. If really true (and I personally doubt it) this would have serious consequences for the attainability of emittances of 10^{-8}.

6. If even lower beam power per luminosity is required. (For a 5 TeV machine, for instance), then we may attempt to obtain an even lower emittance (example H). This does look pretty bad. The sextupoles are 5 times as long as the quads and the acceptance is only 3 microns!

7. The power can be more easily reduced by allowing a higher beamstrahlung energy loss (example B) the resulting higher current in the cooling ring does make the ring larger and more expensive but to no where near the extent of a lower emittance.

8. Finally I give the parameters of a 10^{-7} m radian emittance case. With a radius of only 7 meters it would be a lovely ring to try and build. Note, however, that this would not be suitable for the SLC. The number of particles per bunch is far too low.

I would like to thank Bob Siemann for starting me on this study, and Albert Hoffman for his frequent help.

REFERENCES

1. (a) W. K. H. Panofsky, Limiting Technologies for Particle Beams and High Energy Physics, SLAC-PUB-3735 (1985).

(b) R. B. Palmer, Collider Scaling and Cost Estimation, SLAC-PUB-3849 and Proc. SLAC Summer Inst. 1985.

(c) P. B. Wilson, Linear Accelerators for TeV Colliders, SLAC-PUB-3674; and Laser Acceleration of Particles (Malibu 1985) AIP Conference Proc. # 130, p. 560.

2. M. Sands, The Physics of Storage Rings, SLAC-PUB-121 (1979), p. 110.

3. G. E. Fischer *et al.*, A 1.2 GeV Damping Ring Complex for the SLC, SLAC-PUB-3170 (1983).

4. L. S. Teng, Minimum Emittance Lattice for Synchrotron Radiation Storage Rings, FNAL Report LS-17 (1985).

5. K. Steffen, The Wiggler Storage Ring, Internal Report, DESY PET 79/05 (1979).

6. Matt Sands has pointed out that this relation is more generally true, since

$$H = \alpha R/Q_x$$
$$\alpha = 1/Q_x^2$$

thus $\qquad H \approx \beta_x^3/R^2$

see M. Sands, The Physics of Storage Rings, SLAC-PUB-121 (1979), p. 134.

7. See Ref. 1(c) who quotes: H. Wiedemann, 11th Int. Conf. on High Energy Accelerators (1980), p. 693.

8. The approximation used here was taken from J. Bisognano *et al.*, Feasibility Study of Storage Ring for a High Power XUV Free Electron Laser, LBL-19771 (1985). For a more basic reference see J. LeDuff, Orsay Report LAL 1134 (1965).

DISCUSSION

TAZZARI:

It seems to me that no account is taken, in the expression for admittance, of the very strong wiggler field that is assumed in the calculation and that is instrumental to obtaining a short damping time.

As a comment, the present state of the art has problems with 700 m long rings at ≈ 5 GeV with Q values of ≈ 35. A 700 m long ring with a Q of ≈ 400 has to be designed in detail in order to be believable. Also, a ring with a normalized emittance of 10^{-8} m rad would most probably be longitudinally unstable. But of course one needs an actual lattice to check that.

COURANT:

In estimating ultimate emittance of damping ring, you used the approximation $H \approx \eta^2/\beta x$. If wigglers are present, there is a second term $(\eta')^2 \beta$ which may be much larger. Has this been included in calculations?

PALMER:

I did check it. It is discussed in the paper.

LINEAR COLLIDERS DRIVEN BY A SUPERCONDUCTING LINAC-FEL SYSTEM

U.Amaldi and C.Pellegrini[*]

CERN
1211 Genève 23, Suisse

ABSTRACT

In this paper we discuss linear colliders in the TeV energy region, based on a two beam accelerator scheme. The low energy beam is used in a Free Electron Laser to produce short wavelength radiation, in the range from one cm to a fraction of a mm. The energy lost by this beam is restored by a superconducting linac. The short wavelength radiation is fed to a high frequency, linac-type structure, where the high energy beam is accelerated. We give a review of the scaling laws for a linear collider and use them to find some possible set of parameters for our system. We then discuss some of the accelerator technical problems and the beam physics problems encountered in the design of such a system.

1. INTRODUCTION

Particle accelerators have been built and used for about 50 years. During this time their energy has increased from the MeV to the TeV region, using at each energy step new and innovative ideas and techniques. Of the next electron positron colliders, like Tristan, LEP and SLC, now being built, LEP (100 GeV, 4.3 km radius) is probably the last member of a family of accelerators, the electron-positron storage rings, which starting from AdA (250 MeV, 1 m radius) in the early 60's, has evolved through ACO, Adone, Vep2 and 3, Spear, DC1, CESR, PEP and Petra and has given many important contributions to high energy physics.

---

[*] On leave from Brookhaven National Laboratory, Upton, N.Y., USA

It was pointed out by Richter[1] that, because of synchrotron radiation losses, the radius of an electron storage ring scales like the square of its energy. This makes difficult and uneconomical the construction of rings at energy larger than that of LEP.

The next family of electron-positron colliders is that of "linear colliders"[2], of which SLC (50 GeV, length 3 km) is the first member. We hope that an extension of the ideas, physical principles and techniques that are under test in the SLC will allow us to build colliders in the TeV region, with luminosity much larger than $10^{32}/cm^2/s$, as required by the expectation of the point-like cross-section which drops as $1/E^2$.

In this paper, we discuss some of the problems facing us in the design and construction of a high luminosity linear collider in the TeV region. The scaling laws for this system will be discussed in section 2. This will help us in understanding what are the characteristics needed for the electron and positron beams. As has already been discussed by many authors[2-6], the two most important effects limiting the collider performance are the "disruption"[7] and the "beamstrahlung"[8-10]. In particular the beamstrahlung determines two different regimes of operation[8], the "quantum" and the "classical" regime, defined by the condition that the radiation critical energy is larger or smaller than the beam energy. To reduce the beam power, and the total power needed to operate the system, is necessary to work in the quantum regime. This becomes more so when going to energies of several TeV or luminosities in the 10^{34} cm^{-2} s^{-1} range.

In sections 3 and 4 we relate these beam characteristics to the accelerator characteristics, like operating frequency and acceleration rate. In these sections we show that there is a relationship between the choice of the regime of operation, quantum or classical, and the choice of the accelerator frequency. In particular the quantum regime requires the use of accelerators operating at wavelength in the millimeter or submillimeter range, while in the classical regime one has to use wavelength around one centimeter. We believe this to be very important because the development of colliders in the range of 5 TeV and 10^{34} cm^{-2} s^{-1} will force us in the quantum regime and so in the millimetric waves.

To illustrate these results we will discuss, in sections 5, 6 and 8, a two-beam accelerator scheme based on a low frequency superconducting linac, which feeds energy to an electron beam used to operate a Free Electron Laser. The radiation produced in the Free Electron Laser is then used to accelerate a high energy beam to the TeV energy range, in a high frequency Linac-type structure, following an idea proposed by Sessler[11]. We will consider two examples, at (1+1) and (5+5) TeV. For each one we consider a collider in the classical regime using a linac operating at about 30 GHz, and one in the quantum regime operating at about 700 GHz. We will show that our scheme allows to reach accelerating fields in the range 0.1 to 1 GeV/m.

The damping rings needed to produce the electron and positron beams with the emittance required for high luminosity are discussed shortly in section 7. In the last section we discuss our conclusion and point out the many problems that need to be studied to obtain a full understanding of this collider.

2. SCALING LAWS

The main formulae relating the beam current and current density to the collider luminosity and other important parameters, have been discussed by several authors (see, for instance references 3 to 6). For the convenience of the reader we will summarize them again, limiting ourselves for simplicity to the case of electron and positron beams with cylindrical symmetry. The discussion of the case of a flat beam and of its possible advantages can be found in ref.4.

For cylindrical bunches of N electrons or positrons, having transverse and longitudinal gaussian distribution with r.m.s. radius σ_t, and length σ_1 respectively, the luminosity is

$$\mathscr{L} = \frac{fN^2 H}{4\pi\sigma_t^2} \tag{1}$$

where f is the number of bunch crossing per second and H is an "enhancement factor", describing the self focusing of the electron and positron bunches during the collision.

When crossing, each bunch produces a focusing force on the particles of the other bunch. The resulting deviation is described by the disruption parameter, D, given by[2]

$$D = \frac{r_e N \sigma_1}{\gamma \sigma_t^2} \equiv \frac{\sigma_1}{F} \tag{2}$$

where, in the small D case, F is the focal distance of a focusing lens equivalent to the bunch, r_e is the classical electron radius and γ is the electron energy, E, in rest mass units, $E = \gamma mc^2$. This focusing of the bunches produces a pinch effect, which increases the bunch density and hence the luminosity. On the other hand if the focusing becomes too large, it can lead to a beam filamentation and a decrease in luminosity. This effect is described by the enhancement factor, H, which is a function of the disruption parameter D. This focusing and pinch effect has been studied by Hollebeek[7]. His results show that up to values of the disruption parameter, D, of the order or smaller than one, the luminosity can be calculated ignoring this effect ($H \simeq 1$); for D increasing from one to two the luminosity is enhanced by the pinch effect and one can gain up to a factor of 6 ($H \simeq 6$); for D larger than two, the luminosity tends to remain constant and then to decrease.

A quantity very important in determining the construction and operation cost of a collider is the average beam power, P, related to the particle energy, E, by

$$P = fNE . \tag{3}$$

The next important effect is beamstrahlung, describing the energy loss of an electron caused by the emission of synchrotron radiation in crossing the high magnetic field produced by the other bunch. Following the work of references 4,8,9,10 we can characterize this effect with two of three related parameters, namely the ratio, Υ, of the critical photon energy, E_c, to the beam energy $\Upsilon = 2E_c/3E$); the fluctuation parameter Γ; and the beamstrahlung parameter δ, which equals the average fractional energy loss in the limit $\Upsilon \ll 1$ [12]. In the limit $\Upsilon \ll 1$ the beamstrahlung spectral intensity can be described using the "classical" synchrotron radiation spectrum which is nearly flat up to the critical photon energy and then decreases exponentially. When $\Upsilon \gg 1$ quantum corrections to the synchrotron radiation spectrum become important and the photon spectrum is cut off at the beam energy. This is the so called quantum regime.

The synchrotron radiation critical energy E_c is such that[8]

$$\Upsilon = \frac{r_e \hbar c \gamma N \sqrt{H}}{\sqrt{2} mc^2 \sigma_t \sigma_1} \quad .$$ (4)

The fluctuation parameter is

$$\Gamma = \frac{r_e \gamma}{\alpha_{em} \sigma_1}$$ (5)

where $\alpha_{em} = 1/137$. The quantity Γ is equal, in the electron rest frame, to the ratio DE/mc^2, DE being the uncertainty in the electron energy produced by the finite interaction time $\Delta T = \sigma_1/2\gamma c$. The last parameter is

$$\delta = \frac{8 r_e^3 N^2 \gamma H}{21 \sqrt{\pi} \sigma_1 \sigma_t^2} = \frac{2^4 \alpha_{em} \Upsilon^2}{21 \sqrt{\pi} \Gamma} \quad .$$ (6)

The last equality shows that only two of the three parameters Υ, Γ, and δ, are independent.

The average fractional energy loss, $\langle \varepsilon \rangle = \langle \Delta E/E \rangle$, is given by

$$\langle \varepsilon \rangle = \delta, \quad \text{for} \quad \Upsilon \ll 1$$ (7)

$$\langle \varepsilon \rangle = (2/3)^{1/3} \delta / \Upsilon^{4/3}, \quad \text{for} \quad \Upsilon \gg 1 \quad .$$ (8)

Note that a factor 1/2 was missing in the equation of ref.9 corresponding to eq.8, as pointed out by Yokoya[10].

There are two other quantities that are of interest in evaluating the beamstrahlung effect; one is the average number of photons emitted per crossing[10]

$$N_\gamma = 8 \langle \varepsilon \rangle / 3 \Upsilon, \quad \text{for} \quad \Upsilon \ll 1$$ (9)

$$N_\gamma = 4 \langle \varepsilon \rangle, \quad \text{for} \quad \Upsilon \gg 1$$ (10)

and the other is the r.m.s. electron energy spread[9]

$$\sigma_\varepsilon = (\langle \varepsilon \rangle / \sqrt{2\pi}) \left\{ 1 + 31.3/N_\gamma \right\}^{1/2}, \quad \text{for} \quad \Upsilon \ll 1$$ (11)

$$\sigma_\varepsilon = 0.19 \langle \varepsilon \rangle \left\{ 1 + 64.8/N_\gamma \right\}^{1/2}, \quad \text{for} \quad \Upsilon \gg 1 \quad .$$ (12)

Of the two terms appearing in σ_ε, the first represents the energy spread due to the change of the energy loss with the particle trajectory (a particle crossing the other bunch on axis has zero energy loss, while one crossing on the edge see the largest magnetic field and has the largest loss), the second describes the effect of the fluctuation in the number of emitted photons. Note that the energy distribution due to beamstrahlung has typically a peak very close to the beam energy E, and a long tail at smaller energy. In these circumstances the r.m.s. spread σ_ε has a doubtful meaning.

Although we do not have any analytical result for the average energy loss in the intermediate cases, $\Upsilon \simeq 1$, we will use in our calculations simple interpolating formulae, as done in ref.4,

$$<\varepsilon> = \delta F(\Upsilon) \tag{13}$$

with the function $F(\Upsilon)$ given by[4]

$$F(\Upsilon) = \frac{1}{(1+(3/2)^{1/4}\Upsilon)^{4/3}} \quad . \tag{14}$$

Similarly we take

$$N_\gamma = 4 <\varepsilon> (2+3\Upsilon)/3\Upsilon \quad . \tag{15}$$

In designing a collider we want to obtain a certain luminosity for a given beam power, disruption and average beamstrahlung energy loss. The system is described by ten parameters (E, $\mathscr{L}$, P, D, $<\varepsilon>$, Υ, f, N, σ_t, σ_1) and the five equations (1),(2),(3),(4),(13), relate them. Hence one can specify five parameters and compute the others. If we use the five parameters E, $\mathscr{L}$, P, D, $<\varepsilon>$, (and remember that H is determined by D), we obtain the relationships

$$\sigma_1 = (1/4\pi r_e mc^2)(HPD/\mathscr{L}) \tag{16}$$

$$f = (1/C_1)(E\mathscr{L}^2 F(\Upsilon)/PHD <\varepsilon>) \tag{17}$$

$$N = C_1(P/E\mathscr{L})^2(<\varepsilon>/F(\Upsilon))HD \tag{18}$$

$$\sigma_t^2 = (C_1/4\pi)(P/E\mathscr{L})^3(<\varepsilon>/F(\Upsilon))DH^2 \tag{19}$$

where

$$C_1 = 21/(2^7\pi^{3/2}r_e^4) \tag{20}$$

and the ratio of critical to beam energy is obtained by solving the equation

$$\Upsilon^2 F(\Upsilon) = (7\pi^{3/2}r_e^2/2a_{em}^2)(E\mathscr{L}<\varepsilon>/PHD) \tag{21}$$

obtained from (1),(5),(6),(13),(16). This equation can be solved by introducing the "quantum parameter"[4]

$$q = (r_e/2\alpha_{em})(294\pi^3)^{1/4}(E\mathscr{L}\!\!<\!\varepsilon\!>/PHD)^{1/2} \qquad (22)$$

so that it becomes $F^{3/4}+qF^{1/4}=1$. For all values of q an approximate solution, to better than 5%, is

$$F(\mathsf{T}) = F(q) = \left\{1+4q(1+4q/5)/3+q^4\right\}^{-1} \qquad (23)$$

which allows to express directly Eqs.(16-19) as function of E, $\mathscr{L}$, P, D, $<\varepsilon>$.

Eq.(23) displays very clearly the two regimes. For $q\ll1$, classical regime, the q^2 and q^4 terms in (23) are negligible, and $F\simeq1-4q$. For $q\gg1$, quantum regime, the q^4 term is dominant, and $F\simeq1/q^4$. Using (18) and (23) one can then obtain a simple expression for the number of electrons, N_Q, in the quantum limit

$$N_Q \approx (7\sqrt{\pi})^3(3/32\alpha_{em}^2)^2(<\varepsilon>^3/HD) . \qquad (24)$$

We will use (24) later on in discussing the accelerator system.

Instead of σ_t it is often useful to introduce the normalized beam emittance ε_n

$$\varepsilon_n = \gamma\sigma_t^2/\beta^* \qquad (25)$$

where β^* is determined by the optical system focusing the beam at the collision point. Not to unnecessarily reduce the luminosity, we must require that $\beta^*\geq3\sigma_l$. Once β^* is given, eqs. (19),(23) and (25) can be numerically solved for D as a function of ε_n and of the other main parameters E, $\mathscr{L}$, P, $<\varepsilon>$. This is relevant in the quantum regime, where D is usually small and well below its upper limit $D\approx10$, while the emittence becomes very small and ε_n plays the role of a limiting parameter.

To illustrate the scaling laws we have considered two colliders, one with E=1 TeV, $\mathscr{L}=10^{33}$ cm^{-2}s^{-1}, the second with E=5 TeV, $\mathscr{L}=10^{34}$ cm^{-2}s^{-1}. In figures 1 to 4 we give the collision frequency, number of electrons/ /bunch, transverse and longitudinal r.m.s. bunch size versus beam power. In all cases we have assumed the same average energy loss, $<\varepsilon>$=0.08, and have changed the disruption parameter from a high, D=1.3, to a low, D=0.05, value.

The choice of one particular set of parameter must take into account the technical limitations existing for some quantities, like for instance the minimum transverse beam radius. This cannot be made smaller than a value determined by the beam emittance and by the final focusing system. Some of these limitations are not yet fully understood, so that any choice of parameters done now is necessarily only tentative.

In this spirit we consider now in more details two colliders, one for (1+1) TeV, $\mathscr{L}$=10^{33}/cm^2/s, another for (5+5) TeV, $\mathscr{L}$=10^{34}/cm^2/s. Even larger values of the luminosity might be needed to compensate the decrease in e$^+$-e$^-$ reaction cross section with increasing center of mass energy[4], but we shall see later that this implies very difficult conditions for some of the accelerator parameters. A low value of the average beamstrahlung

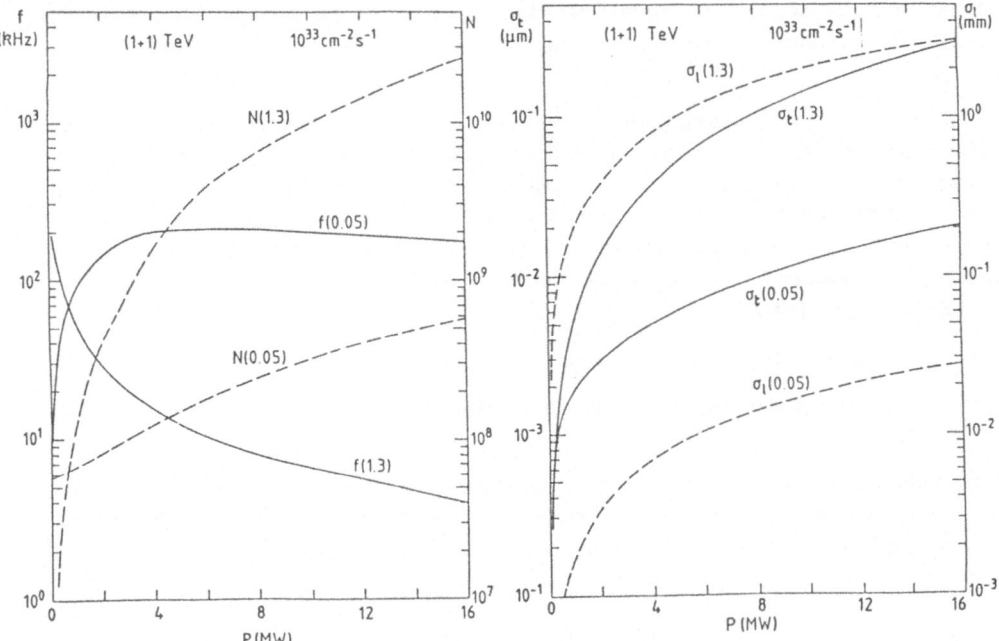

Fig. 1. Number of particles, N, and collision frequency, f, vs beam power, P, for two (1+1) TeV colliders, with $\langle\varepsilon\rangle=0.08$, for D=1.3 and D=0.05.

Fig. 2. R.m.s. bunch length, σ_1, and radius, σ_t, vs beam power, P, for two (1+1) TeV colliders, with $\langle\varepsilon\rangle=0.08$, for D=1.3 and D=0.05.

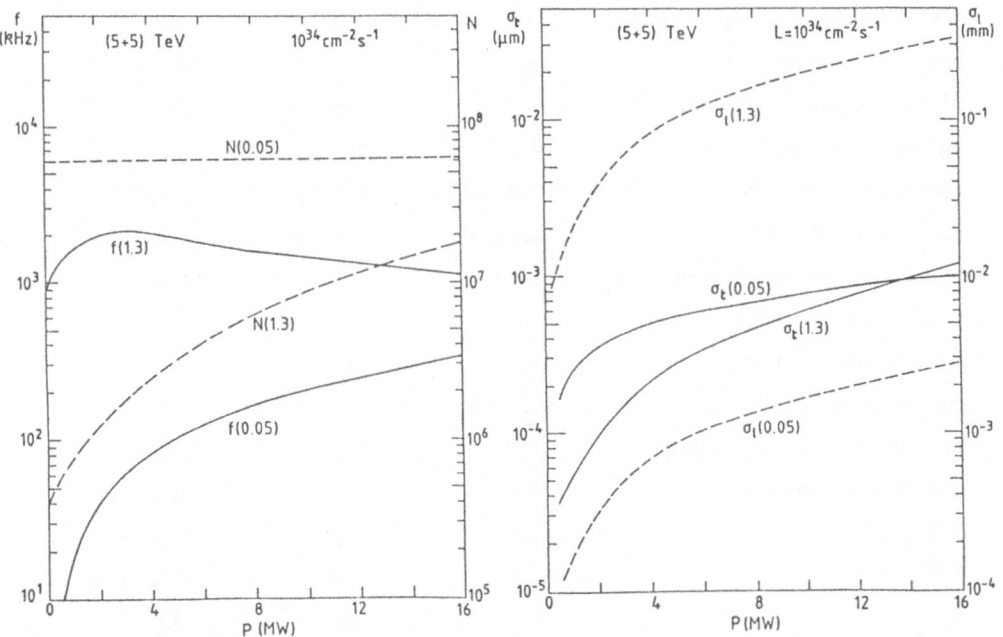

Fig. 3. Number of particles, N, and collision frequency, f, vs beam power, for two (5+5)TeV colliders, with $\langle\varepsilon\rangle=0.08$, for D=1.3 and D=0.05.

Fig. 4. R.m.s. bunch length, σ_1, and radius, σ_t, vs beam power, P, for two (5+5) TeV colliders, with $\langle\varepsilon\rangle=0.08$, for D=1.3 and D=0.05.

energy loss is chosen in order to limit the particle energy spread in the initial state and not to throw away too much of the energy given to the beam.

The sets of parameters given in Table 1 describe examples of possible colliders, and are not intended to represent optimized systems. They can however be useful to illustrate some of the problems that will be encountered in the construction of such systems. For these reasons we have considered colliders working in the classical regime, and in the quantum regime. These numbers will also be used to evaluate the characteristics of our FEL-driven linac.

We have chosen these examples using the following criteria:

a) among the five input parameters discussed earlier we keep constant E, $\mathscr{L}$, $\langle\varepsilon\rangle$;

b) when we reduce the power for constant luminosity, we have to compensate the reduced number of particles by squeezing down the beam radius, i.e. decreasing the emittance; since at the present stage of development

Table 1. Colliders Parameters.

	"Classical"	"Quantum"	"Quantum"
luminosity(cm^{-2}s^{-1}), $\mathscr{L}$	10^{33}	10^{33}	10^{34}
energy (TeV), E	1	1	5
fractional energy loss, $\langle\varepsilon\rangle$	.08	.08	.08
beam power (MW), P	10	3	6
disruption parameter, D	1.3	.05	.015
enhancement factor, H	4.5	1	1
electrons/bunch, N	9.8×10^9	1.1×10^8	2.0×10^8
frequency, Hz, f	6.4×10^3	1.7×10^5	3.7×10^4
r.m.s. bunch length (mm), σ_l	2	5.1×10^{-3}	3.1×10^{-4}
r.m.s. bunch radius (μm), σ_t	.15	4.0×10^{-3}	1.1×10^{-3}
final focus, β^* (mm)	7.0	.1	.1
emittance (m rad), ε_n	6.1×10^{-6}	3.1×10^{-7}	1.2×10^{-7}
beamstrahlung parameter, δ	.09	1.3	5×10^3
quantum parameter, q	.09	1.75	15.8
$\Upsilon=2E_c/3E$	.087	6.4	3.6×10^3
$F(\Upsilon)$	.89	.062	1.6×10^{-5}
N_γ	2.8	.35	.32
rms energy spread, σ_E	.11	.21	.22
Γ	3.7×10^{-4}	.15	12.2
electron density (cm^{-3})	1.4×10^{19}	8.6×10^{22}	3.4×10^{25}

we do not know how small a beam emittance can be produced we have chosen our three cases with an emittance smaller by about a factor of four than that obtained in SLC damping rings, for case 1, and smaller by about two order of magnitudes in cases 2 and 3; a smaller emittance would lead to an even smaller beam power;

c) when reducing the emittance we are also forced by (2) and (4) to go in a regime with smaller disruption and larger ratio of critical to beam energy; this fact is also reflected in Table 1, where one can see that case 1 is "classical" ($\Upsilon < 1$) and cases 2 and 3 are "quantistical";

d) we have not given a "classical" (5+5) TeV collider because it would require either an emittance smaller by another order of magnitude, and/or much larger beam power, so that we consider it impractical (the problems are apparent from Figs. 3 and 4).

The numbers appearing in Table 1 require some further comments. The first point is on the beam power; the real power needed to operate a system is obtained from P by dividing by the efficiency of energy transfer from the energy stored in the accelerator to the beam, η_3 and by the efficiency of energy transfer from the main energy source to the accelerator, η_2. The product $\eta_2 \eta_3$ can easily be small, in the range 10^{-1}-10^{-2}, and this can bring the power needed to operate the collider in the GW range. This points to the importance of reducing not only the beam power but also to develop acceleration methods and accelerator power sources with high efficiency. In other words, the efficiency is as important as the maximum accelerating field in comparing different accelerator schemes.

The normalized emittance values appearing in the Table must be compared with the present state of the art, based on the use of electron storage ring optimized for producing a small emittance. These rings have been developed in connection with SLC[13,14,15] and also as synchrotron radiation sources[16] and Free Electron Laser sources[17]. The emittance obtained in the SLC rings is on the order of 10^{-5} m, much larger than the value needed in the low power "quantum" collider of Table 1. In one synchrotron radiation source an emittance of about 10^{-7} m has been obtained in the vertical plane[16]. Preliminary designs for 10^{-7} in both planes are being considered.

It is also interesting to notice the very high values of the electron density during the crossing. Again one must compare with values presently obtained in storage rings, of the order of 10^{13} cm^{-3}, and in SLC[14], of the order of 10^{18} cm^{-3}. The "classical" collider of Table 1 is close to SLC also in this respect.

The bunch frequency is also much higher than that of existing linacs; SLAC's frequency is a few hundred Hertz. One can also notice that the accelerating field used in SLC[14] would lead to a length of 60+60 km for a 1 TeV collider.

The final focus designed for SLC[14] provides a β^* of about 5 millimeters. Again the β^* value of the second and third collider of Table 1 is much smaller. All this shows that the construction of high energy, high luminosity colliders, requires improvements over the present state of the art in many areas, like the final focus, the emittance, the accelerating field and frequency of the accelerator, and the understanding of very high density electron bunches. All this is particularly true for parameters which lead to the quantum regime of beamstrahlung.

Figs. 3 and 4 show, on the other hand, that the beam parameters and collision frequency needed for a (5+5) TeV, $\mathscr{L}=10^{34}$ cm^{-2}s^{-1} or larger, are less challenging in the quantum regime than in the classical regime. Thus the very short and small bunches considered in the (1+1) TeV "quantum case" of Table 1, have to be studied if one wants to move in the direction leading to colliders with even larger energy and luminosity.

3. ACCELERATION TECHNIQUES

A number of new acceleration teshniques have been studied recently and a description of the most recent work can be found, for instance, in the Proceedings of the Los Angeles Workshop on New Acceleration Techniques[18]. One of these techniques is the Two Beam Accelerator (TBA), proposed by A.M.Sessler[11]. In the proposed TBA a low energy electron beam produces radiation, at around 1 cm wavelength, in a Free Electron Laser (FEL). The energy lost by the low energy beam is periodically restored by an induction linac. The one centimeter radiation is fed to a linac structure where accelerating fields of a few hundred MeV/m can be obtained.

Recently an experiment by a Berkeley-Livermore group has demonstrated the possibility of using Free Electron Lasers as power sources at a wavelength around 1 cm, obtaining accelerating fields up to 250 MeV/m in a linac structure[19].

The possibility of going to still shorter wavelengths, in the mm or μm range, using structures like gratings, droplet arrays or an inside-out linac, has also been actively studied[20,21]. In these cases one could work at around 1 mm, using again an FEL as a power source or at optical wavelengths, where lasers like an FEL or a CO_2 laser could be used. In these last cases, and using very short (picosecond) pulses, one can reach accelerating fileds of the order of 1 GeV/m, before the accelerator walls break down or melt[22].

The recent developments in design and construction of superconducting cavities[23], which have established the possibility of reaching fields on the order of 5 to 10 MeV/m, with quality factors of about 5x10^9, at temperature of about 4K, has also renewed the interest in the use of superconducting linacs[2,24], including the energy recovery option[25,26], for high energy colliders. Although the maximum accelerating field for these systems is not high, less than 60 MeV/m for Nb cavities[23], they might offer the possibility of high efficiency and reduced power consumption, for not too large final energies.

In this paper we want to study the possibility of substituting in a TBA the induction linac with a superconducting linac. The possible advantages are: 1) large efficiency of energy transfer to the low energy beam; 2) possibility of obtaining easily the high e$^+$-e$^-$ collision rate needed for high luminosity.

We believe that these possible advantages, and the established availability of superconducting cavities, make this scheme worth studying. Since the superconducting cavities for increasing the LEP beam energy from 50 to about 100 GeV work at 350 MHz, we choose this frequency for our driving linac. Low frequencies, as this, have the advantage of larger stored energy per unit length.

4. ENERGY BALANCE AND COLLIDER PARAMETERS

The basic module of our system is given in fig. 5, and consists of an undulator of length L_u, and a length L_c of superconducting cavities. The radiation produced in the undulator is fed to a high frequency linac (HFL) by a number of waveguides. The total module length is L_m, and we make the simplifying assumption that $L_m = L_u + L_c$.

We now want to establish the energy balance between the superconducting cavities and the high frequency structure. We will use a subscript 1 to describe the superconducting cavity characteristics, like frequency ω_1, field E_1 etc., and subscript 2 for the HFL. The average energy stored in the superconducting cavities per unit length is [22,27]

$$U_1 = E_1^2 / (R_1/Q_1) \, \omega_1 \tag{26}$$

where R_1 is the shunt impedance per unit length and Q_1 the quality factor. Of this energy we can transfer a fraction $\eta_1 \eta_2$ to the HFL, where η_1 is the fraction of the energy stored in the SCCs that is given to the low energy electron beam powering the FEL, and η_2 is the fraction of energy produced by the FEL that is transferred to the HFL.

If P_I is the power per unit length in the HFL, and τ_f its filling time, we obtain, by applying energy conservation to one module, the relation [22]

$$\eta_1 \eta_2 U_1 L_c = P_I \tau_f L_m . \tag{27}$$

The HFL filling time and power per unit length are related to the structure attenuation constant, α, quality factor, Q_2, frequency, ω_2, shunt impedance per unit length, R_2, and the electric field, E_2, by [22,27]

$$\tau_f = \alpha (Q_2 / \omega_2) \tag{28}$$

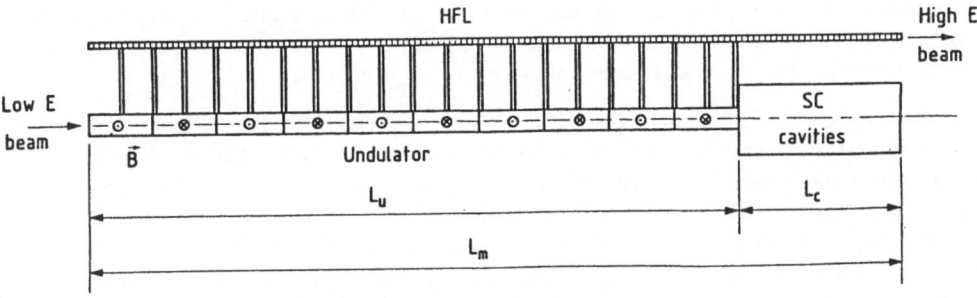

Fig. 5. Schematic layout of SCC-FEL-HFL system.

$$P_\ell = E_2^2/(g^2 \alpha R_2) \tag{29}$$

where g=2(1-exp(-\alpha/2))/\alpha . By combining eqs.(26) to (29) one can introduce[28] an electric field amplification factor, given by

$$\frac{E_2^2}{E_1^2} = \eta_1 \eta_2 g^2 m \frac{(\omega_2 R_2/Q_2)}{(\omega_1 R_1/Q_1)} \tag{30}$$

where m is the ratio of SCC to HFL length, $m = L_c/L_m$. Since R/Q scales like
ω , eq.(30) shows that the ratio of the fields increases with the ratio
ω_2/ω_1 , hence the convenience of going to high frequency in the HFL.

We now make the further step of relating the HFL characteristics to
the high energy bunch parameters determined in section 3. For a given power flux, P_ℓ , in the HFL, and for a given energy transfer to the beam, defined by the efficiency factor η_3 (= energy given to the beam per unit
length/energy stored in the HFL per unit length), we can accelerate N_2
particles in a bunch, with

$$eN_2 E_2 = \eta_3 P_\ell \tau_f \tag{31}$$

and using eqs.(27) and (28)

$$eN_2 \omega_2 (R_2/Q_2) g^2 = \eta_3 E_2 . \tag{32}$$

From the field E_2 we can calculate E_1 using (30). We can also obtain
N_2 from (18) and write directly E_2 and E_1 as function of E, $\mathscr{L}$, P, D (H),
$<\varepsilon>$, and the frequencies ω_1, ω_2

$$\eta_3 E_2 = eC_1 g^2 (R_2 \omega_2/Q_2)(P/E\mathscr{L})^2 (<\varepsilon> DH/F(\Upsilon)) \tag{33}$$

$$\eta_3 E_1 = eC_1 g (\frac{\omega_1 R_1 \omega_2 R_2}{\eta_1 \eta_2 m Q_1 Q_2})^{1/2} (P/E\mathscr{L})^2 (<\varepsilon> DH/F(\Upsilon)) \tag{34}$$

where C_1 is given by (20). Since $R/Q \approx \omega$ we see that $\eta_3 E_2 \approx \omega_2^2/F$ and
$\eta_3 E_1 \approx \omega_1 \omega_2/(F \sqrt{m})$. Eq.(34) shows that if we try to decrease P and D,
thus passing from the classical to the quantum regime, and we want to keep
$<\varepsilon>$, ω_1, η_1, η_2 and $\eta_3 E_1$ constant we have to increase $\omega_2/(F \sqrt{m})$. Since
F^{-1} increases by a factor 15 in the example of Table 1,and $m = L_c/L_m$ should
not be reduced, by necessity the HFL frequency, ω_2, has to increase. This
entails, through (33), a welcome increase of $E_2 \approx \omega_2^2 \sqrt{m}$.

In summary, eqs.(33) and (34) put restrictions on the choice of the
accelerator parameters for given beam and collision conditions. Viceversa
for a given accelerator system we are limited, by the same equations, in
the collider performance. For a fixed, low energy SC structure, and fixed
efficiencies, the frequency and the accelerating field in the HFL come
out to be much larger in the quantum regime than in the classical
one. This puts into light one of the advantages of using a FEL to
produce high frequency fields: one can reduce the wavelength by changing

the undulator system and one can explore the quantum regime.

In the quantum limit, $\Upsilon \gg 1$, the number of particles per bunch is given by the simple expression (24), and one can use this in (33), (34). In this case there is no dependence of the products ηE on the collider energy or luminosity, which makes possible to use the same accelerator system to cover a wide range of beam energy and luminosity. This property makes the quantum regime very attractive and worth exploring in greater depth.

The length of SCC cavities is determined by the electron energy change in the undulator $mc^2(\gamma_o - \gamma_f)$ and the electric field E_1 in the SCC

$$\cdot \, mc^2(\gamma_o - \gamma_f) = eE_1 L_c \, . \tag{35}$$

The electron beam that radiates energy in the undulator can be made of b_1 bunches separated by the SCC wavelength, λ_1. The number of particles, N_1, per bunch is determined again by energy conservation

$$eN_1 b_1 E_1 = \eta_1 U_1 \tag{36}$$

and does not depend on the number of SCC cavities.

The peak power in the low energy beam is given by

$$P_{1,p} = (eN_1 c/L_B)mc^2 \gamma_o \tag{37}$$

where L_B is the bunch length.

Two sets of accelerators parameters for the SCC and the HFL are collected in Tables 2 and 3. Based on the previous discussion we have consid-

Table 2. Superconducting Linac Parameters.

frequency (MHz), f_1	350
shunt impedance (GΩ/m), R_1	1.3×10^3
quality factor, Q_1	5×10^9
electric field (MV/m), E_1	15
stored energy (J/m), U_1	385

Table 3. High Frequency Linac Parameters.

frequency (GHz), f_2	30	700
wavelength (mm), λ_2	10	.42
attenuation parameter, α	.5	.5
filling time (ns), τ_f	11.3	.10
shunt impedance (GΩ/m), R_2	.17	.82
quality factor, Q_2	4100	848

ered for the HFL two cases, λ=10 mm and λ=0.42 mm, with a wavelength ratio of about 25. The second case implies the development of a new technology for the linac structure. We have simply assumed that the characteristics of this linac can be obtained by scaling from lower frequency systems. In our present situation this cannot be fully justified and we will have to wait until more work is done to make more realistic estimates. The value $\lambda \simeq 0.5$ mm, has been chosen also because at this wavelength there exists a 10 KW operating FEL[29], and the relative diagnostics has been developed.

The attenuation parameter has been chosen equal to 0.5 following an argument by Schnell[27].

Using the data of Tables 2 and 3 and assuming η_1=0.1 and $\dot{\eta}_2$=0.5, we have from (26) for λ=1 cm and λ=0.42 mm respectively

$$P_\ell = (1.7m) \text{ GW/m}, \quad \text{for} \quad \lambda_2=10 \text{ mm} \tag{38}$$

$$P_\ell = (193m) \text{ GW/m}, \quad \text{for} \quad \lambda_2=0.42 \text{ mm} \tag{39}$$

where m=L_c/L_m. These are the power levels available in the two cases and from these we can calculate the accelerating field in the HFL.

5. THE FEL

For a complete discussion and review of FEL properties we refer to references 30, 31; here we summarize only the main formulae. We assume that the radiation at wavelength λ produced by the FEL is propagated in a waveguide and that a tapered undulator is used to extract power from the beam. Let A_o be the wave amplitude at the wiggler entrance, ω its frequency, γ the electron energy in rest mass units. The energy change of the synchronous electron is

$$\frac{d\gamma}{dz} = -(eA_o a_u \sin(\psi_r))/mc^2\gamma \tag{40}$$

where the undulator parameter a_u is given, for a planar magnet, by the undulator magnetic field, B_u, and period, λ_u, as a_u=$eB_u\lambda_u/2^{3/2}\pi mc^2$. The phase, ψ_r, in (40) is a constant provided that the condition of synchronism

$$k_u = \delta k+(\omega/c)(1+a_u^2)/2\gamma^2 \tag{41}$$

is satisfied. In (41) k_u=$2\pi/\lambda_u$, and the quantity δk=$(\omega/c)-k$ describes the change in the radiation wavenumber, k, from its value in vacuum. For a rectangular guide of height b and width a, with a≫b, and for the lowest order transverse electric mode, $\delta k \approx \pi\lambda/4b^2$.

To satisfy (41) all along the undulator we have to change λ_u and/or a_u as γ changes according to (40). It is possible to solve (40) and (41) for different types of undulators, i.e. for different choices of λ_u and a_u as a function of z. In the following we will consider for simplicity the case in which a_u is kept constant and λ_u changes.

To integrate (41) we also assume that A_o is constant, so that the electron energy change per undulator is simply

$$\gamma_f^2 = \gamma_o^2 - 2eA_o a_u L_u \sin(\psi_r)/mc^2 . \tag{42}$$

The assumption A_o=constant can be justified since we can remowe the radiation generated along the undulator and feed it to the HFL. To the field A_o is associated a laser power, P_L, propagating in the waveguide

$$P_L = A_o^2 ab/2Z_o \tag{43}$$

Z_o being the vacuum impedance (Z_o=377 Ω). The energy lost by the electrons is restored by the SCC according to (35).

In addition to the synchrotron condition (41) we must consider the effect of the difference in velocity between the low energy electrons used in the FEL, the radiation pulse propagating in the waveguide, and the high energy electrons in the HFL. Let us consider first the radiation group velocity, v_g, and the low energy electron axial velocity, v_ℓ; we have

$$v_g = c/(1-(\pi c/\omega b)^2)^{1/2} \tag{44}$$

$$v_\ell = c(1-(1+a_u^2)/2\gamma^2) . \tag{45}$$

The group velocity (44) has been written again for the lowest order transverse electric mode, and assuming a≫b. The electron velocity, v_ℓ, will change along the module as γ and a_u change. We can define an average longitudinal velocity

$$\langle v_\ell/c \rangle = 1-S_1-S_2 \tag{46}$$

where

$$S_1 = (L_u/L_m)(1+a_u^2)\ln(\gamma_o/\gamma_f)/(\gamma_o^2-\gamma_f^2) \tag{47}$$

$$S_3 = L_c/2L_m \gamma_o \gamma_f \tag{48}$$

and γ_o, γ_f are the electron energies at the entrance and exit of the undulator, and used (35), (42) to evaluate the effect of the electron energy variation.

The difference between the radiation group velocity and the low energy electron velocity is important in our case since we are dealing with very short, a few centimeter long, electron bunches and we want to keep using the same electrons over many modules, each of length L_m. To avoid a deformation of the radiation pulse and the developments of the sideband instability[31] we assume

$$v_g = \langle v_\ell \rangle . \tag{49}$$

By choosing γ_o, γ_f, and the ratio L_u/L_m we can satisfy (49), so that in one module there is no slippage between the electrons and the radiation pulse. If we satisfy the condition (49), we would still be left with a slippage over the undulator length

$$S_u = L_u\left\{(\lambda/2b)^2/2 - S_1\right\}. \tag{50}$$

In practical cases, like the ones we will consider, this turns out to be much smaller than the bunch length, and we will assume that this effect can be neglected.

We can now use the condition (49), and the synchronism condition (41), to determine λ_u and a_u for given γ_o, γ_o/γ_f, $m = L_c/L_m$, and L_u/L_m. In this way we can design a module producing an energy exchange given by (40) and also $v_g = <v_\ell>$. For our estimates we have chosen the particular parameters given in Table 4 for two cases, corresponding to radiation at 10 mm and 0.42 mm wavelength.

The difference between the low energy electron and the high energy electron longitudinal velocity is given to a good approximation by $c - <v_\ell>$. The delay per module is given by

$$S_{HFL} = L_m(1 - <v_\ell>/c). \tag{51}$$

If (49) is satisfied this equation can also be rewritten as

$$S_{HFL} = L_m(1 - v_g/c) \approx L_m(\lambda/2b)^2/2. \tag{52}$$

This shows that this slippage is determined by the characteristics of the waveguide, and that it is less severe at shorter radiation wavelength. The high energy electron slippage can be neglected as long as it is small compared with the low energy electron train length, which will be discussed in the next section. If it becomes of the order of the train length, then one can compensate by making the length of the waveguides joining the FEL and the HFL of variable length, or by increasing the train length. Assuming a train length of about twice the filling time of the structure, given in Table 3, one obtains that the high and low energy beams slip by the train length in about 1.3 and 1.1 Km respectively for the 1 cm, or

Table 4. Undulators Parameters.

radiation wavelength (mm), λ_2	10	.42
electron energy/mc², γ_0	180	360
energy change, γ_0/γ_f	2	2
SCC length/module (m),	3	6
SCC length ratio, L_c/L_m	.2	.2
module length (m), L_m	15	30
waveguide height (cm), b	5	1.5
undulator parameter, a_u	14.8	8.15
undulator period (cm), λ_u	120	120
undulator slippage, S_u/L_u	6.2×10^{-6}	1.6×10^{-6}
HFL slippage, S_{HFL}/L_m	5.2×10^{-3}	9.8×10^{-5}

the 0.42 mm cases. Tapering the waveguide length can increase this distance, hence "supermodules", using the same low energy bunches, longer than 1 km can be envisaged.

Up to now we have discussed the behaviour of the synchronous particle. There are other conditions to be imposed for the stability of the motion of non synchronous particles. The most important is a limitation on the beam energy spread, that must be smaller than[30,31]

$$(\Delta E/E)_M^2 \;=\; \frac{eA_o a_u \lambda_u}{\pi mc^2 \gamma^2} \left\{ \cos \psi_r - (\pi/2 - \psi_r) \sin \psi_r \right\}. \tag{53}$$

The FEL model that we have considered, with a_u and A_o constant, is a simple one. Other possibilities can be consider, like for instance allowing A_o to grow in the undulator region, which can provide more efficient energy extraction, or a reduced undulator length. However this model gives results which are a lower limit to what can be obtained, and we will use it to obtain our simple estimates of the SCC-FEL-HFL system.

6. THE HFL

We can now combine the results of sections 4 and 5 to determine the performance of the HFL. In addition we also have to consider the condition for matching the HFL filling time to the time structure of the FEL radiation, which consists of pulses with a duration nearly equal to the electron bunch duration, L_b/c, separated by the SCC period, λ_1/c.

In the filling time is much longer than the bunch duration

$$\tau_f \gg L_b/c \tag{54}$$

it is convenient to divide the beam in the FEL in b_1 bunches

$$b_1 = \tau_f c/L_b . \tag{55}$$

When (54) is not satisfied it is convenient to have

$$L_b = \tau_f c . \tag{56}$$

We must also remember that the bunch length determines the energy spread in the low energy beam, and that this must remain below the limit given by (53). If for simplicity we assume that the bunch is travelling on the crest of the wave in the SCC, than the energy spread is related to the bunch length by

$$\Delta E/E = (1/2)(2\pi L_b/\lambda_1)^2 . \tag{57}$$

From the equations (33) and (34) we can calculate E_2 and E_1 and the RF power, P_{RF}, to be fed into the SCC's

$$P_{RF} = \eta_2 \eta_3 P = \eta_1 U_1 m L_T f \tag{58}$$

where L_T is the total single linac length

$$L_T = E/E_2 .$$ (59)

The energy stored in the linac, $U_1 m L_T$, determines also the amount of cryogenic power, P_{cryo}, needed

$$P_{cryo} = U_1 \omega_1 m L_T / Q_1 \eta_4$$ (60)

where η_4 is the efficiency in cooling the SC cavities.

7. THE DAMPING RINGS

To complete the dicussion of our colliders we must give a description of the damping rings producing the electron and positron beams with the required phase space density. We assume that the damping rings are built as a sequence of $N_d/2$ achromatic modules, each with two bending magnets, similar to a Chasman–Green structure[33], and that its normalized emittance, ε_n, and momentum compaction, α_c, are given by[17]

$$\varepsilon_n = 1.2 \times 10^{-11} (\gamma/N_d)^3 \text{ m rad}$$ (61)

$$\alpha_c = (2/3)\pi^2 (\varrho/R)(1/N_d^2)$$ (62)

where ϱ is the bending radius and R the average radius of the damping rings. Both the emittance and the momentum compaction value depend on the particular magnetic structures one assumes, but the values given in (61), (62) are reasonable approximations for a rather large class of storage rings.

The choice of the beam energy in (61) has to be done to make the intra-beam scattering effect[34,35,36] small or of the same order than synchrotron radiation effects, in determining the beam emittance.

Small emittance storage rings, based on a Chasman–Green structure, have been built up to now mainly as synchrotron radiation sources. Normalized emittances of the order of about 10^{-4} m rad, in the horizontal plane, and down to 3×10^{-7} m rad, in the vertical plane, have been obtained. Similar rings with an emittance of about 10^{-5} m rad, have been studied in great details and look feasible[37].

Damping rings based on a FODO structure have been built for SLC reaching an emittance of about 10^{-5} m rad, with more than 10^{10} particles per bunch[15]. The design of rings pushing the emittance down to 10^{-7} or 10^{-8} m rad might introduce new problems, related, for instance, to the strong chromatic and non-linear aberrations characteristic of these systems. This, in turn, might lead to a strong reduction of the useful apertura and thus injection and lifetime problems. It is likely that to solve these problems it will be necessary to introduce much tighter magnetic element tolerances and beam orbit control.

The beam energy spread and bunch length have to be chosen to produce a longitudinal phase space emittance which matches that of the collider.

While the bunch length is given in Table 1, the energy spread is also related to the choice of the accelerator characteristics, in particular the beam loading coefficient, η_3. A larger η_3 produces a larger energy spread and that might make more difficult to design the system to focus the beams at the collision point. To design the storage ring we assume η_3=2%; a larger value would make the storage ring design easier and increase the collider efficiency but it is likely to produce problems with the final focus.

To evaluate the beam energy spread and bunch length in the damping ring we assume that these quantities are determined by radiation effects[38], thus neglecting the microwave instability and the coupled bunch longitudinal instability[39]. This cannot be justified without a full analysis of a damping ring design. We notice however that for the very short bunch length we are considering, the ring impedance driving the microwave instability can be reduced only to the vacuum impedance and that this can be reduced to about 0.1 Ω[37]. The control of the coupled bunch instability poses a greater problem, since this is mainly determined by the low frequency part of the impedance, which in the damping ring is not likely to be reduced respect to exsisting rings, and because to obtain the high collision frequency needed in the collider, we need to fill almost all available RF buckets. Thus in the damping ring we want to achieve a large average current and a small bunch separation. To control the coupled bunch longitudinal instability under these conditions, would require a large bandwidth (of the order of 1 GHz), feedback system, and we assume that this can be done.

With these assumptions[38] we can assume that the energy spread and the bunch length are given by

$$\sigma_\varepsilon = 4.4 \times 10^{-7} \, \gamma / \sqrt{\varrho} \, (m) \tag{63}$$

$$\sigma_1 = \sigma_\varepsilon R \left\{ 2\pi E \alpha_c / heV_o \cos \Phi_s \right\}^{1/2} . \tag{64}$$

The number of bunches per second that can be obtained from a system of N_R damping rings, assuming that each bunch stays in the ring for 5 damping times[4], is given by

$$f = h N_R F_f / 5 \tag{65}$$

where h is the harmonic number, F_f is a filling factor (F_f = ratio of number of bunches to the harmonic number), and τ is the ring betatron damping time[35], given by

$$1/\tau = (1/3)(r_e c \gamma^3 / \varrho R) \tag{66}$$

having assumed a damping partition factor of 2.

To have a large frequency, f, we can increase F_f and N_R or decrease τ. To obtain F_f equal to one we need very fast kicker magnets, with a rise time of the order of the RF period; present technology limits this rise time to tens of nanoseconds, which would give, for f_{RF}=500 MHz a filling factor, F_f, smaller than 0.1. A larger value of F_f can be obtained if a train of b_2 bunches is injected, damped and extracted simultaneously from a damping ring. Such a procedure could be limited by wake field effects in the HFL and has to be investigated.

Table 5 contains possible set of parameters for the damping rings needed for the colliders we are considering. The conclusion is that, for $F_f \approx 0.1$, the total length of the positron damping ring system is about 1.5 km, 30 km, and 9 km. Moreover for the last two cases it is likely that one will need predamping in an ad-hoc fast damping ring which will have to reduce the emittance from about 10^{-3} m, to 10^{-5} m, before injecting in the rings giving 10^{-7} m.

Table 5. Damping Rings.

E, TeV	1	1	5
$\mathscr{L}$, $cm^{-2}s^{-1}$	10^{33}	10^{34}	10^{34}
$\langle \varepsilon \rangle$	.08	.08	.08
D	1.3	.05	
P, MW	10	3	6
Energy, GeV	2.5	2.5	2.5
# dipoles	64	180	240
dipole length, m	.87	.29	.29
Bending Field, T	1	1	1
Bending Radius, m	8.3	8.3	8.3
Avergage Rdius, m	25	25	33
Normalized Emittance, m rad	5.7×10^{-6}	2.6×10^{-7}	1.1×10^{-7}
Momentum Compaction	5.3×10^{-4}	6.8×10^{-5}	2.9×10^{-5}
Energy loss/turn, KeV	111	111	111
Damping time, ms	6	6	8
r.m.s. energy spread	7.6×10^{-4}	7.6×10^{-4}	7.6×10^{-4}
f_{RF}, MHz	500	500	500
V_o, MV	3.5	1.2	1.2
r.m.s. bunch length, mm	2.2	1.1	.96
$f/F_f N_R$, KHz/ring	8.7	8.7	8.6
$N_R F_f$	1	20	4
$2\pi R N_R F_f$, m	157	3100	900

8. RESULTS AND CONCLUSIONS

We now consider the three collider cases illustrated in Table 1, and calculate the accelerator characteristics, using the results of sections 4, 5 and 6. In all cases we have tried both the 30 GHz (10 mm) case and the 700 GHz (0.42 mm) case.

For the classical, 1 TeV collider (case 1 of Table 1), it is not possible to find a satisfactory solution at 700 GHz. At this frequency, the product $\eta_3 E_2$ obtained from (33) is always very large, and gives unrealistic values for the efficiency and the field. The opposite is true for the two quantum colliders at 1 or 5 TeV (cases 2 and 3 of Table 1), where, at 30 GHz, the accelerator length and the power needed to operate the system are always very large.

A set of solutions obtained at 30 Ghz for the 1 TeV classical collider, and at 700 GHz for the 1 and 5 TeV quantum colliders are given in Table 6. Both structures are assumed to be fed by a superconducting linac, with cavities having the characteristics given in Table 2. We have also used the HFL characteristics given in Table 3 for the two cases of 30 and 700 GHz. The undulators characteristics for these two frequencies are given in Table 4. In evaluating the total length, L_T, of each linac we have assumed a 20% empty space, corresponding to the total length of the SC cavities.

Some relevant FEL parameter are given in the Table 7.

As one can see from Table 6, the quantum case and a HFL at 700 GHz are very attractive. The accelerating fields are large compared to SLC, the power levels are reasonable and are balanced between RF power and cryogenic power. One can change η_3, and so change the ratio of RF to cryogenic power; a smaller η_3 leads to a larger accelerating field and larger RF power, and viceversa. The 30 GHz case, used for the classical collider, is certainly a much smaller extrapolation with respect to today techniques. However it gives a smaller accelerating field, and a larger power.

Table 6. HFL Parameters.

E, TeV	1	1	5
$\mathscr{L}$, cm^{-2}s^{-1}	10^{33}	10^{33}	10^{34}
D	1.3	.05	.015
$\langle\epsilon\rangle$	.08	.08	.08
P, MW	10	3	6
f_2, GHz	30	700	700
E_1, MV/m	15	15	15
η_3, %	10	10	10
E_2, MV/m	96	574	1080
L_T, km	12.6	2.1	5.5
P_{RF}, MW	200	60	120
P_{cryo}, MW	181	30	80
$b_1 N_1/10^{13}$	.62	.04	.14
b_1	4	1	1
η_1, %	3.8	.25	.9

Table 7. FEL Parameters.

E, TeV	1	1	5
$\mathscr{L}$, cm^{-2}s^{-1}	10^{33}	10^{33}	10^{34}
$\langle\varepsilon\rangle$	.08	.08	.08
D	1.3	.05	.015
P, MW	10	3	6
f_2, GHz	30	700	700
E_1, MV/m	15	15	15
η_1, %	3.8	.25	.9
$b_1 N_1 / 10^{13}$	.62	.041	.14
b_1	4	1	1
λ_2, mm	10	.42	.42
γ_0	180	360	360
γ_f	90	180	180
L_c, m	3	6	6
L_u, m	12	24	24
m	.2	.2	.2
a_u	14.8	8.1	8.1
λ_u, cm	120	120	120
Ψ_R	30°	30°	30°
A_0, MV/m	70	250	250
P_L, GW	32.5	38.5	38.5

Furthermore a classical 5 TeV collider at 30 GHz requires totally unrealistic characteristics. Based on these results, the 700 GHz, quantum case seems worth of further studies.

A number of important issues have not been addressed in this paper. The most important is certainly a detailed study of an accelerating structure operating at this frequency, and which can be built with the required tolerances, and used to accelerate a high power beam. Other important issues concern the phase stability of the accelerating field, and its sensitivity to errors in beam energy and undulator magnetic field. This problem has been studied for the TBA[39] at 30 GHz; it needs to be studied for our configuration and also at 700 GHz. Still in connection with the HFL structure one has to study the radiation extraction from the FEL waveguide, its propagation through the SC cavities, and its transmission to the HFL. Again a study has been done for TBA[30] but not for the 700 GHz case. Transverse wake field effects in the HFL and the low energy electron beam dynamics in the FEL also need much work, before any definite statement on the viability of the system considered in this paper can be made.

To be able to evaluate if a system like the one discussed in this paper can be used as a competitive TeV collider requires that many difficult problems be studied, understood and solved; we feel however that the level of performance that the system can offer would justify this effort.

REFERENCES

1. B.Richter, Nuclear Instr. and Meth. 136:47 (1976).
2. E.Keil et al., e^+-e^- colliders, in: Proc. of the second ICFA Workshop on Possibilities and Limitations of Accelerators and Detectors, U.Amaldi, ed., CERN, Geneva (1980).
3. B.Richter, in: Laser Acceleration of Particles, C.Joshi and T.Katsouleas, eds., American Institute of Physics, Conf. Proc., no. 130, p.8 (1985).
4. U.Amaldi, Nuclear Instr. and Meth. A243: 312 (1986).
5. J.Lawson, Linear collider constraints: some implications for future accelerators, CERN report 85-15 (1985).
6. C.Pellegrini, in: Proc. of the 1985 Intern. Symp. on Lepton and Photon Interactions at High Energies, Kyoto (1985).
7. A.M.Sessler, in: Laser Acceleration of Particles, American Institute of Physics, Conf. Proc. vol. 91, p.154 (1982); also A.M.Sessler, IEEE Trans. Nucl. Sci., NS-30:3145 (1983).
8. R.Hollebeek, Nuclear Instr. and Meth. 184:333 (1985).
9. T.Himmel and J.Siegrist, in ref.2, p.602 (1980).
10. K.Yokoya, KEK report 85-53 (1985).
11. R.Noble, Simulation of beamstrahlung for colliding e^++e^- beams with negligible disruption, AAS-note 3, SLAC (1985).
12. In ref.4 the beamstrahlung parameter is defined to be $2\sqrt{2}$ times smaller.
13. W.K.H.Panofsky, Limiting Technologies for Particle Beams and High Energy Physics, SLAC report, SLAC-Pub-3735 (1985).
14. SLC Design Handbook, Stanford Linear Accelerator Center, Stanford (1984).
15. A.M.Hutton et al., IEEE Trans. Nucl. Sci. NS-32:1659 (1985).
16. M.Barton, IEEE Trans. Nucl. Sci. NS-32:3350 (1985).
17. S.Krinsky, in: Free Electron Generation of Extreme Ultraviolet Coherent Radiation, J.M.J.Madey and C.Pellegrini, eds., American Institute of Physics, Conf. Proc., vol. 118, p.44 (1984).
18. C.Joshi and T.Katsouleas, eds., Laser Acceleration of Particles, American Institute of Physics, Conf. Proc., vol.130 (1985).
19. D.B.Hopkins et al., IEEE Trans. Nucl. Sci. NS-32:3476 (1985).
20. R.B.Palmer et al., ref.18, p.234 (1985).
21. N.M.Kroll, ref.18, p.253 (1985).
22. See for instance: P.B.Wilson, ref.18, p.560 (1985).
23. H.Piel, IEEE Trans. Nucl. Sci. NS-32:3565 (1985).
24. U.Amaldi, H.Lengeler and H.Piel, Linear colliders with superconducting cavities, CLIC Note-15, CERN/EF 86-8 (1986).
25. U.Amaldi, Phys. Letters 61B:313 (1976).
26. M.Tigner, Nuovo Cimento 37:1228 (1956).
27. W.Schnell, Dissipation versus peak power in a classical linac, CERN report LEP-RF/WS/PS (1985).

28. W.Schnell, Consideration of a two beam twin RF scheme for powering an RF linear collider, CERN report LEP–RF/WS/PS (1985).

29. L.R.Elias et al., Nuclear Instr. and Meth. A237:203 (1985).

30. N.M.Kroll, P.L.Morton and M.W.Rosenbluth, IEEE J. Quantum Electronics QE–17:1436 (1981).

31. D.B.Hopkins, A.M.Sessler and J.S.Wurtele, Nuclear Instr. and Meth. A228:15 (1984).

32. E.J.Sternbach and A.M.Sessler, A steady state FEL: Particle dynamics in the FEL portion of a two beam accelerator, Lawrence Berkeley Laboratory report, LBL–19939 (1985).

33. R.Chasman and K.Green, Brookhaven National Laboratory report, BNL 50505 (1980).

34. H.Bruck, Accelerateurs Circulaire de Particules, Presse Universitaire, Paris (1966).

35. A.Piwinsky, in: Proc. 9th Intern. Conf. on High Energy Accelerators, Stanford (1974).

36. J.D.Bjorken and S.E.Mtingwa, Particle Accelerators 13:115 (1983).

37. J.Bisognano et al., Feasibility study of a storage ring for a high power XUV free electron laser, Lawrence Berkeley Laboratory, report LBL–19771 (1985).

38. M.Sands, The physics of electron storage rings. An introduction, in: Physics with Intersecting Storage Rings, B.Touschek, ed., Academic Press, New York (1971).

39. C.Pellegrini, IEEE Trans. Nucl. Sci. NS–28:2413 (1981).

40. R.W.Kuenning and A.M.Sessler, Nuclear Instr. and Meth. A243:263 (1986).

LOW TECHNOLOGY FELs AND IFELs: HOW CAN A SMALL LABORATORY CONTRIBUTE?

G. Dattoli and E. Sabia

ENEA, Dip. TIB, Divisione Fisica Applicata
C.R.E. Frascati
C.P. 65 - 00044 Frascati, Italy

INTRODUCTION

It is well known that the basic FEL Compton mechanism is a genuine example of radiation–matter interaction. The process can be indeed described as the two-wave interaction shown in Fig. 1. Therefore the most simplified view of the process is

a) initial state: two photons moving from opposite directions with approximately the same frequency head on a nonrelativistic electron with momentum p_0.

b) final state: the electron scatters the photon (1) in the backward direction losing $2\hbar k$ of momentum; the electron scatters the photon (2) in the forward direction gaining $2\hbar k$ of momentum.

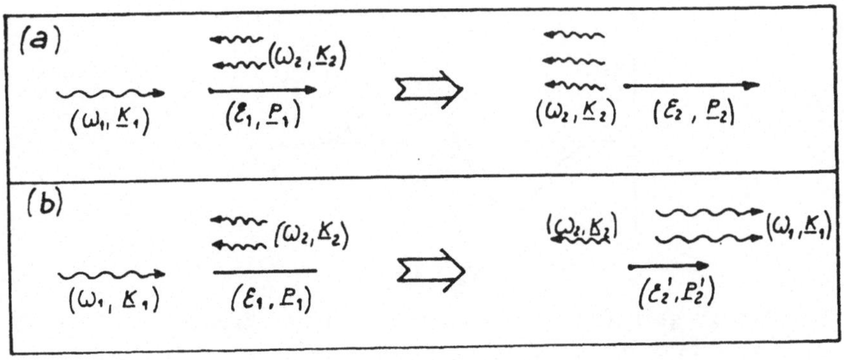

Fig. 1

(a) Stimulated Compton back-scattering;
(b) Stimulated Compton forward-scattering

Suitable competition between the two processes may give rise to the well-known FEL gain curve shown in Fig. 2 and thus a net "laser" or electron energy gain may result according to the value of the detuning frequency. This simple example shows that FEL gain and electron acceleration are, within this framework, the obvious counterparts of the same mechanism.

The simplest scheme to get electron beam (e-beam) acceleration via a laser beam is the radiation pressure mechanism. This process shown in Fig. 3 is a particular case of that discussed in Fig. 1.

A photon beam characterized by a wavenumber K_1 and frequency ω_1 copropagates together with an e-beam with momentum p_1 and energy ε_1. The photons can be backscattered by the electrons (final state) and the electron energy increases. From the laws of conservation of energy and momentum it follows that the frequency of the backscattered photon is down-shifted by Doppler effect thus getting in the extreme relativistic case

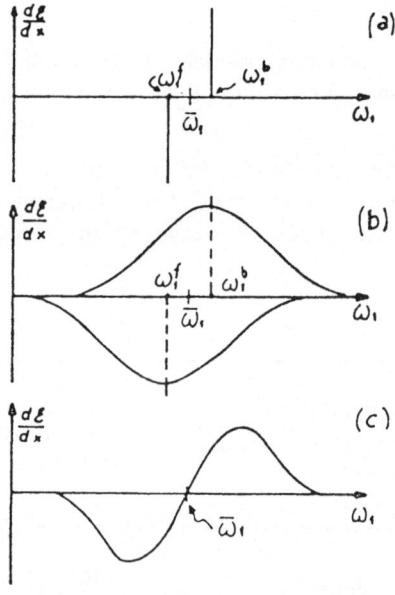

Fig. 2

Energy gain (arbitrary units) vs laser frequency
(a) Infinitely long interaction time
(b) Finite interaction time (acceleration and
    deceleration processes)
(c) Finite interaction time (net energy gain).

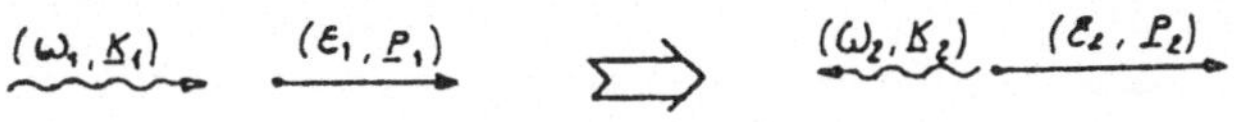

$$(\omega_1, \vec{k}_1) \quad (\varepsilon_1, \vec{p}_1) \quad \Longrightarrow \quad (\omega_2, \vec{k}_2) \quad (\varepsilon_2, \vec{p}_2)$$

Fig. 3.

Radiation pressure effect (Compton back-scattering).

$$\omega_2 \approx \frac{1}{4} \omega_1 \gamma^{-2} \tag{1}$$

The electron energy gain is given by the quantity

$$\delta\varepsilon \approx \hbar\omega_1 \tag{2}$$

If the laser beam contains n_1 photons, the energy change per unit time is $\delta\varepsilon$ multiplied by

$$\dot{\mathcal{N}} = \sigma_{Th} \cdot \frac{n_1}{V} \cdot c \tag{3}$$

where σ_{Th} is the Thomson cross section

$$\sigma_{Th} = \frac{8\pi}{3} r_0^2 \quad , \quad r_0 = \frac{e^2}{m_0 c^2} = \text{classical electron radius}$$

and

$$\frac{n_1}{V} = \text{photon density}$$

The energy gain per unit length[1] reads

$$\frac{d\varepsilon}{dx} = \frac{\dot{\mathcal{N}}\delta\varepsilon}{c} = \sigma_{Th} \cdot \hbar\omega_1 \cdot \frac{n_1}{V} \tag{4}$$

where the term $\hbar\omega_1 \cdot n_1/V$ can be immediately recognized as the laser energy density, which for a linearly polarized wave can be linked to the electric field by the relation

$$\hbar\omega_1 \frac{n_1}{V} = \frac{1}{4\pi} |E_L|^2 \tag{5}$$

Therefore, combining (4) and (5)

$$\frac{d\varepsilon}{dx} = \eta \cdot e E_L \tag{6}$$

where the efficiency η is defined by the relation

$$\eta = \frac{2}{3} \frac{E_L}{m_0 c^2/(e \cdot r_0)} \tag{7}$$

In practical units the efficiency reads

$$\eta = 3.6 \times 10^{-21} E_L [V/m]$$

therefore, if one requires an efficiency at least 1% one should have a laser electric field of

$$2.7 \times 10^{18} V/m$$

about eight orders of magnitude larger than those available with the most powerful CO_2 laser.

It is clear that the simplest way to enhance this efficiency is to realize a stimulated scattering, i.e., the backscattering of Fig.3 in the presence of a number of photons n_2 thus getting the same process of Fig. 1.

Let us now discuss in more detail the mechanism which may lead to an FEL or IFEL.

In an FEL device an ultrarelativistic e-beam interacts with an undulator magnet (Fig. 4) in which it undergoes transverse oscillations and emits radiation at a fixed wavelength. In this way the e-beam can amplify a copropagating laser beam, or once the radiation is stored in an optical resonator and reinteracting with the e-beam it is reinforced and the system works as a self-sustained oscillator.

At first glance this process has nothing to do with a Compton scattering, but according to an improperly defined Weizsäcker-Williams approximation of the FEL the modulator field can be treated as a radiation field with wavelength

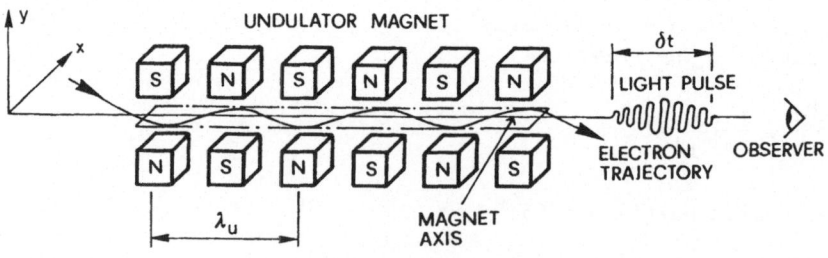

Fig. 4

Undulator magnet

$$\lambda^* = 2\lambda_u \tag{8}$$

and photon density

$$\bar{n} = \frac{\alpha}{4} \frac{\bar{K}^2}{\lambda_u r_0^2} \tag{9}$$

where

$$K = e\bar{B}\lambda_u / 2\pi m_0 c^2$$

α = fine structure constant

r_0 = classical electron radius

λ_u = undulator wavelength

$\bar{B}$ = B_0 for helical undulator

$\qquad B_0/\sqrt{2}$ for linear undulator

Within this framework the interaction of the electron with the magnetic undulator can be understood as the interaction with a very intense electromagnetic wave. To give an example for K=1 (which is a typical working parameter) and λ_u=5 cm we have

$$\bar{n} = 4.6 \times 10^{21} \ [cm^{-3}]$$

which is three orders of magnitude larger than that available from a powerful CO_2 laser.

It is now very easy to evaluate the wavelength of the scattered light (see Fig. 5). According indeed to the well-known formula of Compton scattering we get

$$\lambda = 2\lambda_u \frac{1-\beta\cos\theta}{1+\beta} \tag{10}$$

which for small angles and ultrarelativistic energies reads

$$\lambda = \frac{\lambda_u}{2\gamma^2} (1+K^2+\gamma^2\theta^2) \tag{11}$$

The corrective term K^2 due to the transverse motion of the electron is the photon density dependent Compton wavelength shift suggested by

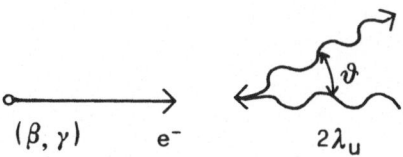

Fig. 5

Compton scattering diagram

Brown and Kibble in the analysis of the electron motion in an intense laser wave[2].

We have so far clarified the basic electron undulator interaction within the framework of Compton scattering of pseudo-photons and we have seen that the details of the process are almost similar to those of the interaction of an electron with an intense laser wave. In the hypothesis in which the scattering takes place in the presence of another electromagnetic field copropagating with the electron, the stimulated scattering may be visualized as the interaction of an electron with two counterpropagating waves, the same depicted in Fig. 1.

In discussing the process of stimulated Compton scattering we referred to a nonrelativistic electron while the main assumption of the electron undulator interaction was that the e-beam is ultrarelativistic. The apparent contradiction may be however bypassed treating the process of laser-electron-undulator pseudo radiation field in a frame of reference where the electron motion can be treated nonrelativistically. Furthermore, assuming that the electromagnetic wave copropagating with the electron is a single mode, a convenient choice is the frame where both the laser and the undulator field have the same frequency.[3] In this connection the process can be described by a Hamiltonian of the type:

$$H = \frac{p^2}{2m} + \hbar\Omega(a_L^+ a_u e^{-2ikz} + a_u^+ a_L e^{2ikz}) +$$

$$+ \hbar(\omega+\Omega)(a_L^+ a_L + 1/2) + \hbar(\omega+\Omega)(a_u^+ a_u + 1/2) \tag{12}$$

where

$\Omega = 2\pi c^2 r_0/\omega V \quad (<< \omega)$

$p \equiv$ electron momentum

$z \equiv$ longitudinal electron coordinate

$a^+(a) \equiv$ creation (annihilation) operator of electromagnetic field

$k \equiv$ laser (undulator) wave number $(\underline{k}_L = -\underline{k}_u)$

$\omega = |\underline{k}|c =$ laser (undulator) frequency

$V \equiv$ interaction volume

The Hamiltonian (12) is the mathematical translation of the process qualitatively described in Fig. 1. It has also been shown that from the Hamiltonian (12) two laws of conservation follow:

$$\begin{cases} n_L + n_u = \text{const.} \\ \\ p + \hbar k(n_L - n_u) = \text{const.} \end{cases} \tag{13}$$

which allow the FEL process to be understood as a photon transfer from the undulator to the laser via the electron momentum. Viceversa we can also describe the IFEL process. At any rate, the above laws of conservation allow the following state vector representation of the electron field interaction:

$$|\psi> = \sum_{\ell=-n_L}^{n_u} C_\ell |\ell> \qquad (14)$$

where ℓ is the number of the exchanged photons between the laser and the undulator and C_ℓ are time dependent coefficients accounting for the amplitude probability of exchanging ℓ photons.

The equation of motion of the coefficients C_ℓ can be derived from the Schrödinger equation which once solved in the small signal approximation yields for the variation of the number of photons after one passage the following expression

$$\Delta n_L = \Omega_R^2 [(\frac{\sin W_0/2}{W_0/2})^2 - (2|\alpha_0|^2+1)\varepsilon \frac{\partial}{\partial W_0} (\frac{\sin W_0/2}{W_0/2})^2] \qquad (15)$$

where

$$\Omega_R = \Omega\sqrt{\overline{n}_0}\,\Delta t$$

$$W_0 = 2(\omega\Delta t)p_0/mc \qquad p_0 \equiv \text{initial electron momentum}$$

$$\varepsilon = \tilde{\omega}\Delta t \qquad\qquad \tilde{\omega} = 2\hbar k^2/m$$

and $|\alpha_0|^2$ is the average number of photons of the laser field assumed to be initially coherent Glauber states.

The first term in (15) is the spontaneous emission while the second is the stimulated term. We stress that the "+1" term is the stimulated emission induced by the vacuum field fluctuations.

The well-kown gain formula can be deduced from (15) when the spontaneous emission can be neglected with respect to the stimulated one thus getting in the laboratory frame the following expression

$$g(\omega) = -\frac{4\pi^2}{\gamma} \frac{\lambda L}{\Sigma_E} \frac{\hat{I}}{I_0} \mathscr{F} \frac{K^2}{1+K^2} (\frac{\Delta\omega}{\omega})_0^{-2} \frac{d}{d\nu} (\frac{\sin\nu/2}{\nu/2})^2 \qquad (16)$$

where

$$\Sigma_E = \text{e-beam cross section}$$

$$\mathscr{F} \equiv \text{filling factor} = \begin{cases} 1 & \text{if } \Sigma_E > \Sigma_L \\ \Sigma_E/\Sigma_L & \text{if } \Sigma_E < \Sigma_1 \end{cases}$$

$$\Sigma_L \equiv \text{laser beam cross section}$$

$$L \equiv \text{undulator length}$$

$\hat{I}$ ≡ e-beam peak current

$I_0 = ec/r_0 (\equiv 1.7 \times 10^4 A)$ Alfvén current

$(\Delta\omega/\omega)_0 \equiv 1/2N$

$\nu = 2\pi N(\omega-\omega_0)/\omega_0$ (17)

Since emission at higher harmonics occurs in undulator magnets on and off axis, an analogous gain formula for higher harmonics can be derived (see Ref. 5 for a general formulation).

In particular linear undulators allow odd harmonic emission on axis and the relevant gain can be written as

$$g_n(\omega) = - g_n^\circ \frac{\pi}{n} \frac{d}{d\nu_n} \left(\frac{\sin\nu_n/2}{\nu_n/2}\right)^2 \, , \qquad n=1,3,\ldots$$

$$\nu_n = 2\pi N \frac{n\omega_0-\omega}{\omega_0}$$

$$g_n^\circ = \frac{4\pi}{\gamma} \frac{\lambda_n L}{\Sigma_E} \frac{\hat{I}}{I_0} \; \mathscr{F}_n \; F_n(\xi)\left(\frac{\Delta\omega}{\omega}\right)_0^{-2} \, , \qquad \lambda_n = \lambda/n$$

$\mathscr{F}_n$ ≡ n-th harmonic filling factor

$$F_n(\xi) = \xi[n(J_{n+1/2}(n\xi) - J_{n-1/2}(n\xi))]^2 \, , \qquad \xi = (1/2)K^2/(1+K^2)$$

$(J_n(\cdot)$ ≡ n-th cylindrical Bessel function) (18)

In this section we have discussed the basic FEL process from a very heuristic point of view and in the next section we will discuss more technical details relevant to the specific aspects of what we define "low technology FEL".

LOW TECHNOLOGY FEL

So far many FEL experiments have successfully operated.[6] An up-to-date chart of the operating and proposed experiments is shown in Fig. 6, where the FEL sources are characterized by a wavelength-energy plot. The energy is relevant to an e-beam accelerator while the wavelength is the operating laser one. The continuous line refers to the tunability FEL curve which for an undulator with K=1 and λ_u=5 cm ranges from ultraviolet to microwave. This last statement could be erroneously interpreted as the claim of the large FEL tunability. Such a wide tunability range would in turn require an accelerating machine with an e-beam whose energy could be continuously tuned from MeV to GeV region. The ideal accelerator for a "radio-like" FEL has not yet been designed but its performances can be easily listed:

a) easy energy tunability

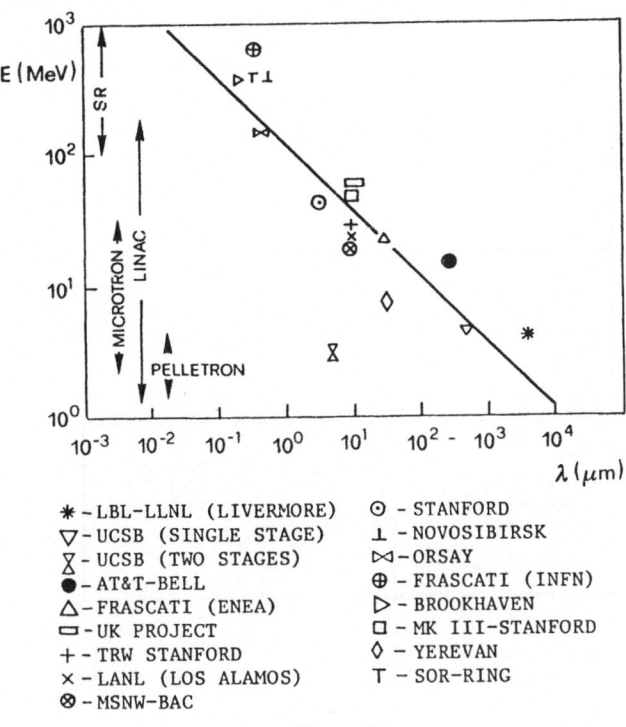

Fig. 6

FEL Scenario

b) modest size
c) high beam power (average and peak)
d) good beam quality (small energy spread and emittances)
e) low cost both of construction and management

In Fig. 7 we have summarized in a current-energy plot the accelerating devices which can be used for FEL. The energy and current ranges are impressive from MeV and KA to GeV and hundreds of mA and what is clear is that an accelerator capable of this flexibility is far from the present technological possibilities.

By "low technology FEL" we mean FELs operating with low energy single passage accelerators with well-established technology and without particularly sophisticated performances. Therefore referring to Fig. 7 we restrict our attention to RF linacs, race-track and circular microtrons.

An FEL grounded on low technology will be characterized by modest size, moderate power and limited to the IR, sub-millimeter region of the spectrum. Its cost will also be modest and according to the plot of Fig. 8. will be less than 1 M$.

In designing the FEL device one must take into account all the effects which can severely affect the gain such as the inhomogeneous broadening coming from emittance and energy spread. It is well known that the parameters which control[6] the deviation from the homogeneous broadening operation are the following:

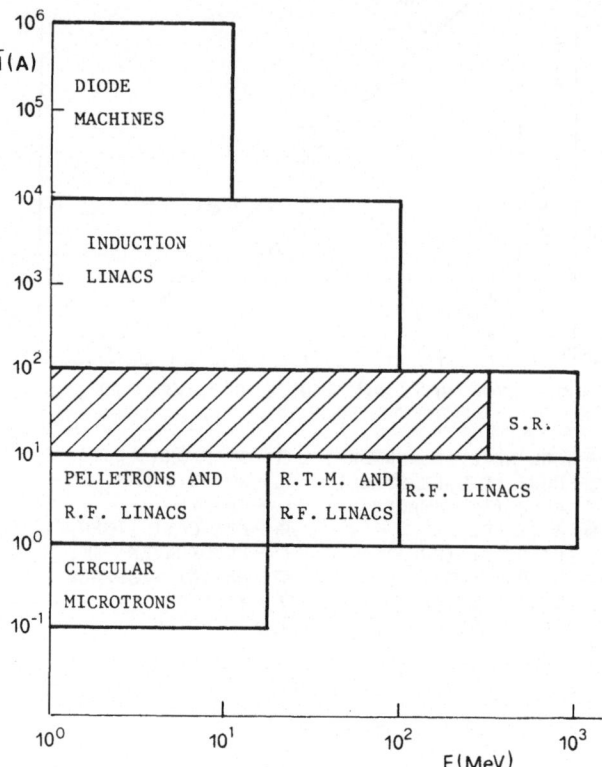

Fig. 7

Current-energy plot for existing electron accelerators

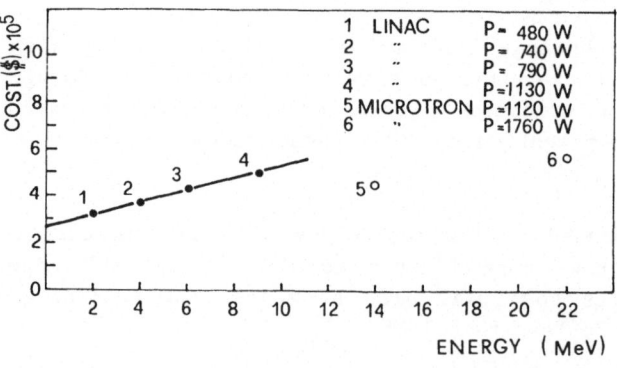

Fig. 8

Costs vs energy for commercial Linacs and microtrons

$$\mu_\varepsilon = 4N\sigma_\varepsilon \qquad \sigma_\varepsilon \equiv \text{r.m.s. energy spread}$$

$$\mu_{x,y} = \frac{4N\gamma^2}{1+K^2}\left\{ \frac{1}{2\sigma_{x,y}^4}\left(\frac{\varepsilon_{x,y}}{2\pi}\right)^4 + 2\left(\frac{K\pi}{\gamma\lambda_u}\right)^4 h_{x,y}^2 \sigma_{x,y}^4 \right\}^{1/2} \tag{19}$$

where $\varepsilon_{x,y}$ are the radial and vertical emittances, $\sigma_{x,y}$ the transverse e-beam dimensions, $h_{x,y}$ are coefficients depending on the undulator geometry, namely, $h_x = h_y = 1$ for helical undulators and $h_x = -\delta$, $h_y = 2+\delta (\delta \ll 1)$ for the linear case, with polarization along the y-axis. Physically δ is the magnitude of the sextupolar term along the x-direction.

The expression (19) suggests that one can choose an optimum $\sigma_{x,y}$ to minimize the effect of the emittance inhomogeneous broadening, i.e.,

$$\sigma_{x,y} = \frac{1}{\pi}\left(\frac{1}{2|h_{x,y}|}\right)^{1/4}\left(\frac{\lambda_u\gamma\varepsilon_{x,y}}{2K}\right)^{1/2} \tag{20}$$

Therefore combining (20) and the second of (19) we get

$$\mu_{x,y} = 2N\sqrt{2|h_{x,y}|}\,\frac{K}{1+K^2}\frac{\gamma\varepsilon_{x,y}}{\lambda_u}$$

The μ coefficients can be understood as the ratio between the inhomogeneous broadening due to the energy spread and emittance and the homogeneous one; their effect on the gain is shown in Fig. 9 and it is clear that large values of μ significantly reduce the gain. To give an example, an FEL operating with the following parameters: energy spread of 5%, N=50, K=1, λ_u =5 cm, γ=40, $\varepsilon_{x,y}$=2πmm mrad, $h_{x,y}$=1 we have

$$\mu_\varepsilon = 1 \qquad \mu_{x,y} \simeq 0.4 \tag{21}$$

In these conditions, which are the operating ones of a low energy accelerator, the gain is more than halved with respect to the homogeneous case.

Another important parameter is the longitudinal bunch length which plays an important role in producing a kind of mode-locking in the laser operation.[6] It can indeed be shown that the strength of the coupling between longitudinal modes is given by the parameter

$$\mu_c = \frac{N\lambda}{\sigma_z} \tag{22}$$

where σ_z is the r.m.s. bunch longitudinal length. The larger μ_c, the greater the number of the coupled longitudinal modes. The bunched e-beam structure (see Fig. 10) is also responsible for the so-called lethargic FEL behaviour, i.e., the slow down of the light pulse due to the interaction, and the necessity of shortening the cavity length with

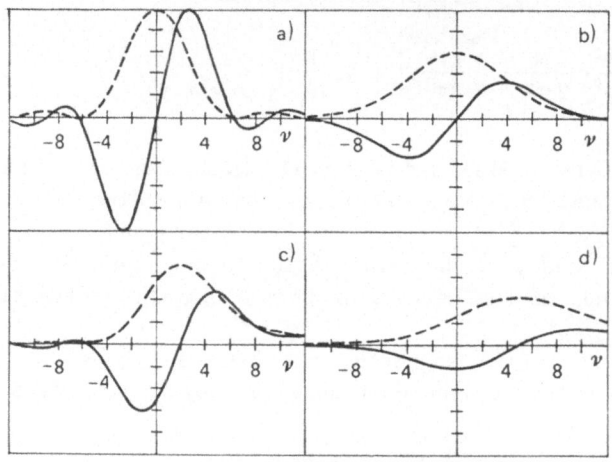

Fig. 9
Inhomogeneous broadened gain
(a) $\mu_\varepsilon = \mu_x = \mu_y = 0$, (b) $\mu_\varepsilon = 1$, $\mu_x = \mu_y = 0$
(c) $\mu_\varepsilon = 0$, $\mu_x = 1$, $\mu_y = 0$, (d) $\mu_\varepsilon = \mu_x = \mu_y = 1$

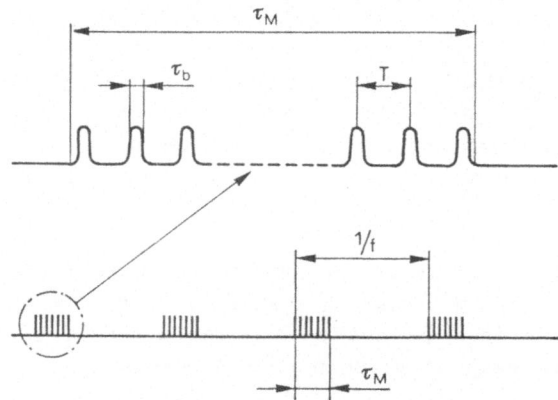

Fig. 10
e-beam structure from a RF machine
$\tau_b \equiv$ microbunch time duration;
$\tau_M \equiv$ macropulse time duration;
$T \equiv$ microbunch time separation
$f \equiv$ repetition frequency

respect to the nominal round-trip period to have timing between light and e-bunches.

From the above very brief description of the parameters which can affect the gain of an FEL, it seems clear that the optimized design of this kind of laser device is a rather complicated game involving the simultaneous interplay of parameters of different nature such as those relevant to the machine, to the cavity, to the undulator, and to the RF accelerating field. It is however possible to comprise all the above effects in a single gain formula which can be written as follows:

$$g_h = g_h^\circ \ |Req_\gamma[\theta; \mu_c; \mu_x; \mu_y, \mu_\varepsilon] \qquad h \equiv helical$$

$$g_{\ell,n} = g_{\ell,n}^\circ \ |Req_\gamma[\theta_n; \mu_c; n\mu_x, n\mu_\gamma, n\mu_\varepsilon], \qquad \ell \equiv linear$$

(23)

$$g_h^\circ = 88 \times 10^{-4} N^2 \ (\frac{k}{1+k^2}) \ \frac{P[MW]}{\delta} \frac{\lambda}{\lambda_u} \quad ,$$

$$g_{\ell,n}^\circ = 88 \times 10^{-4} n^2 N^2 (\frac{k}{1+k^2}) \ \frac{P[MW]}{\delta} \frac{\lambda}{\lambda_u} [J_{(n-1)/2}(n\xi) - J_{(n+1)/2}(n\xi)]^2 \ ,$$

where δ is the machine duty cycle, P[MW] is the e-beam power in megawatts, θ and θ_n are the "delay parameters" and are given by

$$\theta = -\frac{1}{\pi N} \frac{\omega_0 \delta T}{g_h^\circ} \ , \qquad \theta_n = -\frac{1}{\pi} \frac{\omega_0 \delta T}{g_{\ell,n}^\circ}$$

(24)

$\delta T = T_c - T$, $T_c \equiv$ cavity round trip period, $T \equiv$ bunch-bunch time distance (see Fig. 10).

The quantity $|Req_\gamma$ represents the maximum value of the multimode gain function.

The typical behaviour of $|Req_\lambda$ vs θ is shown in Fig. 11, together with the dimensionless laser power χ. It is to be remarked that the maximum gain and the maximum output laser power do not correspond to the same value of θ. Therefore the maximization of the gain does not mean the maximum output laser power.

The average laser power can be evaluated according to the following formula.

$$P_L[MW] = P[MW] \ f[Hz] \ \chi(\theta) \ (\tau_M[\mu s] - \tau_R[\mu s])$$

(25)

where f is the machine repetition frequency, τ_M is the e-beam macropulse duration (see Fig. 10) and τ_R is the pulse rise time linked to the gain by the following expression

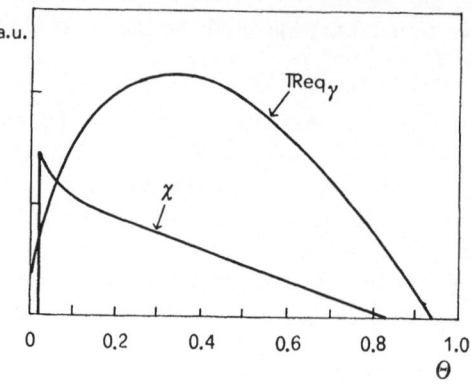

Fig. 11

Gain function and dimensionless laser power vs θ.

$$\tau_R[\mu s] \simeq 0.14 \; \frac{L_c[m]}{g-\gamma_T} \tag{26}$$

g is the gain as a function of the above parameters, γ_T are the cavity losses and L_c is the length of the cavity. In Figure 12 the curves represent the average output power of an FEL operating at the 1st and 3rd harmonics resepctively, with an e-beam of 20 MW. It is evident that in the region 10÷100 μm a low technology FEL may play a significant role yielding indeed larger power than the existing conventional laser sources.

So far we have mainly discussed an FEL process with the main emphasis on low technology devices fixing the limits within which a laboratory not involved in high energy and without a large budget can direct its experimental research. Therefore an experimental program on laser acceleration is essentially a side line activity of the leading program on FEL physics and technology and might be useful to deduce scaling laws. We have described the FEL process as a stimulated Compton scattering and within this framework according to Figs. 1 and 2 (see also Ref. 1) the energy gain per unit length reads

$$\left(\frac{d\varepsilon}{dx}\right)_{ST} = \eta_{ST} \; eE_L \tag{27}$$

$$\eta_{ST} = \eta \cdot n_2 \cdot \frac{\delta p}{p} \frac{\omega_2}{\omega_1} \tag{28}$$

ST ≡ stimulated

where η is the previously defined radiation pressure efficiency (see Eq.(7.)). The efficiency enhancing factor is almost immediately undestood; indeed n_2 is the number of stimulating photons in a volume $L^2\lambda_1$ and L is the interaction length; the last term (the phase-space term) $\delta p/p(\omega_2/\omega_1)$ takes into account the fact that the gain mechanism is just due to two disentangled processes shifted in frequency just by a factor

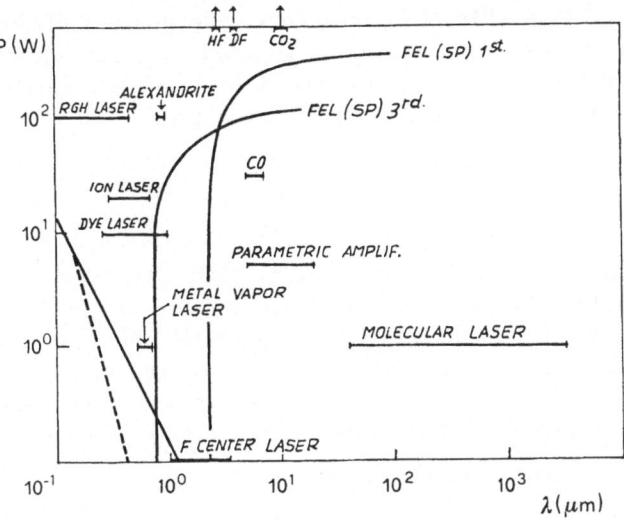

Fig. 12

Comparative chart between FEL and conventional coherent sources. Average power vs λ. Curves FEL (SP); single passage FEL average power vs λ 1st and 3rd harmonic respectively, maximum e.b. power 20 MW, K=1, λ_u=5 cm, N=50, L_c=6 m, τ_M=12 μs, δ=5%. Straight curves: FEL storage-ring average power vs λ without (continuous) and with (dashed) Touschek effect respectively. I=100 A and operating parameter of LEDA-F, see Ref.2. [$L_c \equiv$ cavity length, $\delta \equiv$ duty cycle].

$\delta p/p$. In this connection, expressing (28) in terms of the power density of stimulating field (2) we can write the following expression:

$$\eta_{ST} = \eta \ \frac{dP_2/dS}{e\varepsilon/(L^2\lambda_1)} \tag{29}$$

The obvious condition to get an efficiency enhancement with respect to the previous case is

$$\frac{dP_2}{dS} > \frac{e\,\varepsilon}{L^2\lambda_1} \tag{30}$$

To give an example for ε=20 MeV, L=1 m, λ_1=10 μm, we get

$$\frac{dP_2}{dS} > 9.6 \ \frac{mW}{cm^2}$$

thus indicating that, within this framework, the process of laser acceleration can take place with a very low power stimulating field. This example shows that in principle one could have a very large efficiency enhancement indeed with the parameters given above.

CONCLUSIONS

In the previous section we have emphasized that even though interesting in principle, the laser acceleration is out of the low technology devices. It could be interesting to speculate about the possibility of using a real stimulated Compton scattering to get shorter wavelengths with a low energy FEL device. It is well known that a number of tricks have been proposed to reach short wavelengths with a low energy e-beam.

(a)   the use of a higher harmonic mechanism;

(b)   the use of a device operating with a real e.m. wave in the milli-meter or infrared region as pump wave rather than an undulator provided by a static magnetic field.

Point (a) has been discussed in a number of papers[7]; here we recall that the possibility of operating at higher harmonics requires increasingly good e-beam qualities. In this paper we briefly discuss case (b). According to what has been discussed about the undulator wave approximation, we need a few changes to discuss this kind of FEL operation. We will limit the discussion to accelerating devices without radiation damping and to pump field wavelength in the millimeter range. The equivalent undulator parameter K in this connection reads[8]:

$$K^2 = 3.7 \times 10^{-11} \frac{P[W]\lambda_p^2}{\Sigma_p} \tag{31}$$

where Σ_p is the mode area of the pump field, P its power in watts, and λ_p its wavelength.

The gain may be rewritten as

$$g \sim 3.7 \times 10^{-14} \frac{N^2 \cdot \lambda_p^2[cm]}{\gamma^3 \delta} \cdot \alpha \cdot P[W] f(\mu_c; \mu, \mu_\varepsilon) \tag{32}$$

where

   N ≡ number of optical undulator periods

   δ ≡ duty cycle

   α ≡ brightness parameter

   f ≡ the inhomogeneous part of the gain function (see Fig. 13).

To give and example for $\lambda_p = 1$ mm, $N = 2.5 \times 10^3$, $\sigma_\varepsilon = 10^{-4} (\mu_\varepsilon = 1)$, $\delta = 5\%$, $\mu = 1$,

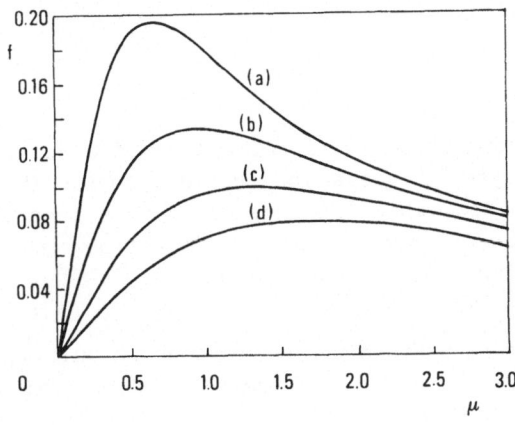

Fig. 13

Gain function vs μ for different values of μ_ε;
(a) $\mu_\varepsilon = 0.5$, (b) $\mu_\varepsilon = 1.0$, (c) $\mu_\varepsilon = 1.5$, (d) $\mu_\varepsilon = 2.0$.

we can rewrite Eq.(32.) as

$$g \approx 05.5 \times 10^{-9} \alpha P[W]/\gamma^3$$

If one wants to operate in the sub-micrometer region, $\gamma = 20$ is sufficient ($\lambda_0 = 0.6 \ \mu m$), furthermore requiring a gain of the order of 1% we find $\alpha P[W] \sim 1.4 \times 10^{11}$, which corresponds, for $\alpha = 1.1 \times 10^4$, to a power of the pump field $P \approx 1.3$ MW. This power is within the limits of the present microwave technology.

REFERENCES

1. A. Renieri, Proc., "Seminar on new trends in particle acceleration techniques", S. Stipcich, S. Tazzari, V. G. Vaccaro, eds. 135 (1982).
2. L. S. Brown and T. W. Kibble, Phys. Rev. 133A:705 (1964).
3. G. Dattoli and A. Renieri, Experimental and Theoretical Aspects of the Free Electron Laser, in: "Laser Handbook", vol. IV, M. L. Stich and M. S. Bass, eds., North Holland, Amsterdam, (1985).
4. G. Dattoli and A. Renieri, ENEA Report RT/TIB/85/42, to be published in: "FEL - Handbook", W. B. Colson, C. Pellegrini, A. Renieri, eds., North Holland, Amsterdam.
5. W. B. Colson, G. Dattoli and F. Ciocci, Phys. Rev. 31A:828 (1985).
6. U. Bizzarri, F. Ciocci, G. Dattoli, A. De Angelis, E. Fiorentino, G. P. Gallerano, T. Letardi, A. Marino, G. Messina, A. Renieri, E. Sabia, A. Vignati, to be published in: "La Rivista del Nuovo Cimento".
7. See Ref. 5 and references therein.
8. F. Ciocci, G. Dattoli and J. E. Walsh, Nucl. Instrum. Methods A237:401 (1985).

BOSCOLO:

As far as I know in your ENEA-FEL experiment you did not reach saturation. Is it true? If yes, I guess you should increase the accelerator
current. I do not know how much the current must be increased. Anyway,
do you think that the increased current microtron you are looking
for can still be considered a low-tech device?

SABIA:

We have not reached the saturation regime because the micropulse duration
is not enough to reach it. It's a high-tech microtron; we want 200
mA of average current.

DEVELOPMENTS IN THE PHYSICS OF
HIGH CURRENT LINEAR ION ACCELERATORS

T.P. Wangler

Los Alamos National Laboratory
Los Alamos, NM 87545, USA

I. INTRODUCTION

Linear ion accelerators have been proposed for both high peak and high average current applications.[1] The design of high-intensity linacs is strongly influenced by the requirement of providing sufficient focusing (confinement) for the beam to balance the effects of the space-charge (self-field) forces, which become important for high beam current. Longitudinal focusing is obtained by operation on the rising rf electric field, where both acceleration and phase stability are obtained. Strong transverse focusing results from electric or magnetic quadrupole lenses arranged in a quasi-periodic structure; in the drift-tube linac (DTL), these individual quadrupole elements are installed within the drift tubes. In a smooth approximation, the quadrupole lenses produce an equivalent linear continuous focusing force upon which is superimposed the local forces of individual lenses that create a flutter in the beam envelope.

The self-field forces between beam particles in high-current linear accelerators produce two generally undesirable effects: (a) defocusing and (b) rms emittance growth.[2,3] Charged particle beams exhibit properties similar to plasmas. Although only one charge species is present, the external focusing forces play a role analogous to the background charge of opposite sign in a neutral plasma. Plasma oscillations are observed in numerical simulation studies of intense beams at the usual plasma frequency $\omega_p^2 = e^2 n / e_0 m$. An effective temperature can be defined in terms of squared rms velocity of the beam particles (even though the velocity distribution may not be Maxwellian), and plasma shielding of the external fields is observed, characterized by the Debye length. For linear external focusing, exact shielding implies a uniform charge density, which results in the approximate property of homogenization (uniformity) of space-charge dominated beams.

These self-field effects in a linear accelerator impose limits for both the peak and the average beam current.[4] The peak current is limited by several effects,[5,6] including (a) collective instabilities driven by periodic focusing lenses,[7] (b) attainable external focusing fields, and (c) higher multipoles and other nonlinear aberations at large beam radii. In addition, the average beam current may be limited by many engineering issues, some of which are related to emittance growth associated with the formation of an

---

*This work was performed under the auspices of the U.S. Dept. of Energy.

outer beam halo. This low-density halo can lead to particle losses, which for high average current linacs, can produce an increased heat load, vacuum degradation, peak surface field reduction, and radioactivation of the structure that would make routine maintenance very difficult.

The emittance (phase-space area occupied by the beam) is a critical beam property that must be controlled in the design of a high-intensity linear accelerator. Generally, small emittance is desirable for two reasons: (a) for fixed aperture and fixed focusing strength, smaller emittance implies a reduction in the beam halo and a larger beam current capacity and (b) some applications (for example, heavy-ion fusion) place severe constraints on the output beam optics and require a high-current beam with a very small emittance. But even when Liouville's theorem is satisfied (when a collisionless system exhibits continuity of flow in phase-space, and no dissipative forces), irreversible changes in the emittance contour can cause an increase in an effective emittance.[8] A useful measure of effective emittance is the rms emittance ε, which can be defined for a phase-space projected area x–x' in terms of the second moments $\overline{x^2}$, $\overline{x'^2}$, and $\overline{xx'}$ of the particle distribution as

$$\varepsilon = 4\left(\overline{x^2}\ \overline{x'^2} - \overline{xx'}^2\right)^{1/2} .$$

(1)

The factor of 4, frequently introduced, results in an emittance that contains all particles for a 2-D continuous uniform beam and contains about 85–90% of the particles for nearly all other distributions, whether for continuous or bunched beams.

Generally, most emittance growth occurs at low velocities, where beam densities are higher, focusing is weaker, and where the beam becomes bunched before injection into the linac. Some known causes of emittance growth include (a) mismatch of the beam in the presence of nonlinear external forces, (b) coupling between degrees of freedom that makes accurate matching difficult in practice, and (c) space-charge effects.

The suggestion of space-charge-induced emittance growth associated with a kinetic-energy exchange mechanism was made many years ago to explain the numerical studies for the CERN and Brookhaven linac injectors.[9-12] The law of equipartition of energy was invoked, which asserts that in thermal equilibrium, the same average kinetic energy is associated with each degree of freedom. But further study was required to establish this principle, in a charged particle accelerator, where collective fields dominate over particle collisions. More recent numerical simulation studies of high-current, asymmetric continuous beams[2,3] and bunched linear accelerator beams[13-15] reaffirmed the importance of the kinetic-energy exchange mechanism. Detailed analysis of the Kapchinskii-Vladimirskii (K-V) distribution for 2-D asymmetric beams[2,3] resulted in a prediction of coherent mode instability thresholds for asymmetric beams, which established a collective field mechanism for kinetic-energy exchange. The predicted threshold values even agreed closely with numerical simulation results for bunched beams.[15,16] Simulation studies clearly showed that equipartitioning does occur when the space-charge forces become large.[16] However, theoretical predictions for the magnitude of the emittance growth from equipartitioning were still not available.

An interest in high-current, low-emittance beams for heavy ion fusion stimulated work to understand the transport of intense beams in periodic quadrupole focusing systems. The instabilities associated with the K-V distribution were identified analytically and confirmed by numerical simulation.[7] A mechanism of emittance growth associated with excitation of coherent modes by a periodic focusing system was clearly established by this work.

I will concentrate on problems associated with the self-field or space-charge-induced growth of rms emittance. I will discuss recent work in which a relationship has been established between the rms emittance and a quantity I call the nonlinear field energy.[17-23] This relationship will be used to derive general equations for rms emittance growth for continuous focusing channels that include two mechanisms: (1) charge-density redistribution and (2) kinetic energy exchange between different degrees of freedom. The equations contain two final-state parameters, the final nonlinear field energy, and the final value of a quantity called the partition parameter. I will invoke two hypotheses to characterize the final state of space-charge dominated beams: (1) homogenization (charge-density uniformity) and (2) equipartitioning. For very intense beams, these hypotheses result in values for the two final-state parameters and equations for emittance growth that depend only on the initial beam properties. In addition to the emittance growth equations, I will present equations for the minimum final emittance for beams at the extreme space-charge limit and will show how the minimum final emittance depends on beam current, focusing strength, and the initial charge-density profile. The dimensionless scaling parameters will be identified, which give emittance growth as a function of the parameters of the beam and accelerator. I will present results both for a 2-D continuous beam (a smoothed representation for an unneutralized beam in a quadrupole transport channel) and for an axially symmetric bunched beam (a smoothed representation for a well-bunched linac beam). In a modern ion linear accelerator with a radio-frequency quadrupole (RFQ) accelerating structure,[24] both continuous and bunched beams are present and both must be considered.

II. RMS EMITTANCE AND NONLINEAR FIELD ENERGY

Consider the problem of a continuous, azimuthally symmetric 2-D beam that propagates in the +s-direction at constant velocity v. Assume that the beam is confined radially by a continuous, linear, external focusing force and that the paraxial approximation is valid in which all particles have the same longitudinal velocity $v \gg v_t$, where v_t is the transverse velocity component. Also consider the characteristics of a steady-state solution, where the charge density, current density, and fields have no explicit dependence on time. Allow for initial phase-space distributions that are not necessarily stationary, and in general, expect that the charge density $\rho(r,s)$ will evolve from an initial state at $s = 0$ to some final state, which may be stationary or independent of s. Using these assumptions, Eq. (1) can be differentiated and written in the form[17,18]

$$\frac{d\varepsilon^2}{ds} = - \frac{X^2 K}{2} \frac{d}{ds}\left(\frac{U}{w_0}\right) , \tag{2}$$

where $X = 2\sqrt{\overline{x^2}}$ is interpreted as the total beam radius of the equivalent uniform beam (uniform beam with the same second moments as the real beam). The quantity K is the generalized perveance given by

$$K = \frac{eI}{2\pi\varepsilon_0 mv^3\gamma^3} ,$$

where I is the beam current, m is the mass, γ is the relativistic mass factor, and ε_0 is the permittivity of free space. The quantity U is defined as

$$U = W - W_u ,$$

where W is the self-electric field energy per unit length, and W_u is the self-electric field energy per unit length for an equivalent uniform beam. The quantity w_0 is defined as

$$w_0 = (eN)^2/16\pi\varepsilon_0 , \tag{3}$$

where N is the number of particles per unit length, related to the beam current by $I = Nev$.

The quantity U is the residual self-electric field energy possessed by beams with nonuniform charge distributions. Because nonuniform beams have nonlinear self-fields, U is called the nonlinear field energy. Equation (2) implies that a decrease in nonlinear field energy U corresponds to an increase in rms emittance. Both the electric and magnetic-field contributions are contained in Eq. (2) by including the factor γ^3 in the definition of K (γ^2 accounts for the magnetic field and γ accounts for relativistic mass). Although the total electromagnetic stored energy is the sum of electric plus magnetic terms, it is the difference that is related to changes in transverse rms emittance because the transverse electric and magnetic self-force terms have opposite signs. Therefore, rms emittance growth, induced by space charge, is inherently a nonrelativistic effect; it is most important when γ is near unity.

The quantity w_0 is the self-electric field energy per unit length within the beam boundary of an equivalent uniform beam. The quantity U/w_0 is zero for a uniform charge distribution and is positive both for peaked and hollow distributions, increasing as the distribution becomes more nonuniform. Furthermore, U/w_0 is independent of both beam current and rms beam size and is a function only of the shape of the distribution. Thus, Eq. (2) shows that rms emittance changes are associated with three separate factors: rms beam size, perveance, and changes in shape of the charge distribution. The quantity $U_n = U/w_0$ is defined as the normalized nonlinear field energy, and values of U_n are given for example distributions in Table I. These distributions are listed in order from most peaked to most hollow.

TABLE I

$U_n = U/w_0$ FOR SOME COMMON 2-D DISTRIBUTIONS

Distribution Function	Charge Density $\rho(r)$		U/w_0
Gaussian	$\exp(-r^2/2\sigma^2)$		0.154
Parabolic	$1 - (r/R)^2$	$r \leq R$	0.0224
Uniform	1	$r \leq R$	0.000
Hollow (n = 2)	r^2	$r \leq R$	0.0754
Hollow (n = 10)	r^{10}	$r \leq R$	0.245

The basic relationship between field energy and rms emittance is not restricted to a continuous azimuthally symmetric 2-D beam. A more general result was derived by Hofmann[20] to include asymmetric 2-D beams and bunched 3-D beams, which can be written as

$$\frac{1}{\overline{x^2}} \frac{d\varepsilon_x^2}{ds} + \frac{1}{\overline{y^2}} \frac{d\varepsilon_y^2}{ds} + \frac{1}{\overline{z^2}} \frac{d\varepsilon_z^2}{ds} = \frac{-32}{mv^2\gamma^3 N} \frac{d}{ds} (W - W_s) \qquad (4)$$

where ε_x, ε_y, and ε_z are the x-, y-, and z-plane emittances, and s is the coordinate along the beam direction. This equation is not exact in the general case, but is a good approximation for practical beam distributions. For a general 2-D continuous beam (the relationship is exact for elliptical beams), the equation can be written as[21]

$$\frac{1}{\overline{x^2}} \frac{d\varepsilon_x^2}{ds} + \frac{1}{\overline{y^2}} \frac{d\varepsilon_y^2}{ds} = -K \frac{dU_n}{ds} \quad , \qquad (5)$$

where x and y are directions transverse to the beam axis s, and where

184

$X = 2 \sqrt{\overline{x^2}}$ and $Y = 2\sqrt{\overline{y^2}}$ are the semiaxes of the equivalent uniform beam. The values of U_n, given in Table I for the symmetric beam, are also valid for the more general elliptical beams. For a symmetric beam, Eq. (5) reduces to Eq. (2).

For an axially symmetric bunch (smooth approximation for a bunched beam in a linac) with rms semiaxes $a = \sqrt{\overline{x^2}} = \sqrt{\overline{y^2}}$ and $b = \sqrt{\overline{z^2}}$ in the beam rest frame, Eq. (4) can be written as[21]

$$\frac{2}{a^2} \frac{d\varepsilon_x^2}{ds} + \frac{1}{b^2} \frac{d\varepsilon_z^2}{ds} = \frac{16K_3}{bw_3} G_0(b/a) \frac{dU}{ds} \quad , \tag{6}$$

where x and y are transverse to the beam direction s, and K_3 is a perveance-like parameter with dimensions of length given by

$$K_3 = \frac{Ne^2}{20\sqrt{5}\pi\varepsilon_0 \, mv^2\gamma^3} \quad . \tag{7}$$

The quantity U is the nonlinear field energy and w_3 is the field energy normalization parameter given by

$$w_3 = \frac{(Ne)^2 G_0(b/a)}{40\sqrt{5}\pi\varepsilon_0 b} \quad . \tag{8}$$

The function $G_0(b/a)$ is given by

$$G_0(b/a) = (1 - M) + M(b/a)^2 \quad , \tag{9}$$

where M is the ellipsoid form factor,[25] which is approximately $M = 1/3(b/a)$ for a nearly spherical bunch, and $M = 1/3$ for an exact sphere. The number of beam particles per bunch is N, and the average beam current for a string of bunches in an rf linac with one bunch per rf period is given by $I = Nec/\lambda$, where c is the speed of light, and λ is the rf wavelength. Also, $U_n = U/w_3$ depends only on the shape of the charge-density profile, is independent of rms beam size and beam current, and is a measure of charge-density nonuniformity, having a minimum value of zero for a uniform bunch. Table II shows values of U_n for some common, spherical, charge-density distributions.

$$\frac{\varepsilon_{xf}}{\varepsilon_{xi}} = \left[1 - \frac{(P_i - P_f)}{P_i(1 + P_f)} - \frac{P_f}{P_i(1 + P_f)} G_2(X/Y) \left(\frac{k_{0y}^2}{k_{yi}^2} - 1 \right) (U_{nf} - U_{ni}) \right]^{1/2} , \tag{13a}$$

and

$$\frac{\varepsilon_{yf}}{\varepsilon_{yi}} = \left[1 + \frac{(P_i - P_f)}{(1 + P_f)} - \frac{1}{(1 + P_f)} G_2(X/Y) \left(\frac{k_{0y}^2}{k_{yi}^2} - 1 \right) (U_{nf} - U_{ni}) \right]^{1/2} , \tag{13b}$$

where the subscripts i and f refer to the initial and final states.

The function $G_2(X/Y)$ is defined as

$$G_2(X/Y) = \frac{1}{2} (1 + X/Y) \quad . \tag{14}$$

The quantity k_{yi} is the initial single-particle wave number (tune) of the equivalent uniform beam including space charge. The betatron tune ratio in the x-plane can be easily expressed in terms of the y-plane ratio[29] and could equally well have been used instead. The betatron wave numbers k_{yi}

and k_{0y} measure the effectiveness of the focusing with and without space charge, and as beam intensity increases, k_{yi}/k_{0y} decreases. The relationship between the tune depression ratio k_{yi}/k_{0y} and the beam and channel parameters K, ε_x, ε_y, k_{0x}, and k_{0y} is algebraically complicated and is most easily obtained numerically, but near the space-charge limit, a simple result is[29]

$$\frac{k_y}{k_{0y}} = \left(\frac{\varepsilon_y k_{0y}}{2K}\right)\left(\frac{k_{0x}^2 + k_{0y}^2}{k_{0x}^2}\right) . \tag{15}$$

Also, one can write $P_i = (\varepsilon_x/\varepsilon_y)^2/(X/Y)^2$, where $X/Y \approx k_{0y}^2/k_{0x}^2$ for a space-charge dominated beam.

Equations (13a) and (13b) express the emittance growth ratios in terms of the initial beam variables X/Y, P_i, k_{yi}/k_{0y}, and U_{ni} and two final state variables P_f and U_{nf}. They contain two growth (or decay) terms, one that depends on the change in the partition parameter P and corresponds to kinetic-energy exchange between the x- and y-planes, and the other that depends on the change in nonlinear field energy U_n, caused by charge-density redistribution. For the case of a round symmetric beam, where X = Y and $P_f = P_i = 1$, these equations reduce to the results already presented.

The values of the final-state parameters must be determined either from additional theory or from numerical simulation. I invoke two hypotheses, observed from our simulation studies,[22] to approximately characterize the final state for space-charge dominated beams: (1) homogenization (charge-density uniformity), and (2) equipartitioning. For intense beams, these hypotheses allow us to obtain values for the two final-state parameters, and equations for emittance growth that depend only on the initial beam properties.

Thus, if one assumes that the final charge density is uniform or homogenized ($U_{nf} = 0$) and that the final kinetic energy is equipartitioned ($P_f = 1$), then

TABLE II

$U_n = U/w_3$ FOR COMMON SPHERICAL BUNCH DISTRIBUTIONS

Distribution	Charge Density	U_n
Gaussian	$\exp(-r^2/2\sigma^2)$	0.308
Parabolic	$1 - r^2/R^2$	0.0368
Uniform	1	0.00

Equivalent forms for Eq. (10) were discovered earlier,[26,27] but it appears that the utility of this result for obtaining a better understanding of emittance growth effects in linacs and beam transport systems was not fully recognized.

III. EMITTANCE GROWTH IN 2-D CONTINUOUS BEAMS

Emittance Growth Equations

The rms envelope equations for a 2-D beam in a continuous focusing channel can be written[28]

$$\frac{d^2X}{ds^2} + k_{0x}^2 X - \frac{\varepsilon_x^2}{X^3} - \frac{2K}{X + Y} = 0 , \tag{10a}$$

and

$$\frac{d^2Y}{ds^2} + k_{0y}^2 Y - \frac{\varepsilon_y^2}{Y^3} - \frac{2K}{X + Y} = 0 \quad , \tag{10b}$$

where k_{0x} and k_{0y} are the single-particle wave numbers for transverse motion at zero beam current. When the equivalent uniform-beam semiaxes X and Y are constant, the beam is rms matched, and the first term of Eqs. (10a) and (10b) vanishes. In the limit where the space-charge term greatly exceeds the emittance term, the beam is said to be space-charge dominated, and X and Y become independent of emittances ε_x and ε_y. In this approximation one can easily integrate Eq. (5), resulting in

$$\frac{\Delta\varepsilon_x^2}{X^2} + \frac{\Delta e_y^2}{Y^2} = -K \ \Delta U_n \quad . \tag{11}$$

For an rms matched beam, write $\varepsilon_x = XX'$ and $\varepsilon_y = YY'$, where X' and Y', equivalent uniform-beam (in velocity space) divergences are related to rms beam divergences $\sqrt{x'^2}$ and $\sqrt{y'^2}$ by $X' = 2\sqrt{x'^2}$ and $Y' = 2\sqrt{y'^2}$. It is convenient to introduce a new parameter P, the partition parameter, defined by

$$P = X'^2/Y'^2 \quad . \tag{12}$$

The quantity P is a measure of the kinetic-energy asymmetry. Using the partition parameter, Eq. (11) can be re-expressed in the two forms that give emittance growth equations in each plane, obtaining

$$\frac{\varepsilon_{xf}}{\varepsilon_{xi}} = \left[1 - \frac{(P_i - 1)}{2P_i} + \frac{G_2(X/Y)}{2P_i}\left(\frac{k_{0y}^2}{k_{yi}^2} - 1\right)U_{ni} \right]^{1/2} \quad , \tag{16a}$$

and

$$\frac{\varepsilon_{yf}}{\varepsilon_{yi}} = \left[1 + \frac{(P_i - 1)}{2} + \frac{G_2(X/Y)}{2}\left(\frac{k_{0y}^2}{k_{yi}^2} - 1\right)U_{ni} \right]^{1/2} \quad . \tag{16b}$$

Minimum Final Emittance

A general result for a minimum final emittance can be derived. This is done using the results,[29] near the the extreme space-charge limit, that $X^2 = 2K(k_{0y}/k_{0x})^2/(k_{0x}^2 + k_{0y}^2)$ and $Y^2 = 2K(k_{0x}/k_{0y})^2/(k_{0x}^2 + k_{0y}^2)$. Then, using the same assumptions of final-state homogenization and equipartitioning one obtains

$$\varepsilon_{xf}^2 = \varepsilon_{xi}^2 \frac{(1 + P_i)}{2P_i} + \frac{(k_{0y}/k_{0x})^2}{k_{0x}^2 + k_{0y}^2} K^2 U_{ni} \quad , \tag{17a}$$

and

$$\varepsilon_{yf}^2 = \varepsilon_{yi}^2 \frac{(1 + P_i)}{2} + \frac{(k_{0x}/k_{0y})^2}{k_{0x}^2 + k_{0y}^2} K^2 U_{ni} \quad . \tag{17b}$$

The minimum final emittances correspond to initial emittances $\varepsilon_{xi} = \varepsilon_{yi} = 0$ (the extreme space-charge limit). Then

$$\varepsilon_{xf,min} = \frac{k_{0y}/k_{0x}}{\sqrt{k_{0x}^2 + k_{0y}^2}} \, K \, U_{ni}^{1/2} \quad , \tag{18a}$$

and

$$\varepsilon_{yf,min} = \frac{k_{0x}/k_{0y}}{\sqrt{k_{0x}^2 + k_{0y}^2}} \, K \, U_{ni}^{1/2} \quad . \tag{18b}$$

Equations (18a) and (18b) predict that the minimum final emittance depends on the initial nonlinear field energy U_{ni} but not on the initial partition parameter P_i. The minimum final emittances are linearly proportional to beam current through the parameter K.

Scaling With Beam and Channel Parameters

Using the pair of matched rms envelope equations for the x- and y-planes and the definitions of the equivalent uniform beam tunes k_x and k_y, it is straightforward to show that the three variables P, X/Y, and k_y/k_{0y} that determine the emittance growth can be expressed as functions of three new dimensionless variables that depend directly on the beam and channel parameters K, ε_x, ε_y, k_{0x} and k_{0y}.[29] These three new variables consist of two current-dependent parameters $u_x = K/2\varepsilon_x k_{0x}$ and $u_y = K/2\varepsilon_y k_{0y}$, and the zero-current beam aspect ratio $X_0/Y_0 = (\varepsilon_x k_{0y}/\varepsilon_y k_{0x})^{1/2}$. Thus, for a given initial charge-density profile, constant values of u_x, u_y, and X_0/Y_0 should produce the same emittance growth from charge-density redistribution and kinetic-energy exchange. Note that the ratios I/ε_x and I/ε_y enter into u_x and u_y to determine the emittance growth.

IV. EMITTANCE GROWTH IN AXIALLY SYMMETRIC BUNCHED BEAMS

Emittance Growth

The rms envelope equations for an axially symmetric bunched beam are[28]

$$\frac{d^2 a}{ds^2} + k_{0y}^2 a - \frac{\varepsilon_y^2}{16a^3} - \frac{K_3 G_y(b/a)}{ab} = 0 \quad , \tag{19a}$$

and

$$\frac{d^2 b}{ds^2} + k_{0z}^2 b - \frac{\varepsilon_z^2}{16b^3} - \frac{K_3 G_z(b/a)}{ab} = 0 \quad , \tag{19b}$$

where k_{0y} and k_{0z} are the single-particle wave numbers for transverse and longitudinal motion, respectively, at zero beam current. The functions G_x and G_z are given in terms of the ellipsoid form factor M by $G_y = 3(1 - M)/2$ and $G_z = 3 M b/a$, and both equal unity for a spherical bunch. As was true for the 2-D beam, the rms semiaxes a and b will become independent of the emittances for an rms matched beam in the extreme space-charge limit. In this approximation, Eq. (6) can be integrated and

$$\frac{2\Delta\varepsilon_y^2}{a^2} + \frac{\Delta\varepsilon_z^2}{b^2} = \frac{16K_3}{b} \, G_0(b/a) \, \Delta U_n \quad . \tag{20}$$

As for the 2-D problem, introduce the partition parameter P, defined by

$$P = b'^2/a'^2 \quad , \tag{21}$$

where $a' = \sqrt{\overline{x'^2}}$ and $b' = \sqrt{\overline{z'^2}}$, so that P is a measure of the nonrelativistic kinetic-energy asymmetry between the longitudinal and transverse planes in the bunch rest frame. Using the partition parameter, one can re-express Eq. (20) to give emittance growth equations in each plane. The results are

$$\frac{\varepsilon_{zf}}{\varepsilon_{zi}} = \left[1 - \frac{2}{P_i}\frac{(P_i - P_f)}{(2 + P_f)} - \frac{1}{P_i}\frac{P_f}{(2 + P_f)} \; G(b/a)\left(\frac{k_{0y}^2}{k_{yi}^2} - 1\right)(U_{nf} - U_{ni})\right]^{1/2} , \quad (22a)$$

and

$$\frac{\varepsilon_{yf}}{\varepsilon_{yi}} = \left[1 + \frac{(P_i - P_f)}{(2 + P_f)} - \frac{1}{(2 + P_f)} \; G(b/a)\left(\frac{k_{0y}^2}{k_{yi}^2} - 1\right)(U_{nf} - U_{ni})\right]^{1/2} , \quad (22b)$$

where the subscripts i and f refer to the initial and final states. The function G(b/a) is defined by $G(b/a) = 2G_0(b/a)/3(1 - M)$, and the quantity k_{yi} is the initial single-particle wave number of the equivalent uniform beam with space charge for the y-plane. The initial tune depression ratio k_{zi}/k_{z0} can be expressed[30] as a function of k_{yi}/k_{0y}, P_i, and b/a. An approximate relationship for the tune depression ratio k_{yi}/k_{0y}, with respect to the parameters of the beam and channel valid near the extreme space-charge limit, is[30]

$$\frac{k_y}{k_{0y}} = \frac{\varepsilon_y}{4k_{0y}}\left(\frac{k_{0z}^2}{K_3}\right)^{2/3}\left(\frac{1 + 2k_{0y}^2/k_{0z}^2}{3}\right)^{4/3} . \quad (23)$$

Also, one can write $P_i = (\varepsilon_{zi}/\varepsilon_{yi})^2/(b/a)^2$, where $b/a \approx (1 + 2\,k_{0y}^2/k_{0z}^2)/3$ for a space-charge dominated beam. Equations (22a) and (22b) contain two emittance growth mechanisms: (1) kinetic-energy exchange, which depends on the change in the partition parameter; and (2) charge-density redistribution, which depends on the change in nonlinear field energy. As in the 2-D case, if it is assumed that for high beam intensities the beam approaches final-state homogenization ($U_{nf} = 0$) and kinetic-energy equipartitioning ($P_f = 1$), one obtains emittance growth equations given by

$$\frac{\varepsilon_{zf}}{\varepsilon_{zi}} = \left[1 - \frac{2}{3}\frac{(P_i - 1)}{P_i} + \frac{G(b/a)}{3P_i}\left(\frac{k_{0y}^2}{k_{yi}^2} - 1\right)U_{ni}\right]^{1/2} , \quad (24a)$$

and

$$\frac{\varepsilon_{yf}}{\varepsilon_{yi}} = \left[1 + \frac{(P_i - 1)}{3} + \frac{G(b/a)}{3}\left(\frac{k_{0y}^2}{k_{yi}^2} - 1\right)U_{ni}\right]^{1/2} . \quad (24b)$$

Minimum Final Emittance

The assumptions of final-state homogenization and equipartitioning near the extreme space-charge limit result in

$$\varepsilon_{zf}^2 = \varepsilon_{zi}^2 \frac{(2 + P_i)}{3P_i} + \frac{16G_z(b/a)}{3}\left(\frac{K_3^2}{k_{0z}}\right)^{2/3} U_{ni} , \quad (25a)$$

189

and

$$\varepsilon_{yf}^2 = \varepsilon_{yi}^2 \frac{(2 + P_i)}{3} + \frac{16 G_y (b/a)}{3} \left(\frac{K_3^2}{k_{0y}}\right)^{2/3} U_{ni} \quad . \tag{25b}$$

The minimum final emittance occurs when $\varepsilon_{yi} = \varepsilon_{zi} = 0$ (extreme space-charge limit), which results in

$$\varepsilon_{zf,min} = 4 \left(\frac{G_z}{3}\right)^{1/2} \left(\frac{K_3^2}{k_{0z}}\right)^{1/3} U_{ni}^{1/2} \quad , \tag{26a}$$

and

$$\varepsilon_{yf,min} = 4 \left(\frac{G_y}{3}\right)^{1/2} \left(\frac{K_3^2}{k_{0y}}\right)^{1/3} U_{ni}^{1/2} \quad . \tag{26b}$$

Equations (26a) and (26b) depend on the initial nonlinear field energy U_{ni} but are independent of P_i, a conclusion that was found also for the 2-D problem. For the bunched-beam problem, the minimum final emittances are proportional to $I^{2/3}$, through the parameter K_3. This is in contrast to the linear dependence obtained for the 2-D continuous beam.

Scaling With Beam and Channel Parameters

From the matched envelope equations for a bunched beam and the definitions of the equivalent uniform bunch tunes k_y and k_z, one can show[30] that the three variables P, b/a, and k_y/k_{0y} that determine the emittance growth can be expressed as functions of three new variables that depend directly on the beam and channel parameters K_3, ε_y, ε_z, k_{0y}, and k_{0z}. The three new variables are u_y, u_z, and b_0/a_0, where

$$u_y = 4K_3 k_{0z}^{1/2}/\varepsilon_y \varepsilon_z^{1/2} k_{0y} \quad ,$$

$$u_z = 4K_3 k_{0y}^{1/2}/\varepsilon_z \varepsilon_y^{1/2} k_{0z} \quad ,$$

and

$$b_0/a_0 = (\varepsilon_z k_{0y}/\varepsilon_y k_{0z})^{1/2} \quad .$$

Then, constant values of u_y, u_z, and b_0/a_0 should yield the same emittance growth ratios for a given initial charge-density profile, assuming that no other sources of emittance growth are present. For bunched beams, the ratios $I/\varepsilon_y \varepsilon_z^{1/2}$ and $I/\varepsilon_z \varepsilon_y^{1/2}$ enter into u_y and u_z and determine the emittance growth, in contrast to the I/ε ratios that enter for 2-D beams. For spherical bunches, this implies that the emittance growth depends on $I/\varepsilon^{3/2}$.

V. NUMERICAL STUDIES

We have made numerical studies to investigate the effects described above using computer codes written especially for this problem. Our initial studies have included 1-D sheet beams, 2-D continuous beams, and 3-D spherical bunches.[17,18,22,23] In general, we find that our conclusions for the 1-D sheet beams, 2-D azimuthally symmetric beams, and 3-D spherical bunched beams are nearly identical. For these problems, the kinetic-energy exchange is not present; thus, the charge-density redistribution mechanism is isolated. I will discuss the results for 3-D spherical bunched beams, obtained using a code,[31] where the radial self-forces are calculated from Gauss' law. I use 4000 particles with different rms matched distributions, where at each time step, the particles are given an impulse based on both the external and self-forces. First, I present results for a space-charge-dominated spherical bunch with an initial Gaussian distribution, both in position and divergence, truncated at four standard deviations and with an initial tune ratio, $k_i/k_0 = 0.02$. Figures 1a and 1b show $U_n = U/w_3$ and $\varepsilon/\varepsilon_i$ as a function of distance z/λ_p along the beamline, where a plasma length λ is the distance the beam travels during one plasma period. The quantity U/w_3, shown in Fig. 1a, decreases to zero during the first quarter plasma period, then oscillates, and finally settles down to a steady value somewhat larger than zero. Figure 1b shows the emittance ratio $\varepsilon/\varepsilon_i$ as a function of z/λ_p, which grows rapidly during the first quarter plasma period and is in agreement with the emittance differential equation [Eq. (6) with a = b] and the U/w_3 profile shown in Fig. 1a.

Figure 2 shows x,x' phase space and the radial charge density at $z/\lambda_p = 0.0$, 0.25, 0.5, and 10 for an initial Gaussian spherical bunch with $k_i/k_0 = 0.02$. The variation of the phase-space distribution, seen in Figs. 2a through 2c ($z/\lambda_p = 0.0$, 0.25 and 0.5), is qualitatively consistent with the rms emittance curve shown in Fig. 1b. The phase-space distribution at $z/\lambda_p = 10$, shown in Fig. 2d, has evolved to a nearly stationary configuration composed of a central core with an extended diffuse background and corresponds to an emittance growth ratio of about $\varepsilon_f/\varepsilon_i = 16$. Figures 2e through 2h show that, as a result of radial plasma oscillations, the charge-density changes from an initial Gaussian distribution at $z/\lambda_p = 0.0$ to a nearly uniform distribution at $z/\lambda_p = 0.25$ and to a hollow beam at $z/\lambda_p = 0.50$. The charge density at $z/\lambda_p = 10$, after the plasma oscillations have damped, can be seen in Fig. 2h and within statistical uncertainties is consistent with a uniform distribution having a soft edge or finite thickness boundary. A low-density tail or halo remains outside the central core and contributes to the nonzero final value of U/w_3. The phase-space characteristics and the uniform charge density, observed at $z/\lambda_p = 10$, change very little at larger z/λ_p values, indicating the formation (at least approximately) of a final stationary spherical bunch.

Figures 3a and 3b show U/w_3 and $\varepsilon/\varepsilon_i$ versus z/λ_p for an initial semi-Gaussian or thermal spherical bunch (with $k_i/k_0 = 0.25$) corresponding to uniform charge density and a Gaussian distribution in divergence (velocity) space. The quantity U/w_3 increases rapidly, which would imply an emittance decrease according to Eq. (5). Figure 3b, the rms emittance growth ratio $\varepsilon/\varepsilon_i$, shows a corresponding small decrease in rms emittance. (Because rms emittance is a measure of effective rather than the emittance, no violation of Liouville's theorem is implied by this result.) Examination of the charge density, at this and at other tune depressions, shows that the initially uniform beam acquires a tail or soft edge, whose width (consistent with the Debye length) increases with emittance. Even so, the assumption of a final uniform beam for this case will lead to only a small error in emittance growth, because the emittance decrease is so small.

Figure 4 shows numerical simulation results of $\varepsilon_f/\varepsilon_i$ at $z/\lambda_p = 10$, plotted versus k_i/k_0 for an initial Gaussian spherical bunch. The curve obtained from Eq. (22) with $P_f = P_i = 1$ and b = a, using the value $U_{ni} = 0.308$ for a Gaussian beam, is in good agreement with the simulations and shows the steep rise in emittance growth at low tune depressions (high space charge). The good agreement for low k_i/k_0 (high beam intensities) is not unexpected because, for this case, the assumption of final charge-density uniformity is a good approximation. An unexpected result is the agreement of Eq. (22) with the simulation at high k_i/k_0, which is due to the fact that the large error in nonlinear field energy change (by assuming a final uniform beam) contributes little to the calculated emittance growth because of the small value of the tune ratio factor in Eq. (22). Figure 5 shows the same plot for an initial thermal spherical bunch, which has an initial uniform charge density and $U_{ni} = 0.0$. The rms emittance decrease shown in Fig. 3 also can be seen in some plotted points on Fig. 5, but is a small effect. Again the curve from Eq. (22) with $P_f = P_i = 1$ and b/a = 1 is in close agreement with the numerical simulation results. Comparison of Figs. 4 and 5 shows that the strong emittance increase observed at low tune depressions

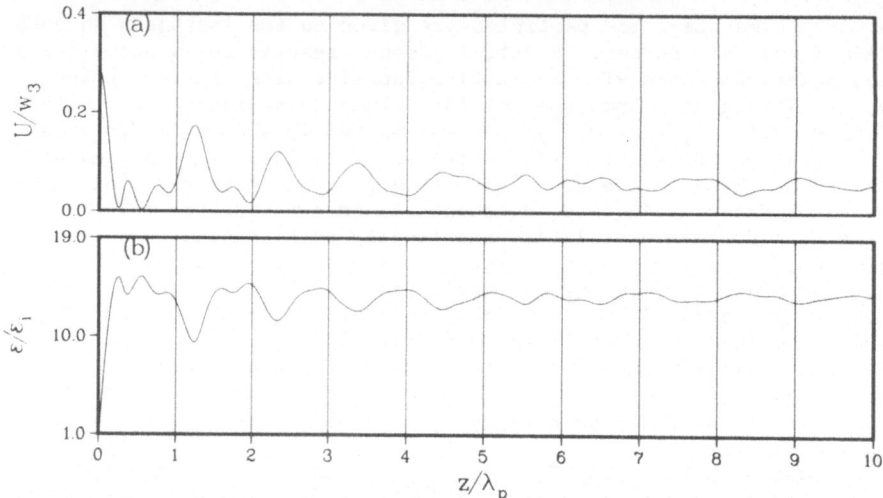

Fig. 1. Results of numerical simulation for 10 plasma periods using an initial Gaussian spherical bunch, truncated at four standard deviations, and initial tune ratio $k_i/k_0 = 0.02$. The abcissa is the
distance z/λ_p along the beamline, where λ_p is the distance the beam
travels in one plasma period.
(a) Dimensionless nonlinear field energy U/w_3.
(b) Emittance growth ratio $\varepsilon/\varepsilon_i$.

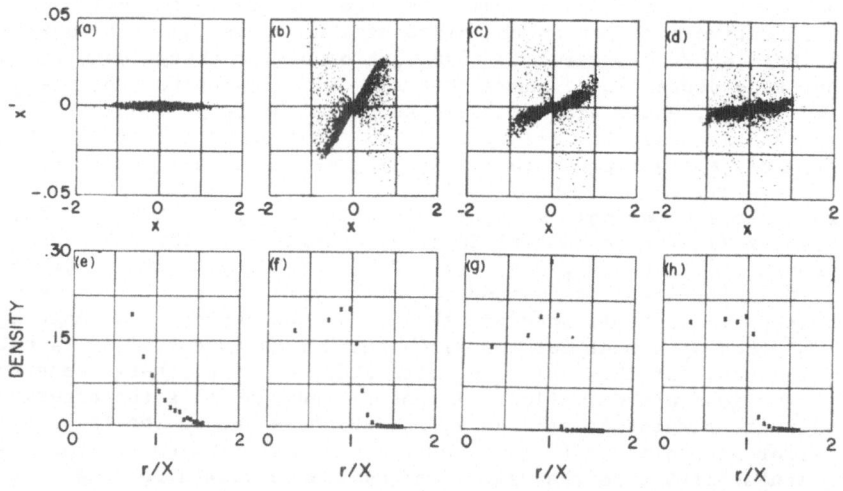

Fig. 2. Phase space (x,x'), and charge density versus normalized radius r/X,
where $X = \sqrt{5}a$, at $z/\lambda_p = 0.00$, 0.25, 0.50, and 10.0 from numerical
simulation for an initial Gaussian spherical bunch, truncated at four
standard deviations, and initial tune ratio $k_i/k_0 = 0.02$. The
quantities x, x', and charge density are shown in relative units.

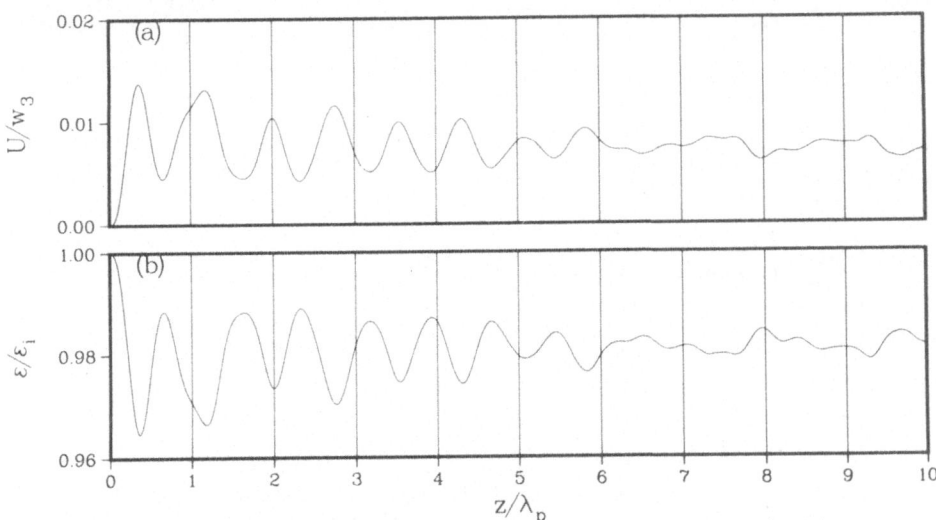

Fig. 3.   Results of numerical simulations for an initial semi-Gaussian
(thermal) spherical bunch, truncated at four standard deviations
in velocity space, and initial tune ratio $k_i/k_0 = 0.25$.   The
quantities plotted are the same as in Fig. 1.

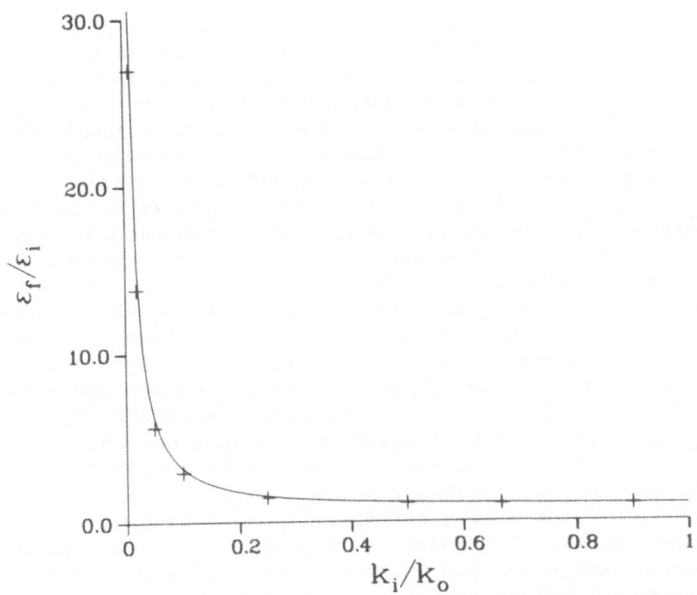

Fig. 4.   Final emittance growth ratio versus initial tune ratio for an initial
Gaussian spherical bunch, truncated at four standard deviations.   The
curve is generated from Eq. (22) and the plus symbols show the results
of the particle simulations after 10 plasma periods.

193

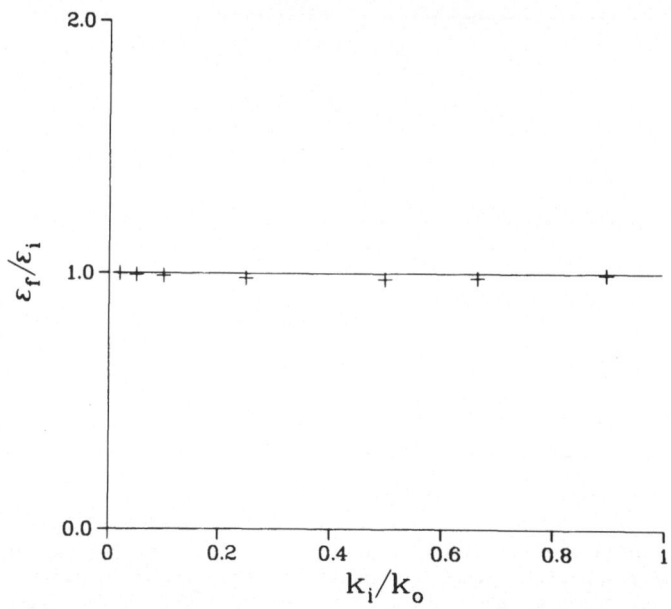

Fig. 5. Final emittance-growth ratio versus initial tune ratio for an initial
semi-Gaussian spherical bunch, truncated at four standard deviations.
The curve is generated from Eq. (22) and the plus symbols show the
results of the particle simulations after 10 plasma periods.

for the initial Gaussian bunch is absent for the initial thermal bunch,
which illustrates the advantage of initially uniform beams for controlling
emittance growth from charge redistribution. Furthermore, from the good
agreement between the curves and the numerical simulation results, shown in
Figs. 4 and 5, the final emittance growth can be closely predicted from
Eq. (22) when $P_f = P_i = 1$, using only the initial tune ratio and the initial
value of U_n. A comparison of Eq. (25) with particle simulations for a spherical
bunch is made in Figs. 6 and 7, where $k_0\varepsilon_f$ versus $k_0\varepsilon_i$ is plotted for initial
Gaussian and thermal distributions, respectively. The results from the simula-
tions and the curves from Eq. (25) are, again, in close agreement. The nonzero
value of the minimum final emittance is evident for the initial Gaussian bunch
in Fig. 6, in contrast to the result for the initial thermal (uniform charge
density) bunch in Fig. 7, where the minimum final emittance is zero.

Studies of 2-D asymmetric beams[22] suggests that kinetic-energy exchange
is a slower process than the charge-density redistribution and typically may
take tens of plasma periods. The final partition parameter, P_f, depends
strongly on k_{yi}/k_{0y}, and for any propagation distance one can identify three
distinct regions: (1) a stable region, where P does not change, (2) a transition region with partial or incomplete equipartitioning, and (3) full equipartitioning, where $P_f = 1$ for sufficiently low tune depressions. As the beam
propagates further, $P_f \to 1$ throughout the transition region. Further studies
of kinetic-energy exchange for bunched beams are needed to confirm these results
and to obtain numerical values for the thresholds.

VI. COMPARISON WITH REAL SYSTEMS AND EXPERIMENTS

It is well known that collective instabilities driven by periodic focusing
systems used in linacs will cause emittance growth.[7] I have derived equations
from space-charge-induced emittance growth from two additional mechanisms:

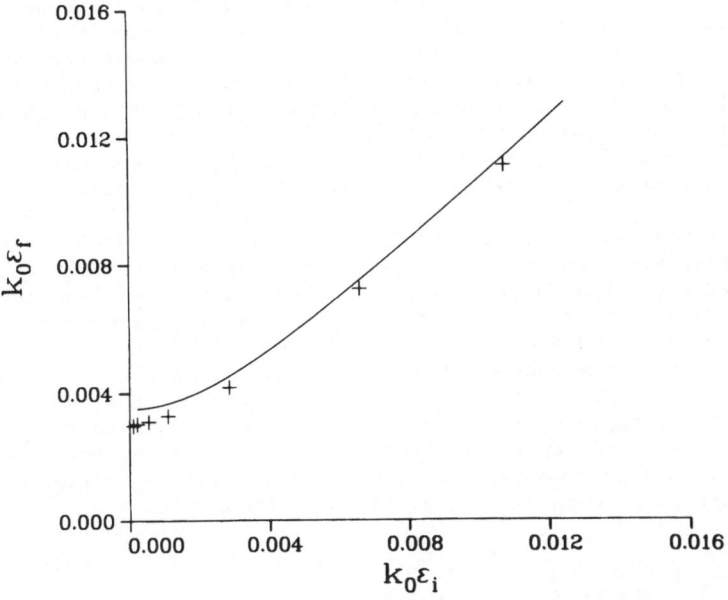

Fig. 6. Final $k_0\varepsilon$ versus initial $k_0\varepsilon$ for an initial Gaussian spherical bunch, truncated at four standard deviations. The curve is generated from Eq. (25) and the plus symbols show the results of particle simulations after 10 plasma periods.

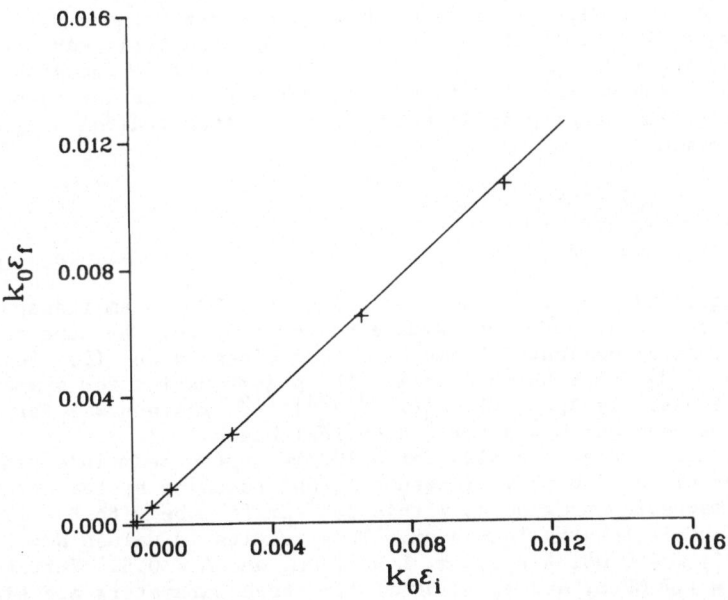

Fig. 7. Final $k_0\varepsilon$ versus initial $k_0\varepsilon$ for an initial semi-Gaussian spherical bunch, truncated at four standard deviations. The curve is generated from Eq. (25) and the plus symbols show the results of particle simulations after 10 plasma periods.

(1) charge-density redistribution and (2) kinetic-energy exchange, obtained by assuming continuous linear external forces. Real beams in transport lines and linear accelerators are focused using periodic systems, usually with quadrupole lenses in an alternating gradient configuration. In a smooth approximation, the periodic system can be replaced by an equivalent linear continuous focusing system with the same average focusing properties. Nevertheless, it is not obvious that the major emittance growth effects in a real beam with periodic focusing can be described with sufficient accuracy by using equations based on an equivalent linear continuous focusing (smoothed) representation for the system.

However, numerical simulation studies for a FODO quadrupole channel by Struckmeier, Klabunde, and Reiser[19] suggest that the emittance growth equations presented are valid for zero-current phase advance per focusing period $\sigma_0 \leq 60°$. Indeed, these authors orginally derived the symmetric beam version ($P_f = P_i = 1$, and X = Y) of Eqs. (13a) and (13b) from an assumption of transverse energy conservation when attempting to explain their numerical simulation results. Therefore, there is strong evidence from numerical simulation work for applying the emittance growth equations to real beams, using a smooth approximation representation for the focusing force.

In the smooth approximation for magnetic quadrupole focusing, the transverse phase advance σ_0 in a linac is approximately[6]

$$\sigma_{0y}^2 = \frac{1}{8\pi^2}\left(\frac{e\chi BL^2}{mc\beta\gamma a_Q}\right)^2 + \frac{\pi e E_0 TL^2 \sin\varphi_s}{mc^2\beta^3\gamma^3\lambda} \quad , \tag{27}$$

where B is the pole-tip magnetic field, a_Q is the radial aperture of the pole tip, E_0 is the average axial accelerating field, T is the transit time factor, φ_s is the synchronous phase, λ is the rf wavelength and L is the focusing period (L = $2\beta\lambda$ for a drift tube linac). The quantity X is given by X = $(4/\pi)$ $\sin\pi\Lambda/2$ for a FODO system, where Λ is the fraction of the focusing period occupied by quadrupole lenses. The first term in Eq. (27) represents the magnetic quadrupole focusing and the second term represents the effect of the rf field (defocusing for a stable synchronous phase in the range $-90° \leq \varphi_s \leq 0°$). Then the zero-current single-particle wave number k_{0y} is given by $k_{0y} = \sigma_{0y}/L$. For an RFQ[32] with rf electric focusing, one must use L = $\beta\lambda$, and replace B with $V/\beta c a_Q$, where V is the intervane voltage and X is the RFQ focusing efficiency. For the longitudinal single-particle wave number[6]

$$k_{0z}^2 = -\frac{2\pi e E_0 T \sin\varphi_s}{mc^2\beta^3\gamma^3\lambda} \quad . \tag{28}$$

When $E_0 = 0$, Eqs. (27) and (28) can be applied to a FODO beam-transport system with no acceleration. Approximate expressions for the tune depression for space-charge-dominated beams have been given in Eq. (15) for a 2-D beam and in Eq. (23) for a bunched beam. The plasma period for symmetric nonrelativistic beams is $\lambda_p = 2\pi L[n\sigma_0^2(1-\sigma^2/\sigma_0^2)]^{-1/2}$, where n = 2 for a round continuous beam and n = 3 for a spherical bunch.

I now calculate three examples for a 100 mA proton beam injected into a 200 MHz Alvarez DTL at 750 keV. I assume a FODO focusing system obtained by installing magnetic quadrupoles within each drift tube with B = 1T, a_Q = 1 cm, and quadrupole lengths l_Q = 3 cm. This results in values N = 3.1 x 10^9, K_3 = 2.7 x 10^{-7}, β = 0.04, λ = 1.5 m, L = 12 cm, and Λ = 0.5. Furthermore, we assume $E_0 T$ = 1.7 MV/m, and φ_s = $-35°$. All these parameters are similar to those that correspond to existing high-current linac injectors. Then from Eqs. (27) and (28), $\sigma_{0x} = \sigma_{0y} = 53°$, and $k_{0x} = k_{0y} = k_{0z} = 8$ m^{-1}. First, consider an initial Gaussian spherical bunch with emittances $\varepsilon_x = \varepsilon_y = \varepsilon_z = 20$ $\pi \cdot$mm$\cdot$mrad. For this example, the charge redistribution mechanism alone will contribute to emittance growth (assuming that non-space-charge mechanisms can be ignored). From Eq. (23), $\sigma_y/\sigma_{0y} = k_y/k_{0y} = 0.24$ or

$\sigma_y = 13°$. Equation (24) gives $\varepsilon_f/\varepsilon_i = 1.65$ or $\varepsilon_f = 33^-$ π•mm•mrad. The final emittance can be compared with the minimum final emittance, which is obtained from Eq. (26) as $\varepsilon_{f,min} = 27$ π•mm•mrad. For the plasma length, $\lambda_p/L = 4.0$ so that a quarter plasma period (characteristic time for charge-density redistribution) corresponds to one focusing period.

As a second example, consider an initial uniform bunch with initial unnormalized emittances $\varepsilon_x = \varepsilon_y = 20$ π•mm•mrad and $\varepsilon_z = 60$ π•mm•mrad, with other parameters kept the same as above. This example approximately corresponds to emittance growth from kinetic-energy exchange alone, and one may assume that the energy exchange is completed to full equipartitioning. Then $P_i = 9$, $U_{ni} = 0$, and Eqs. (24a) and (24b) yield $\varepsilon_{zf}/\varepsilon_{zi} = 0.64$, and $\varepsilon_{yf}/\varepsilon_{yi} = 1.9$. Finally, considering an initial Gaussian beam for the second example, there will be an emittance growth from both mechanisms: (a) charge density redistribution and (b) kinetic-energy exchange. For this case, Eqs. (24a) and (24b) give $\varepsilon_{zf}/\varepsilon_{zi} = 0.77$, and $\varepsilon_{yf}/\varepsilon_{yi} = 2.3$. These numerical examples for emittance growth are of comparable magnitude to measured values in real linac injectors[33,34] and show that the equations presented here predict emittance-growth effects that can be of significant magnitude for real linear accelerators.

Experimental evidence for the validity of the 2-D emittance growth (Eq. 16) with charge-density redistribution alone for an unneutralized beam in real quadrupole transport channels has also been reported.[35,36] Both the data and the curves show significant emittance growth at low σ for the initial Gaussian-like beam from the GSI experiment and no growth for the initial uniform beam from the LBL experiment. Additional experimental observations are reported for unneutralized beams from the GSI experiment, including the minimum final emittance effect and emittance growth for the 2-D asymmetric beams.

VII. CONCLUSIONS

The need for improvements in the design of linear accelerators for high-current, low-emittance beams is providing the motivation to understand the physics of intense charged particle beams and particularly the mechanisms leading to emittance growth from space-charge effects. The emittance growth mechanism caused by collective instabilities excited by the periodic focusing system is already well established in beam-transport systems. I have presented equations for 2-D continuous beams and axially symmetric bunched beams in free space that predict space-charge-induced emittance growth associated with two additional mechanisms: (1) charge-density redistribution and (2) kinetic-energy exchange. I have used the relation between field energy and rms emittance to obtain results, which have been expressed in terms of both the initial beam variables and two dimensionless final-state variables: (1) the final normalized nonlinear field energy, U_{nf}, a measure of the final charge-density nonuniformity, and (2) the final partition parameter, P_f, a measure of the final kinetic-energy asymmetry.

From numerical simulation studies, the assumption of a final uniform charge density ($U_{nf} = 0$), which is expected in the extreme space-charge limit, leads to a good first approximation for the rapid charge-density redistribution component of emittance growth. For 2-D beams, a general characterization of the final partition parameter, which determines the usually slower kinetic-energy exchange process, is more complicated, but final equipartitioning of the beam is observed at low initial tune depressions.

In addition to the emittance growth equations, I have derived formulas for minimum final emittances corresponding to an extreme space-charge limit, when $\varepsilon_i = 0$. I predict that the minimum final emittances should vary with beam current as $\varepsilon_{f,min} \propto I$ for 2-D continuous beams, and as $\varepsilon_{f,min} \propto I^{2/3}$ for bunched beams.

Finally, I have shown how space-charge-induced emittance growth scales with the beam and channel parameters through two dimensionless current-dependent quantities, and the zero-current beam aspect ratio. The emittance growth depends on I/ε_x and I/ε_y for 2-D continuous beams, and on $I/\varepsilon_y\varepsilon_z^{1/2}$

and $I/\varepsilon_z\varepsilon_y^{1/2}$ for bunched beams. These effects are of sufficient magnitude to contribute significantly in real high-current linear accelerators. Further numerical simulations and experimental studies will be needed to test these predictions and to evaluate their utility for improved high-current beam transport and accelerator design. Finally, these results suggest that the challenge for minimizing the emittance growth in high-current, low-emittance linear accelerator beams is to learn how to produce equipartitioned beams, with a uniform charge-density profile.

Acknowledgments

The author acknowledges the work of colleagues F. W. Guy, K. R. Crandall, and R. S. Mills for the numerical simulations described here,
all of which are published in more detail elsewhere. The diligent help of S. R. Watson, and the encouragement of R. A. Jameson and J. E. Stovall are gratefully acknowledged.

References

1. R. A. Jameson, "New Linac Technology - For SSC, and Beyond?" Proc. 12th International Conference on High Energy Accelerators, Fermilab, August 11-16 (1983) 497; and R. A. Jameson, "New Linear Accelerators," presented at the Institute for Nuclear Study Kikuchi Winter School, Fujiyoshida, Yamanashi 403, Japan, Los Alamos National Laboratory document LA-UR-84-25 (1984).
2. I. Hofmann, "Emittance Growth of Ion Beams With Space Charge," Nucl. Instr. and Meth. 187 (1981) 281.
3. I. Hofmann, "Emittance Growth of Beams Close to the Space Charge Limit," IEEE Trans. Nucl. Sci. 28 (3), (1981) 2399.
4. I. M. Kapchinskii, "High Current Linear Ion Accelerators," Uspekhi Fizicheskikh Nauk. 132 (1981) 639.
5. M. Reiser, "Current Limits in Linear Accelerators," J. Appl. Phys. 52 (1981) 555.
6. T. P. Wangler, "Space-Charge Limits in Linear Accelerators," Los Alamos Scientific Laboratory report LA-8388, July 1980.
7. L. Smith, L. J. Laslett, I. Hofmann, and I. Haber, "Stability of KV Distributions in Long Peroidic Systems," Particle Accelerators 13 (1983) 145.
8. J. D. Lawson, "The Physics of Charged Particle Beams," (Clarendon Press, Oxford, 1977), 197.
9. P. Lapostolle, "Round Table Discussion on Space-Charge and Related Effects," Proc. 1968 Proton Linear Accelerator Conf., Brookhaven National Laboratory report BNL 50120 (1968), 433-439.
10. P. M. Lapostolle, C. Taylor, P. Tetu, and L. Thorndahl, "Intensity Dependent Effects and Space Charge Limit Investigations at CERN Injector and Synchrotron," CERN report CERN 68-35 (1968).
11. M. Prome, "Effects of Space Charge in Proton Linear Accelerators," thesis presented at Universite Paris Sud, Centre d'Orsay (1971), Los Alamos Translation LA-TR-79-33, 90-92.
12. R. Chasman, "Numerical Calculations on Transverse Emittance Growth in Bright Linac Beams," IEEE Trans. Nucl. Sci. 16 (3), (1969) 202.
13. R. A. Jameson and R. S. Mills, "On Emittance Growth in Linear Accelerators," Proc. 1979 Linear Accelerator Conf., Brookhaven National Laboratory report BNL 51134 (1979) 231.
14. R. A. Jameson, "Beam Intensity Limitations in Linear Accelerators," IEEE Trans. Nucl. Sci. 28 (3) (1981) 2408.
15. R. A. Jameson, "Equipartitioning in Linear Accelerators," Proc. 1981 Linear Accelerator Conf., Los Alamos National Laboratory report LA-9234-C (1982) 125.

16. I. Hofmann and I. Bozsik, "Computer Simulation of Longitudinal-Transverse Space-Charge Effects in Bunched Beams," Proc. 1981 Linear Accelerator Conf., Los Alamos National Laboratory report LA-9234-C (1982) 116.

17. T. P. Wangler, K. R. Crandall, R. S. Mills, and M. Reiser, "Relationship Between Field Energy and RMS Emittance in Intense Particle Beams," IEEE Trans. Nucl. Sci. 32 (1985) 2196.

18. T. P. Wangler, K. R. Crandall, R. S. Mills, and M. Reiser, "Field Energy and RMS Emittance in Intense Particle Beams," Proc. of Workshop on High Brightness, High Current, High Duty Factor Ion Injectors, San Diego, California, May 21-23, 1985, AIP Conference Proceedings No. 139 (1986) 133.

19. J. Struckmeier, J. Klabunde, and M. Reiser, "On the Stability and Emittance Growth of Different Particle Phase-Space Distributions in a Long Magnetic Quadrupole Channel," Particle Accelerators 15 (1984) 47.

20. Ingo Hofmann, "Emittance Growth," presented at the 1986 Linear Accelerator Conference, SLAC, Stanford, California, June 2-6, 1986, to be published.

21. T. P. Wangler, F. W. Guy, and I. Hofmann, "The Influence of Equipartitioning on the Emittance of Intense Charged Particle Beams," presented at the 1986 Linear Accelerator Conference, SLAC, Stanford, California, June 2-6, 1986, to be published.

22. F. W. Guy and T. P. Wangler, "Numerical Studies of Emittance Exchange in 2-D Charged Particle Beams," presented at the 1986 Linear Accelerator Conference, SLAC, Stanford, California, June 2-6, 1986, to be published.

23. T. P. Wangler, K. R. Crandall, and R. S. Mills, "Emittance Growth from Beam Density Changes in High Current Beams," presented at the International Symposium on Heavy Ion Fusion, Washington, D.C., May 27-29, 1986, to be published.

24. I. M. Kapchinskii and V. A. Teplyakov, "Linear Ion Accelerator With Spatially Homogeneous Strong Focusing," Prib. Tekh. Eksp. 2 (1970) 19.

25. R. L. Gluckstern, "Space Charge Effects," in Linear Accelerators, P. M. Lapostolle and A. L. Septier, Eds. (North Holland Publishing Co., Amsterdam, 1970).

26. P. M. Lapostolle, "Energy Relationships in Continuous Beams," Los Alamos National Laboratory translation LA-TR-80-8, or CERN-ISR-DI/71-6 (1971); and P. M. Lapostolle, "Possible Emittance Increase Through Filamentation Due to Space Charge in Continuous Beams," IEEE Trans. Nucl. Sci. 18 (3) (1971) 1101.

27. E. P. Lee, S. S. Yu, and W. A. Barletta, "Phase Space Distortion of a Heavy-Ion Beam Propagating Through a Vacuum Reactor Vessel," Nucl. Fusion 21 (1981) 961.

28. F. J. Sacherer, "RMS Envelope Equations with Space-Charge," IEEE Trans. Nucl. Sci. 18 (3) (1971) 1105.

29. T. P. Wangler, "Emittance Growth for 2-D Charged Particle Beams Including Equipartitioning and Nonlinear Field Energy Effects," Group AT-1 memorandum AT-1:86-72, March 1986; T. P. Wangler, "Emittance Growth from Charge Redistribution and Energy Repartitioning for 2-D Continuous Beams," Group AT-1 memorandum AT-1:86-73, March 1986.

30. T. P. Wangler, "Emittance Growth of Bunched Axially Symmetric Charged Particle Beams," Group AT-1 memorandum AT-1:86-74, March 1986.

31. K. R. Crandall, R. S. Mills, and T. P. Wangler, "Simulation of Continuous Beams Having Azimuthal Symmetry to Check Relation Between Emittance Growth and Nonlinear Energy," Group AT-1 memorandum AT-1:85-218, June 1985.

32. K. R. Crandall, R. H. Stokes, and T. P. Wangler, "RF Quadrupole Beam Dynamics Design Studies," Proc. 10th Linear Accelerator Conf., Montauk, New York, September 10-14, 1979, Brookhaven National Laboratory report BNL-51134 (1980) 205.

33. C. D. Curtis, "Experimental Results: Experience with Beam Current and Emittance in the Fermilab Linac," Space Charge in Linear Accelerators Workshop, Los Alamos Scientific Laboratory report LA-7265-C, 4 (1978).

34. R. A. Jameson, "Emittance Data from the New CERN Linac," Los Alamos Scientific Laboratory memorandum, January 1979.

35. Denis Keefe, "Summary for Working Group on High Current Beam Transport," Proc. of Workshop on High Brightness, High Current, High Duty Factor Ion Injectors, San Diego, California, May 21-23, 1985, AIP Conference Proceedings No. 139, 165 (1986).

36. J. Klabunde, P. Spädtke, and A. Schönlein, "High Current Beam Transport Experiments at GSI," IEEE Trans. Nucl. Sci. 32 (5) (1985) 2462.

DISCUSSION

WEISS

For the new CERN linacs we tried to have bunches filled as uniformly as possible. Not having had RFQs at that time, we constructed a special buncher system to fulfill this requirement.

The time depression at injection into the linac is about 0.9. This figure gets soon lower, as the particles are accelerated.

The equipartitioning of kinetic energy phenomenon was used for the CERN linac in the sense that we tries to transfer transverse kinetic energy (where we wanted to keep the emittance) into longitudinal motion (where the emittance was of minor concern).

What does a plasma wave length mean when expressed in centimeters?

SCHEMPP

It would be interesting to find a way to design an RFQ able to produce an homogeneous beam which seems to be optimal for low emittance growth.

JOHO

Some years ago Haber showed with numerical simulation that one could avoid emittance growth by adding long beam halos which stabilized the charge distribution. How do you explain this result with your model?

WANGLER

The emittance growth Haber was avoiding was caused by excitation of coherent modes for a K-V beam in a periodic structure. The K-V distribution is very susceptible to such instabilities and adding beam halos changed the distribution and reduced the effect. But such a procedura could also give rise to emittance growth from charge redistribution. A better approach to minimize emittance growth is probably to choose the tune parameters to avoid the instabilities, but still to keep the initial charge density uniform.

RFQ APPLICATION

A. Schempp

Institut für Angewandte Physik der Universität
Frankfurt am Main, Germany

The design of an RFQ accelerator can be done in two nearly independent steps. The specific electrode design determines beam properties and can be shaped to get the output beam current and emittance needed for experiments or further acceleration in other accelerator structures. The rf cavity then has to be designed to provide a uniform electrode potenttial generally as high as possible and with high efficiency in respect to power consumption. In the following paper a new specific beam dynamic design is presented, followed by examples how RFQs are designed and built. In a third part specific high frequency problems are addressed and at last the four-rod-$\lambda/2$ structure developed in Frankfurt is discussed.

INTRODUCTION

The concept of spatial homogeneous focusing[1] has closed the low velocity gap of high frequency ion accelerators. The work has ignited numerous activities starting with the thoroughly work in Los Alamos[2]. While firstly the aim was improving high energy proton accelerators, possible applications for heavier ions were seen very early and research started early too[3,4]. Since numerous RFQs have been built and many new applications have been proposed, some of which are impossible and unrealistic with classical rf structures like decellerating with RFQs[5] and funneling[6,7].

a) Beam Dynamics

The basic structure of an RFQ can be developed going from a chain of static and electric quadrupoles to a system of four long electrodes supplied with rf. There is no acceleration in such a high frequency quadrupole, but a mechanical modulation of the electrodes can now introduce an accelerating field component E_z. The quadrupole of the accelerating field is approximately proportional to the electrode modulation factor m. On the other hand the focusing strength becomes smaller with increasing m resp. E_z. The modulation m can be varied from cell to cell. The flight time per cell is half the high frequency period $T_c = 1/2\nu$. Depending on the design goals like particle energy, beam current, beam emittance, the charge state of the ions, frequency, electrode aperture and voltage will be chosen and the variation of m can be shaped resulting in various functions of m, E_z, particle energy, stable phase and focusing strength along the structure.

All schemes are more or less based on the work of Kapchinskiy[1] at ITEP Moscow and Stokes et al.[2] at LANL.

Each design procedure has some special features to optimize the structure to the specific needs and to apply ideas, which have been investigated since, especially in the field of space charge dominated beams. We have made designs for heavy ion RFQs and for highly charged light ions from an EBIS source. There are some really new applications like the use of the RFQ to decelerate beams[5], to accelerate very heavy clusters[8] and a procedure to funnel beams with RFQs[6,7].

A new design to produce "single bunches" has been made for the specific needs of neutron time of flight experiments[9]. The neutron bunches in this case will be produced by short bursts of protons (6-30 MeV) on a target.

Another interesting application are the separated bunch schemes for EHF, the high current injector for the European Hadron Facility[10]. The 2/8 system planned means that two 400 MHz buckets are injected with a repetition rate of 50 MHz respectively 25 nsec between the microbunches. This is proposed in a straightforward way with a jump in frequency by a factor of 8 and then by forming of two 400 MHz bunches out of one 50 MHz bunch.

For neutron sources the pulse pauses must be even longer, so for a bunch to pause ratio of 1/100 a frequency jump in the normal way is not possible.

A high energy buncher for 1 MHz at 2 MeV beam energy, which should focus several microbunches into one longitudinal spot, would need a voltage of approximately 7 MV and a drift length of 20 m. A main problem would be the transport of this beam to that focal point.

Buncher and chopper schemes are hard to realize for high current beams, because the second important boundary condition of these systems asks for a clean beam pause and good emittance. This is difficult also in the EHF case, in which the reforming of two bunches gives an increase in longitudinal emittance and single particle calculations show that with space charge the adjacent buckets are not totally empty. In addition the low frequency of the first stage is far from optimum, so only 15 mA can be accelerated.

In the following a scheme is presented, which can shape the pulses and pauses with help of asynchronous acceleration of RFQs. The idea uses the well known properties of the RFQ like high beam transport capability and adiabatic bunching as well as the selectivity of the fixed velocity profile to filter the proper beam microstructure.

Basic Simple Bunch System

An accelerator system would consist of an ion source, matching lenses and two RFQs with matching elements in between as sketched in Fig. 2.

At the end of the first RFQ the beam is bunched and has a size which can be expressed in phase width $\Delta\varphi$ of approximataly 3 φ_s (φ_s synchronous phase) Beam dynamics then shows that a stable region exists (bucket Fig. 1), which size depends on particle energy, frequency, accelerating field, specific charge and synchronous phase. The beam pulse now must be matched to the bucket of the second RFQ, which width is different according to the frequency change Asynchronous acceleration now takes the train of bunches and only those, which fit into a bucket for the new frequency, which is not a multiple of f_1, are accelerated stably. After a few accelerating cells all buckets in between will be empty. Fig. 2 illustrates this principle for a frequency ratio of 1.25.

Fig. 3 shows single particle calculations for $f_1/f_2=90/100$ MHz resulting in a bunch repetition rate of 10 MHz. The properties of the synchronous bunch could be calculated with a normal particle code, too. The missing space charge of the nighbouring bunches has to be taken into account in this case.

In the usual design a 10 MHz preaccelerator would have to compress a bunch such that it fits into the approximately 10 times narrower one at the higher frequency. The problems involved in such a system are obvious.

A first attempt to get a lower bunch repetition rate ν_r for a neutron source injector would apply two RFQs with 98 and 100 MHz resonance frequencies (ν_r=2 MHz). Unfortunately the pulse pause would be not empty, because the neighbouring bunch centers are shifted by only 3.6° per bucket. This is illustrated in Fig. 4, which shows the beam current as function of time. The centers of the neighbouring bunches oscillate around the stable phase φ, which deteriorates the beam. To provide a phase shift for the adjacent bunches to be outside of the separatrix the minimum frequency ratio must be $f_2/f_1 > 2\pi/(2\pi-\Delta\varphi)$. A 90/100 MHz system, as illustrated in Fig. 3 gives a 10 MHz repetition rate, when a bunch width of 10% is achieved in the first stage (φ_s=10°, $\Delta\varphi$=36).

IN a next step to reduce ν_r a 1 MHz prebuncher could be applied, but which would partly fill all buckets and would need a long drift length, which is not favourable with high current beams. Another RFQ stage is a better solution. The second stage cleans the neighbouring buckets, the third can use a small frequency ratio like 99/100 to give 1 MHz for ν_r.

A higher average beam current requires higher values of φ_s. The problem can be solved with choice of a subharmonic for RFQ1, which increases the charge per bunch additionally. Fig. 5 shows phase widths along the RFQ system. The injected proton beam current was 25 mA, the output beam (ν_r=5 MHz) is I_{av}=2.2 mA at 1 MeV.

Realization of such a system suggests transition energies as low as possible to minimize power losses in the structures. Frequency tuning of RFQ1, which is possible in a wide range using the four-rod-RFQ structure allows a change of the repetition rate. Beam losses in the structure

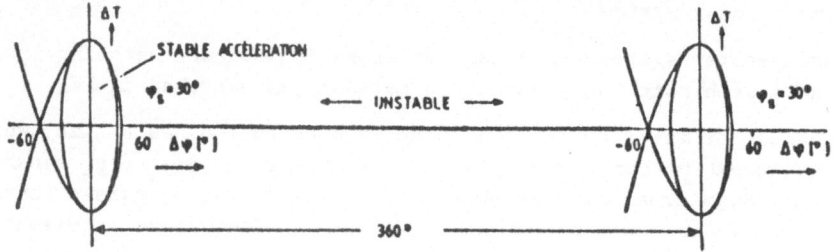

Fig. 1. Buckets for a synchrotron phase of φ_s=30°.

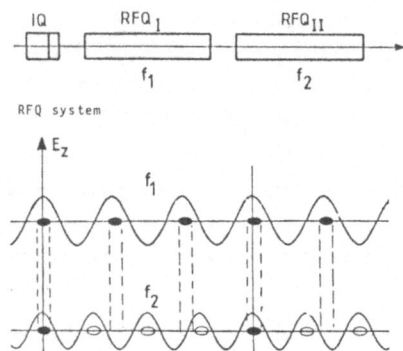

Fig. 2. Scheme of asynchronous acceleration.

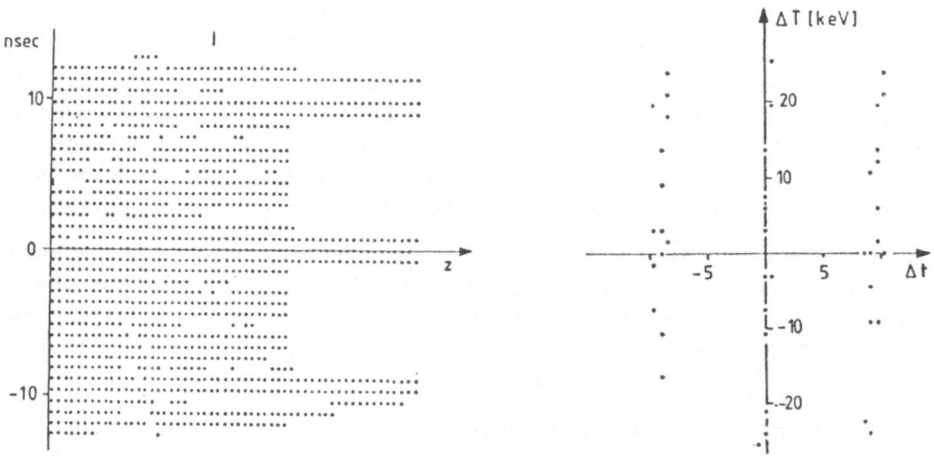

Fig. 3. Single particle calculation for f_1/f_2=90/100 MHz RFQ system.

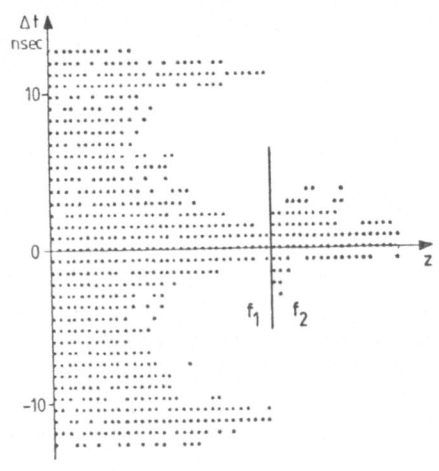

Fig. 4. Beam as function of time
for a 98/100 MHz RFQ system.

Fig. 5. Beam width along a 45/100
MHz RFQ system.

have shown no influence on sparking as long as a good cooling of the electrodes can be preserved, which has been tested with the four-rod-structure, too.

Because only the ratio of frequency detuning determines the repetition rate ν_r, results are valid for applications for 200, 400, 800 and
1200 MHz, too. Even for electron beams in high gradient structures or
wake field accelerators asynchronous acceleration can be used to isolate
single bunches.

Applications

The starting point was the search for a design giving a 1 nsec pulse
with 1 μsec pause (1 MHz repetition rate ν_r), as would be necessary for
a neutron source injector.

With the 45/100 MHz system as a starting point, the bunch repetition
rate ν_r has to be further reduced. The logical solution is another frequency transition: 45/99/100 MHz would give a clean $\nu=1$ MHz. Fig. 6 shows
the longitudinal output emittance for an injected beam of 25 mA at 20
keV and an output beam of I=1.1 mA. This was achieved by adding another
prebuncher with 20 MHz in front of RFQ1.

Tuning the RFQ1 to 47.5 MHz would give a $\nu_r=0.5$ MHz. Now, manipulating the rf pulse of RFQ1 such that the accelerating field amplitude E_o
stays smaller than $E_o \cos \varphi_s$ and then a small spike of less than 1 μsec
length is added with a fast regulation system, pulse repetition rates
of arbitrary low frequency can be produced.

The advantage of this system would be less beam losses in RFQ2. Applying a fast chopper with p.e. 1 MHz (or an asynchronous one with a high-

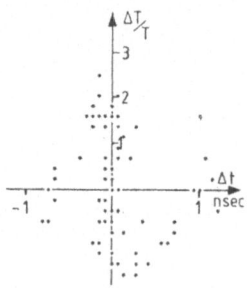

Fig. 6. Output longitudinal emit-tance for a 45/99/100 MHz system with average current of T=1mA.

er frequency, too), the two stage sy-stem 45/100 MHz could be used as well.

As shown in Table 1 the 50 MHz re-petition rate for the EHF injector can be done with a 50 MHz RFQ1 injecting in a 400 MHz RFQ2, forming two bunch-es out of a single 50 MHz bunch by a second adiabatic buncher system.

The use of 150 MHz RFQ1 enables a higher bunch current so that the sec-ond asynchronous RFQ2 (400 MHz) can further accelerate this single bunch to get the high average design current of 100 μA.

In this case asynchronous accele-ration shapes beams to produce empty buckets that an effective high fre-quency rf structure can be used to get low bunch frequencies ν_r.

Table 1. 150/400 MHz EHF injector (ν_r =50 MHz)

	RFQ1	RFQ2
T_i (keV)	50	300
T_F (MeV)	0.3	2.0
f (MHz)	150	400
I_{lim} (mA)	100	200
φ_S (°)	70–30	30
cell number	50	~ 52
length (m)	1.0	0.8
aperture (mm)	2.5	2.5
voltage (kV)	120	150
σ_o (°)	90	60

b) RFQ Structures

We are working on the realization of these schemes, which made use of RFQ structures, which are accepted as unique solution for the low ener-gy part of ion accelerators.

Criteria for RFQ structures are firstly to provide beam dynamics re-quirements like sufficient acceptance, current limits, small emittance growth, possibly high fields and small quadrupole multipole components. With higher fields the structure gets shorter and the current acceler-ated can be higher.

The rf structure must supply the quadrupole voltage on the electrodes. The efficiency described by the shuntimpedance is as important as a high group velocity that means good tolerances, good coupling between different parts resulting in a flat field distribution without dipole components like taken in the beam dynamic designs. Like the shuntimpedance the tolerances are going directly into costs of the system.

Simplicity and reliability mostly come together and are most important criteria for RFQs for injectors in accelerator complexes like HERA at DESY in Hamburg.

The RFQ for HERA, which is being built in Frankfurt, will be taken as example of new developments in RFQ design[11,12].

HERA-RFQ

For the HERA project at DESY an RFQ is being built as injector for the 50 MeV Alvarez linac. Fig. 7 shows the RFQ preaccelerator layout using a FNAL H⁻ ion source and two CERN type solenoids for beam matching into the RFQ. The application of the H⁻ injection allows for a design current of only 20 mA. Beam dynamic calculations show that a short RFQ with an injection energy of only 18 keV (final energy 750 keV) and a modest electrode voltage of 70 kV can be chosen. This has advantages for the mechanical design, the RF properties and the operation reliability.

The beam dynamics design has been done using mostly the standard approach developed in Los Alamos resulting in a high current transmission and a good beam emittance. Table 2 gives some parameters of the HERA-RFQ. Because of the experience with operating RFQs at high energy machines a four-vane-structure has been chosen as rf resonator. It consists of a cylinder, in which four electrodes are mounted symmetrically. In order to provide a proper axial field distribution, the manufacturing of the electrodes and the adjustment has to be done with high precision. In addition the RF properties of this type of resonator require a highly symmetrical structure to avoid dipole components in the axial field, which lead to beam quality deterioration. The mechanical design differs from other 4-vane RFQs. We tried to make a separated function structure. Mechanical adjustement, rf contacts, rf tuning, stabilization and vacuum cooling can be done independently resp. changed independently.

The mechanical adjustment is done with two 3D-positioners per vane, rf contacting between vane and tank is done with a contact bar on each

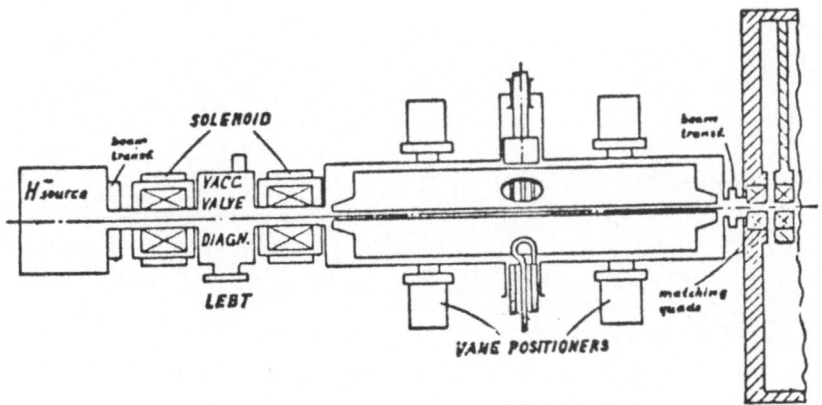

Fig. 7. Layout of RFQ preaccelerator.

Table 2. RFQ parameters

Input energy W_{in}	18	keV
Output energy W_{out}	750	keV
Radio frequency f	202.56	MHz
Beam current I	20	mA
Total length L_{tot}	117.7	cm
Total cell number N	135	
Intervane voltage V^o	70.5	kV
Maximum electric field E_{max}	21.9	MV/V
Vane modulation m	1 to 1.88	
Minimum aperture radius	3.5	mm
Average radius r_o	5.0– 5.2	mm

side of the vane, rf tuning and stabilizing are done in the end cell. Only the outer cylinder is cooled and vacuum can be applied from the outside at the very last.

Fig. 8 shows a cross section of the RFQ, Fig. 9 shows a vane positioner.

The vanes as the main part are made out of CuCr alloy, which has favourable mechanical properties but still very good electrical and heat conductivity. The vanes have been machined out of a solid, forged block, which has been annealed at high temperature, in order to avoid bending by residual internal stresses after milling. An inherent part of the design is the independent control of the critical parts and the mechanical alignment with help of a computerized 3D measuring machine with very good accuracy, 5 μm seem possible.

Careful machining has been done at Pfeiffer (Balzers) at Asslar. The vane tip dimensions and the modulation proved to have deviations below 10 μm from the theoretical values for two test vanes and two final vanes. This was measured independently from the production by Komeg (Zeiss). Measurement was done comparing 2,300 points along the vane with theoretical values and the straightness of the reference edges. Two vanes showed deviations up to 20 μm. A "banana" like bending along the axis of highest momentum indicated residual stresses. After remachining these vanes had the same precision as the other ones.

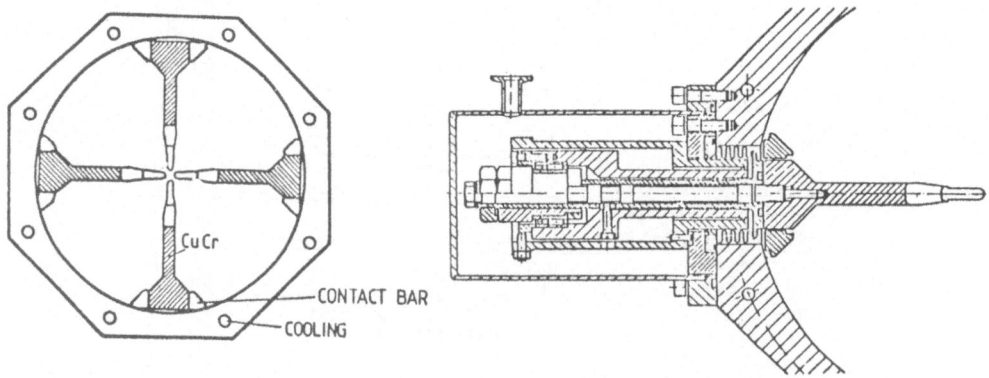

Fig. 8. Cross section of the RFQ.   Fig. 9. Vane positioner.

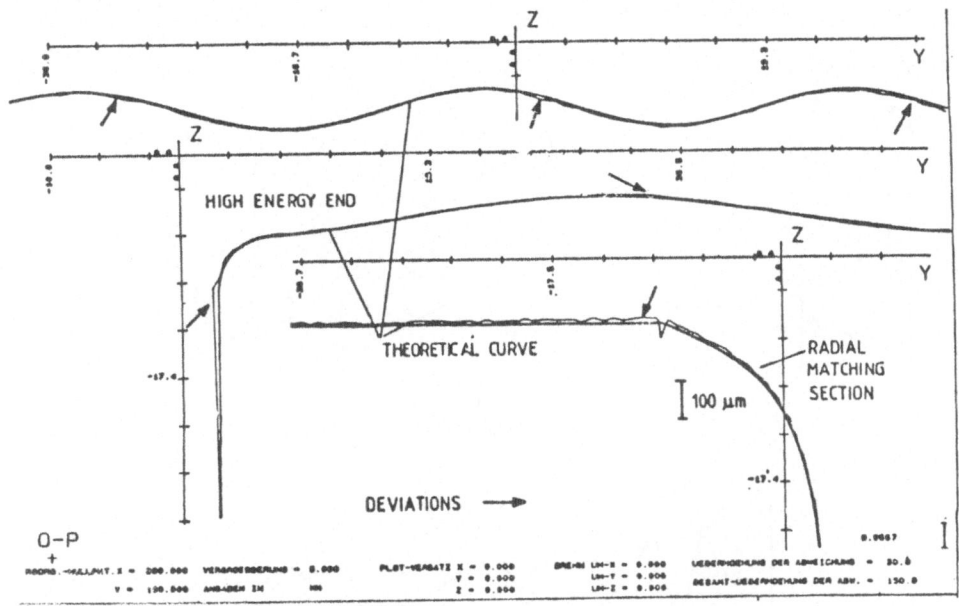

Fig. 10. Measurement of vane tip modulation.

Fig. 10 shows plots for the mechanical measurements along the vanes. The deviations are enlarged by a factor of 150. The maximum deviation occurs at the low energy end at the transition from the radial matching section to the shaper. There the theoretical set of values had not been smoothed out had been done for the milling machine values. The wheel milling tool ($\emptyset_{min}$ =12 mm) cuts with an angle of 15° giving a very clean cut and resembles 150° of a circular arc.

After machining the tank ($\emptyset$ 320 mm±0.01) was copper plated at GSI and then the vanes have been installed and adjusted with help of the Zeiss-3D-machine.

A basic problem for four-vane-structures is the balancing of the four quadrants, which is one reason for the required precision of the manufacturing and tuning. The four quadrants are very weakly coupled with the result that the frequencies (ν_D) of two dipole modes, which have unwanted polarity of the electrode voltages, are approximately 0.5 MHz aside the quadrupole mode ν_o. This gives rise to mode mixing, which makes the voltage distribution in the resonator "unflat". The next longitudinal mode (ν_1) will be approximately 15 MHz higher, because the length L of the structure is smaller than the free space wavelength λ_o. The frequency of ν_1 and the perturbation of the Q-mode is proportional to $(L/\lambda_o)^2$. In general, the flatness is proportional to the mode separation.

We will apply our resonant rings[13,14] to stabilize the HERA-RFQ. There will be one loop coupler (RLC) at each end plate of the RFQ. Fig. 11 shows a scheme of the coupling ring together with two vanes, the end cells of which are tied together. The dipole modes are symmetrically shifted away by the rings, because the stored energy of the RLC is added resp. substracted to that of the dipole modes. The RLC concept allows

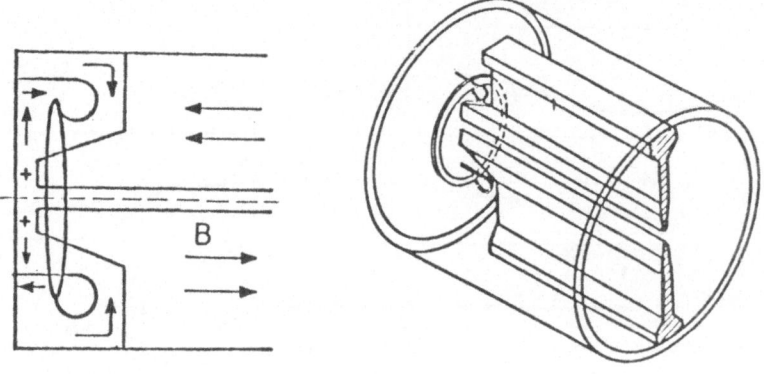

End cell resonant coupler

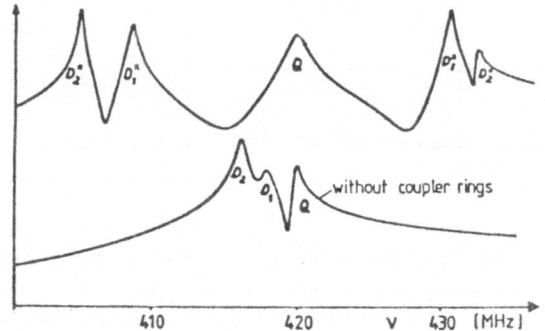

Fig. 11. Mode spectra of RLC coupled RFQ model.

for the use of only one slow-tuner for thermal frequency shift and an adjustable single drive loop to match four different beam currents, because the induced imbalance does not tilt the field distribution.

Fig. 12 shows the end tuner for frequency tuning together with the RLC in the vane cut back area. Fig. 13 shows a drawing of the rf part of the adjustable drive loop featuring a phase shifter near the fixed end, a movement of the total loop and vacuum window.

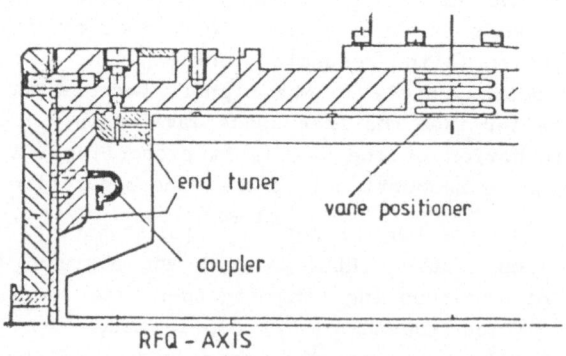

Fig. 12. End tuner and RLC in the vane cut-back.

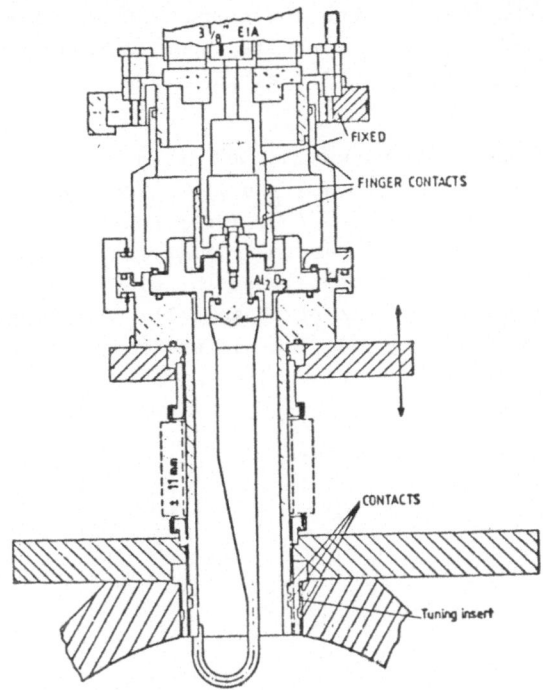

Fig. 13. Adjustable RF drive loop.

There is a tuning insert for balancing the perturbation by the loop. The first measurements showed that the precision manufacturing has resulted in a field, which is azimuthally and longitudinally within 10% flat. Without end plates the resonance frequency was 195 MHz that is 7.5 MHz too low. Adding the endplates the azimuthal symmetry gets worse and the frequency increases to 197 MHz. With additional tuning of the end cells we will be able to get the proper frequency, tune the RLC rings and operate the structure with high-power rf in November.

4-Rod-$\lambda/2$-RFQ

The need for high mechanical precision for a 4-vane-structure is mainly due to rf problems. Balancing the accelerating and focusing fields during tuning is the main problem, because the operating mode is a very sensitive cut-off TE_{210}-like mode. Especially for longer structures (longer than the vacuum wavelength) and high average power operation the technical problems become prohibitive.

From the beginning of our work on RFQ-structures we looked for alternative solutions to bypass these problems, which are still more stringended for the heavy ion structures envisaged.

For heavy ion fusion ions with low e/m, low velocity, and high beam current are accelerated in a tree-like accelerator system. For the first branches RFQ structures will be used, starting as low as 1.0 keV/N.

For protons and light ions the 4-vane-RFQ is the structure mostly used; frequencies as low as 80-100 MHz as chosen in FMIT and TALL seem to be the lower limit, however. For low charged heavy ions a frequency of 10-30 MHz must be chosen, which results in an inhibitory diameter for the 4-vane-resonator. Therefore at GSI a split coaxial heavy ion RFQ is being built for U ions[16,17].

We developed the four-rod-structure, which is a simple alternative to the four-vane-structure for protons and light ions and can be applied for low charged heavy ions as well[5,18,19]. The resonators basic cell consists of two $\lambda/2$ oscillators excited in transversal π-mode to give the proper quadrupole field distribution between the electrodes. The accelerating structure consists of a chain of these cells operating in longitudinal O-mode as indicated in Fig. 14. The difference to the 4-vane-struc-

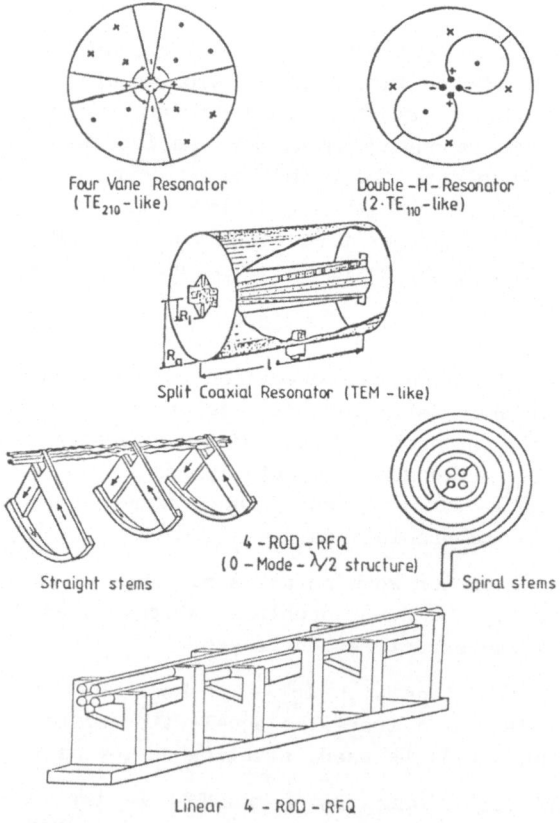

Four Vane Resonator
(TE$_{210}$ – like)

Double – H – Resonator
(2·TE$_{110}$ – like)

Split Coaxial Resonator (TEM – like)

4 – ROD – RFQ
(O – Mode – λ/2 structure)

Straight stems   Spiral stems

Linear 4 – ROD – RFQ

Fig. 14. Scheme of RFQ-resonators.

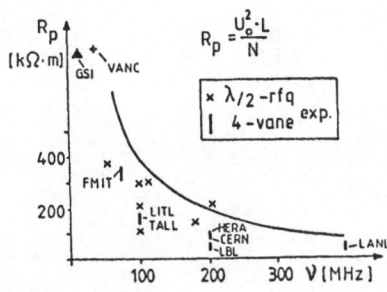

Fig. 15. Optimal shuntimpedance of the four-rod RFQ as a function
of operating frequency compared with four-vane resonators.

ture is clear: There are only four rods or tubes as electrodes machinable
on a turning lathe and there is only one resonator instead of four weakly
coupled ones azimuthally.

Test with such an accelerator structure have been made[20]. With a very
small electrode voltage of 28 kV a 1 mA cw proton beam has been accelerated from 10 to 300 keV.

Fig. 14 shows two cells of a linear version of this structure. Optimizing this structure in respect to shunt impedance results in equidistant arrangement of the stems. The shunt-impedance η of the four-rod-RFQ
is astonishing high and especially favourable for high frequencies. Fig.
15 shows the optimal η as a function of the operating frequency. For
comparison the values of existing 4-vane RFQs are added.

A prototype for light ions has already been built recently. The beam
dynamic parameters of the 4-vane RFQ for HERA[11] had been taken for this
four-rod structure too, which enables a simple comparison of the structure's characteristics and the beam behaviour. Table 3 shows the general

Table 3. Parameters of the four-rod RFQ prototype.

Resonance frequency	202.56 MHz
Beam energy	18-750 kHz
Rod to rod voltage	70.5 kV
rf power	25 kW
R_p-value	180 kΩ
Quality factor	3600
Duty factor	2.5×10^{-4}
Repetition frequency	1 Hz
Tank diameter	25 cm
Structure length	117.7 cm

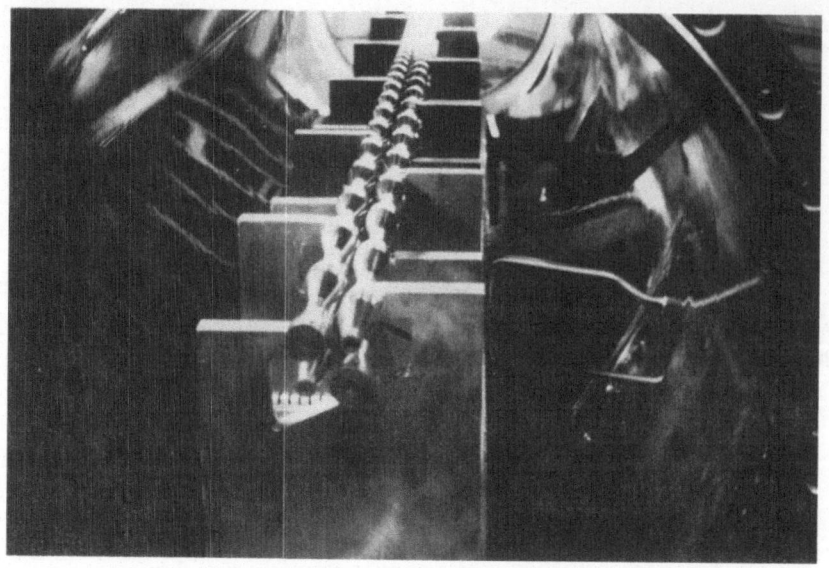

Fig. 16. High power prototype four rod mounted in the vacuum tank.

parameters of this prototype RFQ which has been built at our institute
in Frankfurt (Fig. 16) The rf structure consists of a massive water co
oled ground bar with fourteen specially shaped stems to carry the quadru
pole electrodes The copper plated vacuum tank is provided with a precise
steel base plate with again fourteen bores so that the structure can be
mounted easily from the outside Cooling water is supplied through two
of the screw bolts at both ends Frequency and longitudinal flatness-
-tuning can be achieved in one operation by varying the inductance of
the end cells with adjustable tuning bars between the stems Fig. 17 shows
the effect, which has been measured with unmodulated electrodes. In Fig.
18 a scheme of the modulated electrodes is shown; the sinusoidal shape
was approximated by trapezoidal segments, which can be realized on a lathe
relatively easy.

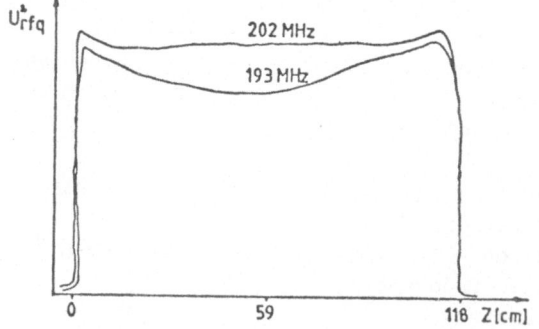

Fig. 17. Effect of the end-cell tuners on longitudinal flatness
and resonance frequency.

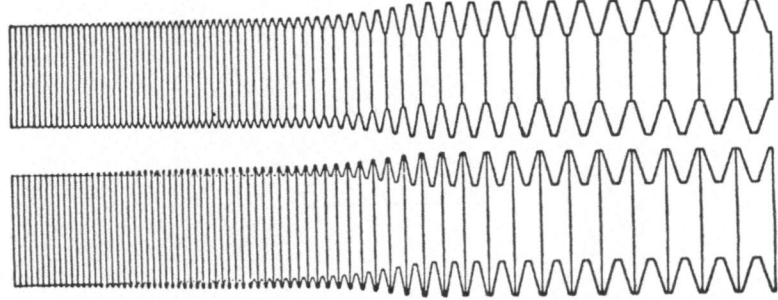

Fig. 18. Trapezoidally modulated electrodes of the four-rod
light ion prototype RFQ.

In rf tests with 25 kW (5% dc) rf power no problems occured. The power was limited by our transmitter. We plan to start with beam tests in
Dezember and there are plans to install it at the HERA beam-line in Hamburg for some time.

Another high power test cavity with a four-rod-structure has been
built by the Chalk River National Laboratory, Canada. Because of the direct cooling of the electrodes and stems the four-rod structure is a promising candidate for the breeder cw accelerator programme[21]. The resonator,
which has been assembled in Canada was sent to Frankfurt recently. Prepa-
rations for the forthcoming high power tests with our 108 MHz power-trans-
mitter up to 200 kW are almost finished (parameters see Table 4).

Table 4. Parameters of the CRNL four-rod RFQ.

Resonance frequency	108.56 MHz
R_p-value	250 kΩ
Quality factor	6000
Tank diameter	50 cm
Length	94 cm
Aperture radius	8.5 mm
Rod radius	6.4 mm

In a first step we will examine properties of cw operation. With a
shorter version we might be able to increase the field strength far over
two times the Kilpatrick limit to study sparking in this configuration.

A spiral loaded four-rod structure with a frequency of 17 MHz is used
in our beam transport experiment[22] and it is in discussion now to employ
such a structure (27 MHz) as a second stage after the first stripper in
the planned high current injector system at GSI[23].

The four-rod is the only RFQ structure, which can accelerate several
beams in parallel in one tank by combining e.g. two identical resonators
to provide two or four beamlines (Fig. 19) and make a direct use of funneling[6,7].

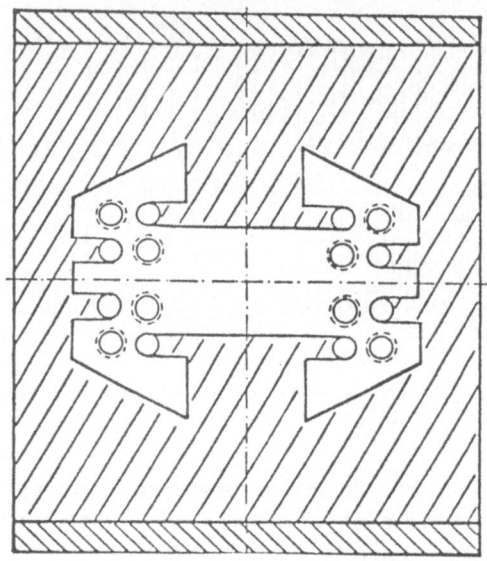

Fig. 19. Possible arrangement of a multi-beam
accelerator using four-rod resonators.

Since it has certain advantages like good efficiency, compact dimensions, flexible rf design, relatively simple manufacture and easy high
duty factor operation, this new structure offers a wide range of possible
application.

REFERENCES

1. I.M.Kapchinskiy and V.A.Teplyakov, Prib. Tekh. Eksp., 4:19 (1970);
   4:17 (1970).
2. K.R.Crandall, R.H.Stokes and T.P.Wangler, Brookhaven National Labo-
   tory, BNL-51143 (1980), p. 20.
3. R.W.Müller, GSI Darmstadt, GSI-Report 79-7 (1979).
4. J.Müller and A.Schempp, Univ. Frankfurt/M., IAP, Int. Report 79-1
   (1979); Engl. Transl. LA-TR-82-28 (1982).
5. A.Schempp et al., Proc. 1984 Linear Accelerator Conference, Seeheim,
   GSI 84-11 (1984), p. 100.
6. R.H.Stokes and G.N.Minerbo, IEEE Trans. Nucl. Sci. NS-32:2593 (1985).
7. A.Schempp, Univ. Frankfurt/M., IAP, Int. Report 85-6 (1985); Engl.
   Transl. LA-TR-85-21 (1985).
8. H.O.Moser and A.Schempp, KFK Katlsruhe, KFK 4090 (1986).
9. A.Crametz et al., Nuclear Instr. and Meth. 242:179 (1986).
10. P.Blüm and A.Citron, Proc. 3rd EHF-Workshop, Karlsruhe 1986.
11. A.Schempp et al., IEEE Trans. Nucl. Sci. NS-32:3552 (1985).
12. DESY Hamburg, HERA-Report 84-12 (1984).
13. A.Schempp, Proc. 1984 Linear Accelerator Conference, Seeheim, GSI-
    84-11 (1984), p.. 339.
14. A.Schempp, Proc. 1986 Linear Accelerator Conference, Stanford, to
    be published.
15. Y.Hirao, Proc. 1984 Linear Accelerator Conference, Seeheim, GSI 84-11
    (1984), p. 490.

16. R.W.Müller et al., Proc. 1984 Linear Accelerator Conference, Seeheim, GSI 84-11 (1984), p. 77.
17. W.Neumann et al., Proc. 1984 Linear Accelerator Conference, Seeheim, GSI 84-11 (1984), p. 80.
18. A.Schempp et al., Nuclear Instr. and Meth. 10/11:831 (1985).
19. A.Schempp et al., Proc. HIF Symposium, Washington 1986, to be published.
20. A.Schempp et al., IEEE Trans. Nucl. Sci. NS-30:3536 (1983).
21. S.O.Schriber, Proc. 1984 Linear Accelerator Conference, Seeheim, GSI 84-11 (1984), p. 501.
22. N.Zoubek et al., Proc. HIF Symposium, Washington 1986, to be published.
23. J.Klabunde et al., Proc. 1986 Linear Accelerator Conference, Stanford, to be published.

DISCUSSION

VRETENAR:

You used the equivalent circuit of the RFQ cavity to calculate the parameters of the structure. How did you compute the numerical values for the equivalent circuit's parameters, L (quadrant inductance) and C (quadrant capacitance)?
I suspect that your transverse stabilizing system, affects the quadrupole mode too, in the sense that the quadrupole frequency may be shifted.

Answer:

I took formulas from standard textbooks for the 4 line capacity and assumed a homogeneous magnetic field in the quadrant.

The dipole modes are shifted by approximately + - 5 MHz for the 202. MHz RFQ. In first order the quadruple mode is not affected at all. A small perturbation of ca. 50KHz caused by the additional copper material in the end cell is negligible compared to frequency shifts of up to 3MHz caused by a single vane coupling ring(VCR) used in other RFQ resonators.

HIGH POWER 35 GHz TESTING OF A FREE-ELECTRON LASER AND TWO-BEAM ACCELERATOR STRUCTURES[*]

D.B. Hopkins, R.W. Kuenning, F.B. Selph, and A.M. Sessler

Lawrence Berkeley Laboratory, University of California, B52B, Berkeley, CA 94720

J.C. Clark, W.M. Fawley, T.J. Orzechowski, A.C. Paul, D. Prosnitz, E.T. Scharlemann, and S.M. Yarema

Lawrence Livermore Laboratory, University of California, L-626, Livermore, CA 94550

B.R. Anderson

Air Force Weapons Center, Kirtland Air Force Base, Albuquerque, NM 87117

Abstract

At the Lawrence Livermore National Laboratory, a pulsed electromagnetic wiggler has been coupled to the Experimental Test Accelerator forming the Electron Laser Facility. This is a single-pass Free-Electron Laser which, because the wiggler excitation can be varied, can operate over a wide frequency range. Efficiency of conversion of electron beam to microwave power is 7%. This new power source is being used in a collateral program for developing and testing structures suitable for a Two-Beam Accelerator. The Two-Beam Accelerator shows much promise for achieving the high average accelerating gradients, e.g., ≥250 MV/m, required in such next-generation electron accelerators as 1 TeV on 1 TeV linear colliders.

_____

[*] This work was supported by the Office of Energy Research, High Energy Physics Division of the U.S. Department of Energy under Contract No. DE-AC03-76SF00098. Performed jointly under the auspices of the U.S. Department of Energy by the Lawrence Livermore National Laboratory under W-7405-ENG-48 and for the Department of Defense under SDIO/BMD-ATC MIPR No. W31-RPD-53-A127. This paper was originally prepared for the 1986 SPIE International Synmposium on Optical and Optoelectronic Applied Sciences and Engineering, June 2-6, 1986, Quebec, Canada.

In this paper, the results of recent tests characterizing the Free-Electron Laser are summarized. Also, progress in the fabrication and testing of key Two-Beam Accelerator hardware components are reviewed. Finally, Two-Beam Accelerator problem areas are discussed.

Description of the Electron Laser Facility

The Electron Laser Facility (ELF)[1] at the Lawrence Livermore National Laboratory (LLNL) is a microwave Free Electron Laser (FEL) consisting of a 3 m pulsed electromagnetic wiggler coupled[2] to the Experimental Test Accelerator (ETA).[3] The latter is an induction linear electron accelerator which typically produces a 3-4 MeV, 10 kA, 20 ns beam at a 1 Hz repetition rate. The beam is transported through an emittance selector. As a result, a 15 ns beam of 800-1000 A having a normalized edge emittance[4] of 0.47 π radian-cm is delivered to the wiggler input. The design of the beamline and matching quadrupole focusing magnets achieves a highly chromatic transport which produces a nearly monoenergetic beam.

A description of ELF and its measured performance has been reported elsewhere.[5] These will be briefly reviewed in this paper. Figure 1 shows a 1 m wiggler module and its pulsed-coil electrical connections. The visible windings are quadrupole coils which produce a field gradient of 30 G/m for horizontal beam focusing. The wiggler windings are inside the quadrupole coils and adjacent to the central interaction region. These can produce an on-axis peak field of 5 kG; additionally, they provide vertical beam focusing. The windings form two series (top and bottom) of alternately opposed solenoids as indicated in Figure 2. Their cusp fields produce a linearly polarized

Fig. 1. ELF 1 m wiggler module.

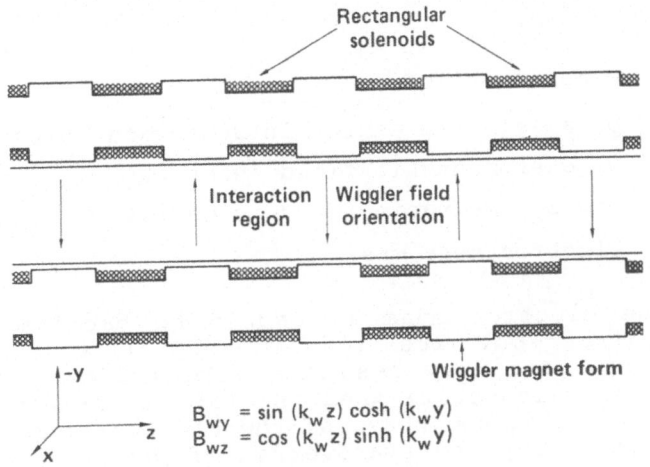

Fig. 2. ELF wiggler magnet design.

$B_{wy} = \sin(k_w z)\cosh(k_w y)$
$B_{wz} = \cos(k_w z)\sinh(k_w y)$

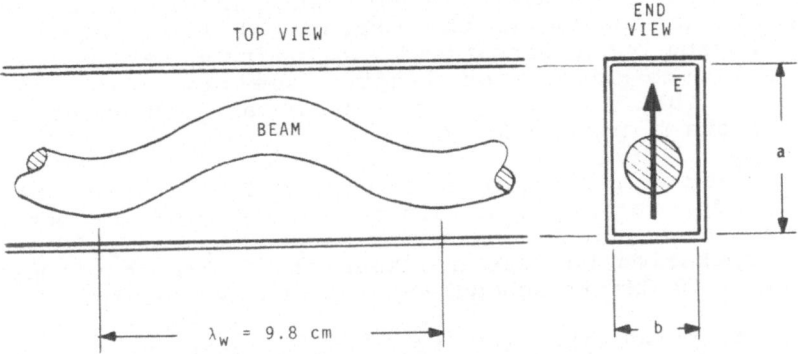

Fig. 3. Wiggler beam orientation.

magnetic field whose polarity alternates and causes the beam to undulate with an axial wiggler period of 9.8 cm. This is indicated in Figure 3.

Each two periods of the wiggler are energized by a separate power supply. These generate half-sinusoidal current pulses of about 1 ms duration. Their output levels are programmable; this permits easy variation of wiggler field strength and the axial intensity profile. For the work reported in this paper, a variable-length "flat" wiggler was used; i.e., while the number of powered wiggler sections was varied, all powered sections were excited equally (except for short beam-matching regions at both ends).

For high-power testing, ELF is operated as a FEL amplifier. (Operation in an oscillator mode is discussed in the following section). A theoretical analysis of such a device is given elsewhere.[6,7] The wavelength of rf radiation, λ_{rf}, for which there is high gain is given by:

$$\lambda_{rf} = \frac{\lambda_w}{2\gamma 2}\left[1 + 1/2\left(\frac{eB_w\lambda_w}{2\pi mc^2}\right)^2\right] \qquad (1)$$

221

where

$$\lambda_w \equiv \quad \text{wiggler period}$$

$$\gamma \equiv \quad \text{parallel beam energy in rest-mass units}$$

$$e \equiv \quad \text{wiggler magnetic field intensity}$$

$$B_w \equiv \quad \text{wiggler magnetic field intensity}$$

$$mc^2 \equiv \quad \text{electron rest mass}$$

Because the electron beam is caused to oscillate in the transverse direction, its longitudinal (i.e., parallel) velocity is significantly less than the velocity of light. Microwave power gain occurs when the FEL is operated so that the beam slips behind the phase of the electromagnetic wave by one wavelength during the transversal of one wiggler period. This is the resonance condition wherein electrons are in step with the field oscillations. The microwave field intensity then grows as it flows through the interaction region, increasing its energy at the expense of the electron beam energy. During other conditions, e.g., with an excited wiggler longer than the saturation length (see next section), the opposite can take place; i.e., the beam can gain energy at the expense of the microwave field.

Unlike nearly all other single-pass FEL's, ours has the capability for varying B_w. This makes possible the operation of ELF over a very large frequency range, typically 30 to >150 GHz. All operation to date has been at 35 GHz, but operation at ~70 and ~140 GHz is scheduled for the near future.

Figure 4 shows the ELF beamline arrangement from the emittance selector through the wiggler, including various steering magnets (SM), quadrupoles (QD), and a beam current monitor (BM). It also shows how the microwave input signal is brought in, reflected by a screen at 45°, and directed into the wiggler interaction region collinearly with the electron beam. The microwave source is a magnetron operated with a 0.4 μs pulse width. The microwave power input at the wiggler is typically 30-40 kW peak.

As indicated in Figure 3, the beam and microwave power flow in a rectangular pipe, i.e., waveguide, in the interaction

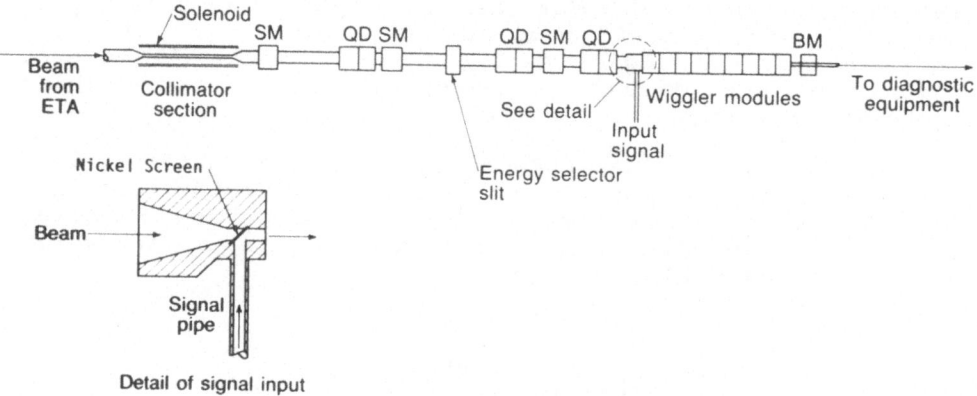

Fig. 4. ELF beamline arrangement.

region. This is made of stainless steel whose poor electrical conductivity minimizes eddy-current shielding effects and ensures that the pulsed magnetic field adequately penetrates to the interior of the guide. In order to accommodate the size of the beam, which is approximately 8 mm diameter, and its undulations, the waveguide is necessarily large compared to the rf wavelength of 8.6 mm at 35 GHz. Most of our work has been with a waveguide having a and b dimensions of 9.83 and 2.91 cm, respectively. Since such an oversized guide can propagate a very large number of modes, care is taken to launch only the desired mode into the interaction region. This is the rectangular TE_{01} mode which interacts with the undulating beam.

Not surprisingly, the undulating beam can couple to other modes and even higher harmonics of the rf frequency in the interaction region. These will be touched upon in the next section. It is possible for an even more subtle effect to occur, namely the generation of a significant amount of power in the upper and lower sidebands, separated from the input frequency by the Lorentz-shifted synchrotron frequency.[6] In the laboratory frame, the sideband separation is typically a few GHz. These three effects can rob power from the desired frequency and/or mode, thereby reducing efficiency. The first two effects (other modes and harmonics) will be shown to be of minor importance. The sideband coupling is presently being analyzed by a number of people. Preliminary studies indicate that the FEL can probably be designed to suppress this effect.

Summarizing the principal features of ELF:

Operates in an amplifier mode

$$\lambda_{rf} = \text{8.6 mm, 4.4 mm, or 2.2 mm}$$
$$P_{in} = \text{1 kW - 100 kW}$$

Pulsed electromagnetic wiggler

$$\lambda_w = \text{9.8 cm}$$
$$B_w = \text{5 kG maximum}$$
Linearly polarized
No axial magnetic guide field
Each two periods independently controlled

Electron beam from Experimental Test Accelerator (ETA)

$$E_b = \text{2 - 4.5 MeV}$$
$$I_b = \text{100 - 1000 Amps}$$
$$\beta_b = \text{2 x } 10^4 \text{ Amps/cm}^2 \text{ - rad}^2$$

The last parameter listed above is the brightness which is proportional to the beam current divided by the square of the normalized emittance. It influences laser gain in direct proportionality.

Small-signal gain measurement[5]

The small-signal gain of ELF has been measured by operating it in a super-radiant mode with no microwave power input, i.e., as an oscillator, as shown in Figure 5. The direction

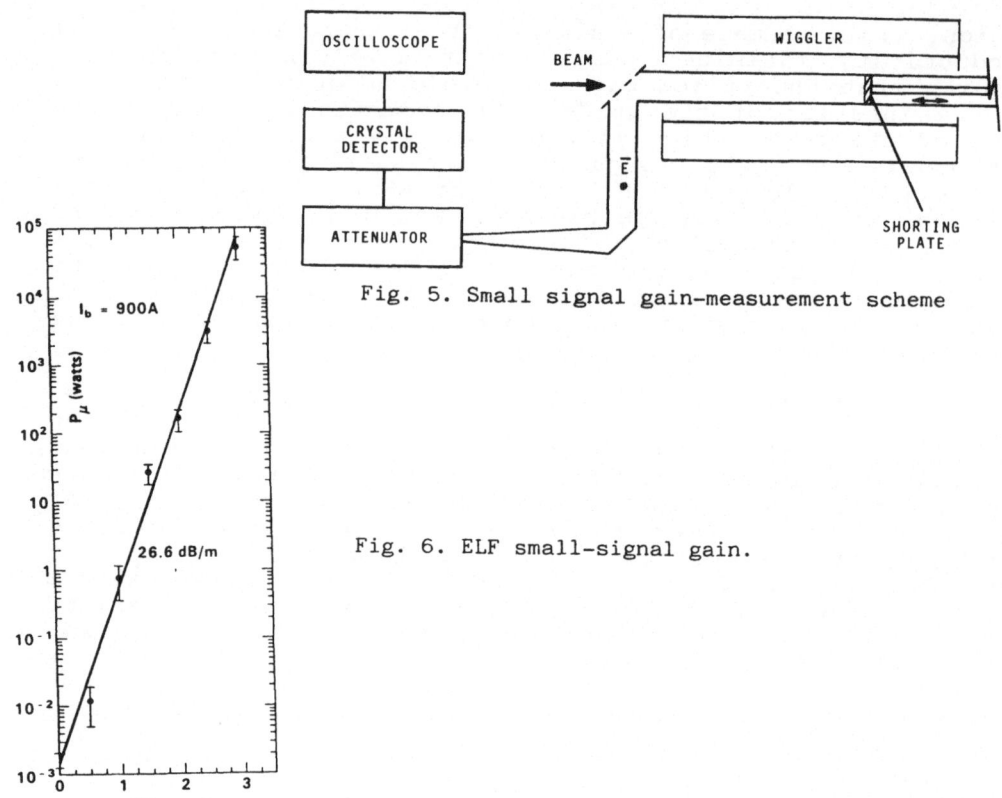

Fig. 5. Small signal gain-measurement scheme

Fig. 6. ELF small-signal gain.

of the microwave electric field vector is into the paper. Power at the resonant frequency builds from noise and increases in magnitude down the length of the wiggler. A shorting plate reflects the power back through a 40 inch linear taper which transforms the TE_{01} mode in the large guide to a TE_{10} mode in WR-28 fundamental guide. The power is appropriately attenuated, then monitored by a calibrated square-law crystal detector terminated in 50 ohms. The length of the interaction region is varied by changing the axial position of the shorting plate. This plate is actually the tantalum target plate of a probe and fiber optics imaging system for measuring the size and position of the FEL beam.[8]

The results of this measurement are shown in Figure 6. For a beam current of 900 A, the microwave power increases exponentially with wiggler length with a gain of 26.6 dB/m. The equivalent noise power level at the wiggler input is seen to be about 1.5 mW. This is in relatively good agreement with the calculated value for single particle radiation at 34.6 GHz into the TE_{01} mode.[9]

High power diagnostics

For measuring ELF performance when operated as a high power amplifier, the diagnostic scheme of Figure 7 was employed. Forward and reverse microwave power are monitored near the magnetron. The WR-28 waveguide components there (0.711 x

224

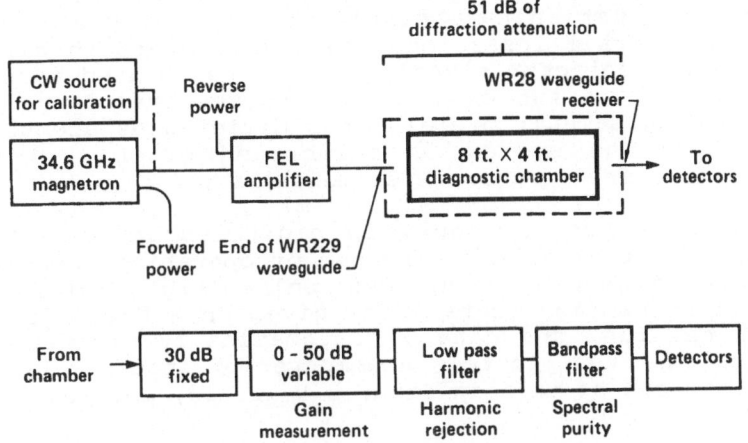

Fig. 7. ELF high-power diagnostics scheme.

0.356 cm inside dimensions) are pressurized to 15 psig with sulphur hexafluoride to prevent sparking at the normal 80-100 kW power level. As mentioned previously, gradual linear tapers are used as transitions between the WR-28 and large waveguides.

At the wiggler output, the interaction waveguide tapers to WR-229 guide (5.817 x 2.908 cm inside dimensions). A 20 foot run of this guide terminates in an open-ended waveguide transmitting antenna at one end of a diagnostic chamber. This is 4 feet in diameter and 8 feet long, is lined with microwave absorber material, and is evacuated to a pressure in the 10^{-5} torr range. A short open-ended WR-28 waveguide stub at the far end functions as a receiving antenna operating in the far field of the transmitting antenna. This arrangement provides 51 dB of diffraction - loss attenuation, reducing the power to a level safely handled by directional couplers and other components operated at atmospheric pressure.

Appropriate attenuation and filtering is introduced at the chamber output to permit operating a calibrated crystal detector in its square law region when terminated in 50 ohms. This is typically at a ~0.5 to 2.0 mW power level. The low-pass filter design results in good rejection of the second and third harmonic frequencies. The bandpass filter has a ~4 GHz passband.

A 2 W cw klystron is first used to calibrate the system in two steps. The insertion loss between the magnetron input location and the diagnostics chamber output is first measured. Then, similarly, the insertion loss between the chamber output location and the crystal detector is measured. The sum of these two loss values is the overall system insertion loss. As a matter of interest, the insertion loss between the magnetron input location and the (unexcited) wiggler output location is typically ~3.0 dB. This value is the sum of loss contributions from waveguide wall losses, vacuum window and flange reflections, and mode conversion in the tapers and reflecting screen region.

As a cross-check, the overall system insertion loss is also measured using the magnetron as the power source. By reducing the amount of variable attenuation before the crystal detector, the overall insertion loss can be measured directly by comparing the detector signal with the calibrated magnetron power detector. The two calibration schemes are always found to be in good agreement; overall loss differences are usually ≤0.5 dB (~12%).The system is calibrated often, certainly before and after every major experimental run.

The crystal detector signal is digitized in a Tektronix model 7912AD digitizer. For each set of operating parameters, during an experimental run, the pulse signal data from typically eight machine shots is archived on a hard disk. At the end of the run, the data is automatically organized and properly scaled, then plotted in computer printouts. Examples of these plots can be seen in Figures 9-11, 13, and 15.

ELF performance[7]

The results of an early study of amplifier gain as a function of wiggler magnetic field intensity, B_w, and wiggler length, L_w, are shown in Figure 8. Chosen values of L_w were 1, 2, and 3 m. The beam energy and current were ~3.3 MeV and 450 A, respectively. For the 1 m and 2 m studies, the remaining wiggler length was excited at a level 30% below that in the interaction region. This permitted the beam to be transported through the remaining wiggler to the current monitor at the wiggler output. However, since the resultant magnetic field was well below the resonant B_w value, this region did not contribute to the FEL interaction.

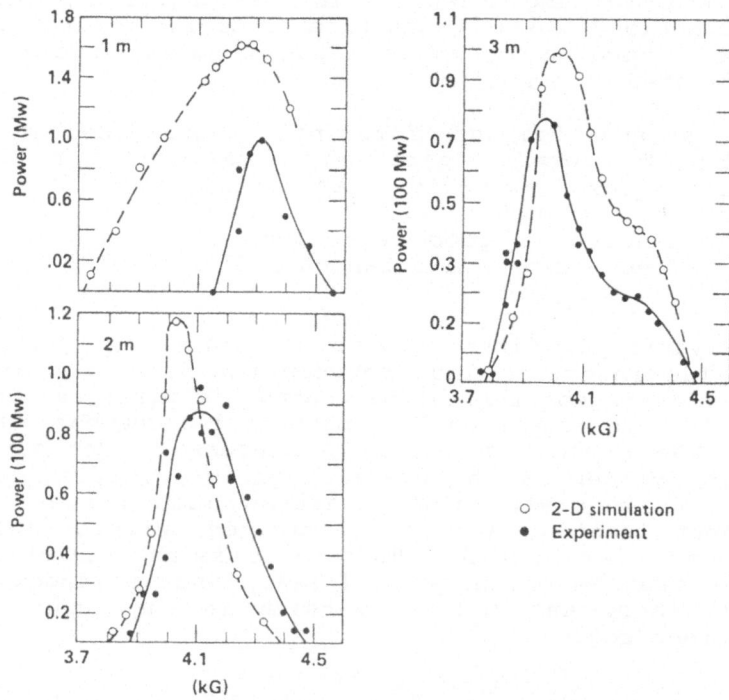

Fig. 8. Early ELF measurement of power vs wiggler magnetic field.

Also shown in Figure 8 are the results of computations using the FRED code.[10] This is a two-dimensional (x-z) numerical particle simulation code developed at LLNL for detailed FEL analysis. It models the electron beam source for the radiation field by 2048 macroparticles executing full betatron motion in the combined wiggler and external focusing fields. In addition to studying basic FEL interaction, it can analyze higher order modes, harmonics, phase front profiles, effects of different initial electron loading distributions, etc. The agreement between FRED simulations and experiment are generally quite good through the exponential gain region up to the saturation length (see below).

More recent measurements of power output vs wiggler magnetic field are shown in Figures 9, 10 and 11 for 1 m, 2 m, and 3 m wiggler lengths. The beam energy was 3 MeV; beam current was 800 A. The total microwave power output vs wiggler length, L_w, is shown in Figure 12. This shows a maximum power level of ~180 MW. The efficiency of conversion of electron beam power to microwave power is ~7%. The output increases exponentially with wiggler length up to ~1.4 m, the "saturation length". Beyond this length, power decreases as the interaction is reversed and electrons gain energy at the expense of the microwave field.

A measurement of the modal power distribution has been made. This determines how much power is produced in the TE_{01} mode in the interaction waveguide and how much is flowing in the two primary competing modes, the degenerate TE_{21} and TM_{21} modes. The measurement is a radiated pattern measurement made in the diagnostics chamber. As indicated in Figure 13, a radially moveable probe assembly positions two mutually perpendicular open-ended WR-28 receiving antennae in the diagnostics chamber. The modal power distribution is obtained by comparing the detected signals from these antennae with

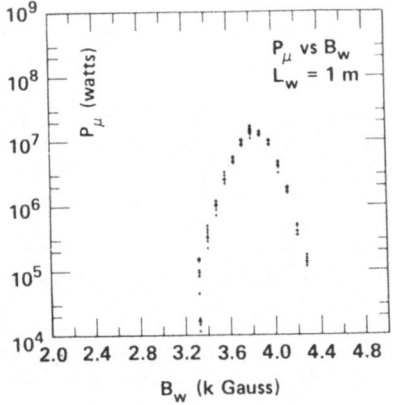

Fig. 9. Output power vs B_w for 1 m wiggler.

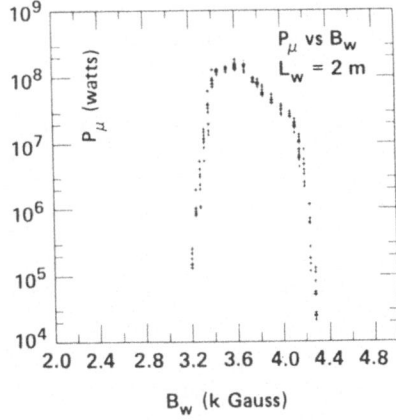

Fig. 10. Output power vs B_w for 2 m wiggler.

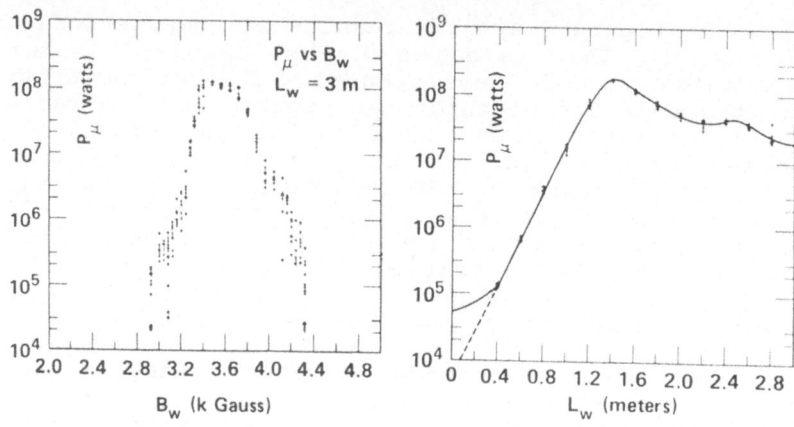

Fig. 11. Output power vs B_w for 3 m wiggler.

Fig. 12. Total output power vs wiggler length.

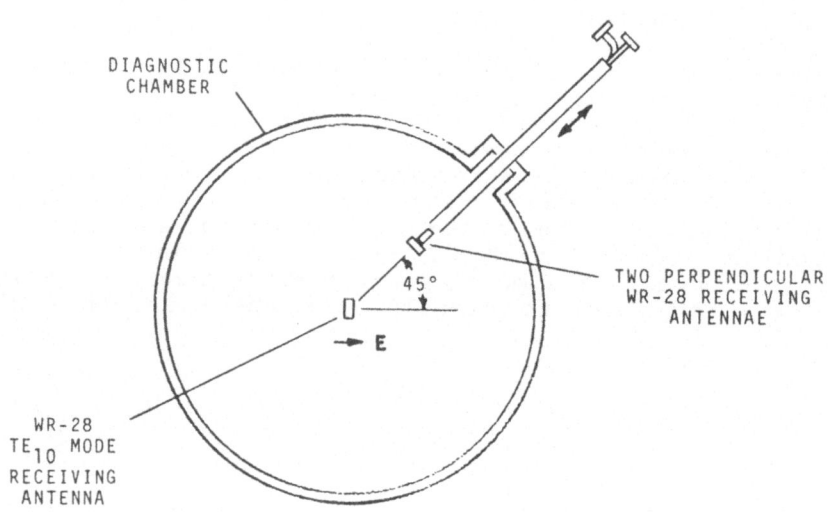

Fig. 13. Antenna probes for modal power distribution measurement.

theoretical radiated antenna patterns and the signal from the fundamental TE_{10} mode pickup antenna. Figure 14 shows the output power in the TE_{01} mode (only) vs L_w. Figure 15 shows relative modal power content vs L_w. This reveals a desirable FEL property, i.e., at the saturation length, the power is essentially totally contained in the single TE_{01} mode. This behavior is in good agreement with that predicted by the FRED code.

As mentioned before, the electron beam undulations also produce radiation at (odd) harmonic frequencies. Electrons are not resonant with the third harmonic, however. Its phase velocity is incorrect for substantial gain to occur. Figure 16 shows the measured third harmonic power output vs L_w. This also agrees well with FRED estimates.

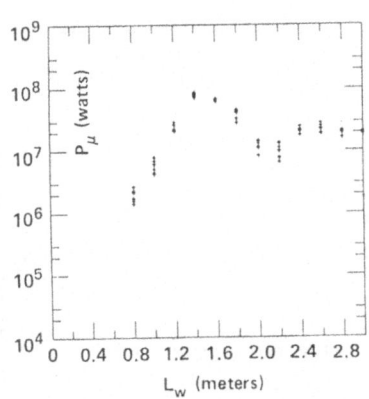

Fig. 14. TE_{01} mode power output vs L_w.

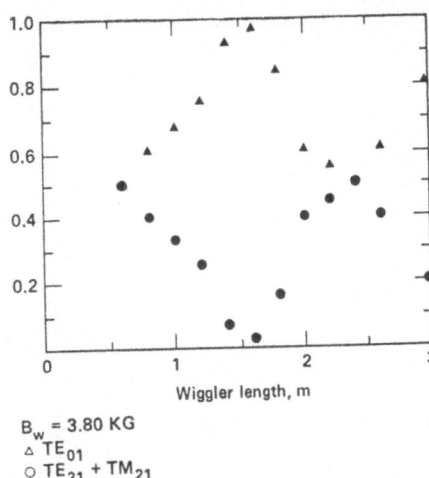

$B_w = 3.80$ KG
△ TE_{01}
○ $TE_{21} + TM_{21}$

Fig. 15. Relative model power content vs L_w.

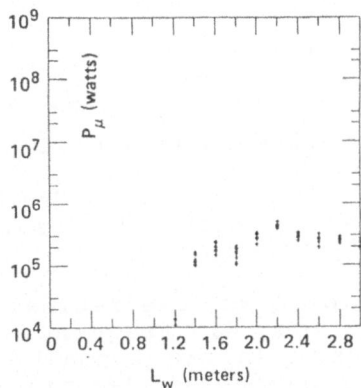

Fig. 16. Third harmonic power output vs L_w.

Motivation for a high-frequency, high-gradient accelerator

In the high-energy physics community, many feel that the next major accelerator development will be a linear collider operating at an energy level of about 1 TeV on 1 TeV. (1 TeV = 10^{12} electron volts). Present-day techniques, such as those employed at the Stanford Linear Accelerator Center (SLAC), produce an average accelerating gradient of ~17 MV/m. A 1 TeV machine of this type would be tens of kilometers long and require thousands of klystron power sources and their

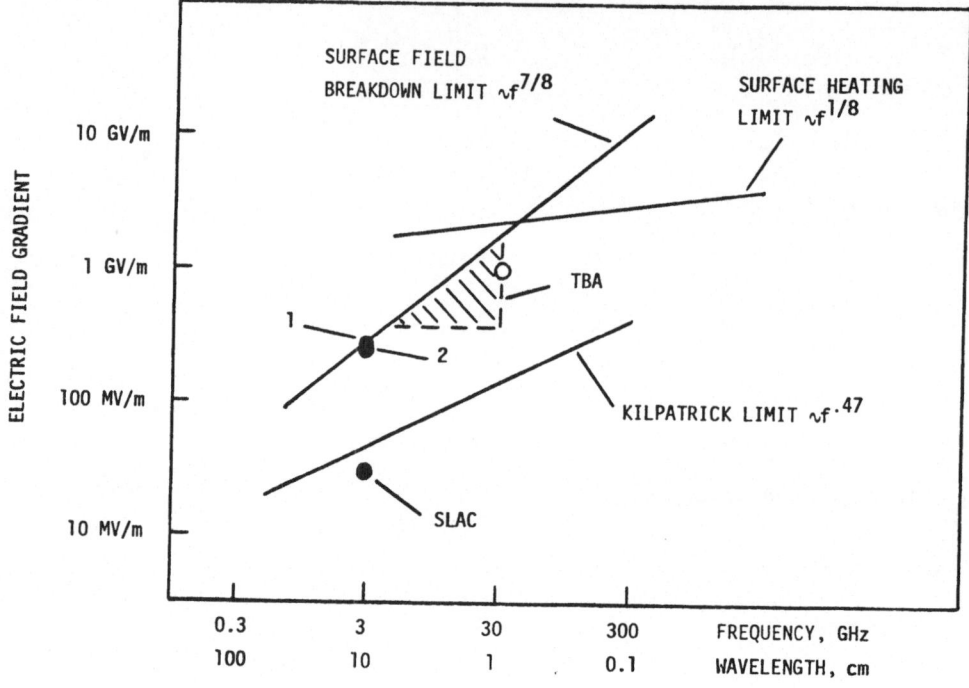

1. WANG & LOEW, SLAC, 2.5 μs, 1985
2. TANABE, VARIAN, 4.4 μs, 1983

Fig. 17. Accelerator maximum surface electric field gradient.

associated modulators. Clearly, such a device is impractical. A new approach is required which will result in dramatically increased accelerating gradients.

Figure 17 shows a plot of theoretical maximum surface electric field gradient vs frequency for copper rf accelerating structures. In the commonly used disk-loaded waveguide geometry, the average accelerating gradient is about half the maximum surface electric field gradient. It is clear from Figure 17 that achievable gradients scale almost linearly with frequency. An average accelerating gradient of at least a few hundred MV/m should be possible at an operating frequency in the 30 GHz range.(The Kilpatrick "limit"[11] is also shown for comparison. This criterion for reliable spark-free operation was often applied to accelerator designs in the past but is now recognized as being overly conservative).

Although the advantage of operating at higher frequency has long been recognized, there have been no suitable high-power sources at, say, 1 cm wavelength until the recent development of gyrotrons[12] and the FEL. Gyrotrons are in an active stage of development and may eventually prove to be a practical answer for long-accelerator power sources. However, their maximum power output at a 1 cm wavelength seems likely to remain below ~200 MW; therefore, thousands of these, all properly phase locked, would be required for a 1 TeV accelerator.

Two-Beam Accelerator

As a promising alternative approach to achieving high accelerating gradients at high frequency, we have introduced the concept of the Two-Beam Accelerator[13] and are engaged in the process of its theoretical and experimental development. Its design is based on the fact that suitably high microwave power is now available from an FEL.

The Two-Beam Accelerator (TBA) concept is shown diagrammatically in Figure 18. As can be seen, the TBA consists of a high - gradient accelerator structure which is periodically coupled to an FEL as a source of high microwave power. To replenish energy given up by the FEL beam to the microwave field, induction accelerator units[14] are placed periodically along the length of the machine. Recent computer studies[15] have shown that the FEL can be several kilometers long without significant degradation of beam quality or current.

Basic TBA analysis and design examples have been published elsewhere.[16-18] These designs are still evolving as analytical work progresses. A sample set of parameters from reference 16 is presented in Table 1.

Table 1 Parameters for a 1 TeV x 1 TeV Two-Beam Accelerator Collider

Low Energy Beam

Average Beam Energy (Units of mc^2)	40
Beam Current	2.15 kA
Bunch Length	6 m
Wiggler Wavelength	27 cm
Average Peak Wiggler Field	2.4 kG
Beam Power	43 GW
Beam Energy	0.8 kJ
Power Production	2.2 GW/m
Number of FEL Injectors	2 x 2
Power From Mains	160 MW

High Gradient Structure

Wavelength	1 cm
Gradient	500 MeV/m
Stored Energy	40 J/m
Fill Time	18 ns

High Energy Beam

Injection Energy	2 GeV
Repetition Rate (f)	0.5 kHz
Final Energy	1 TeV
Length	2 x 2 km
Luminosity	$4 \times 10^{32} cm^{-2} sec^{-1}$
Beam Height (σ_y)	0.14μ
Beam Width (σ_x)	1.4μ
Single Beam Power	8.0 MW
Number of Particles	10^{11}
Disruption (D)	1.3
Beamstrahlung (δ)	0.2
Overall Efficiency (From Mains to HEB)	10%

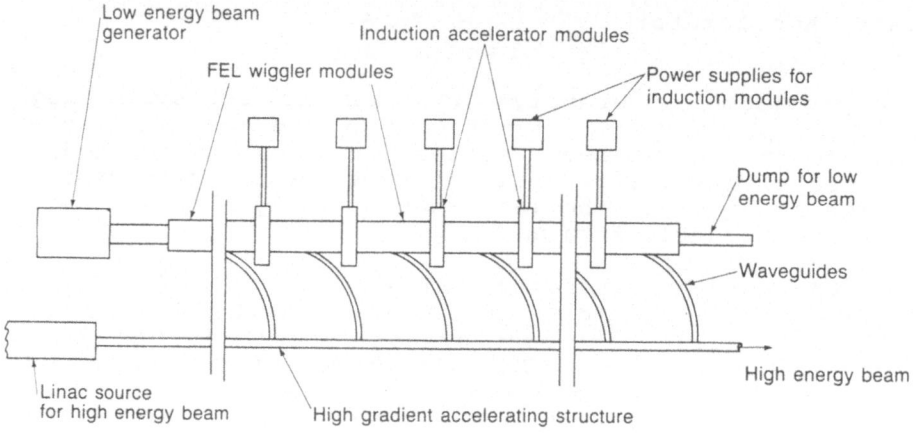

Fig. 18. Two-Beam Accelerator concept.

An initial goal of our TBA program has been to demonstrate ultra-high gradients in an actual accelerating structure in order to increase our confidence in the breakdown gradient scaling shown in Figure 17. The seven-cell high-gradient accelerator test structure (HGS) shown in Figure 19 was constructed for testing at ELF. Its method of fabrication and other details have been reported elsewhere.[19] It is a copper $2\pi/3$ mode structure[20] with all cavity and input/output coupler dimensions scaled from SLAC dimensions by the ratio of 34.6/2.856 (GHz) = 12.11. The center-to-center distance between the input and output coupling cavities is 0.6828 inches. The overall length is 1.6776 inches. The couplers were scaled from the output coupler of a SLAC accelerating section. The cavities are all of equal diameter, scaled from SLAC cavity number 84.

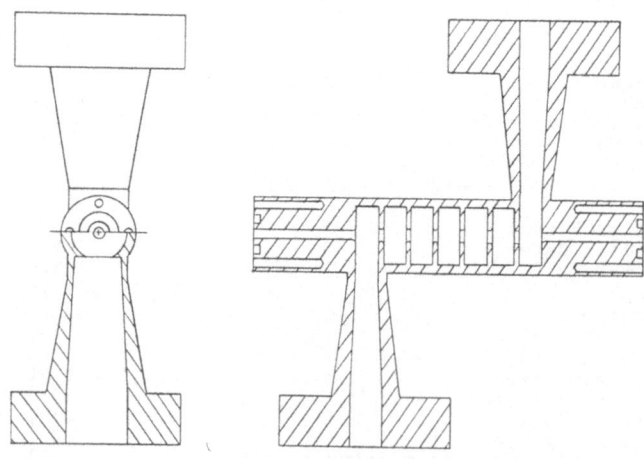

Fig. 19. High-Gradient Accelerator test structure (HGS).

In this structure, the highest electric field gradient is produced in the input cavity. Since the traveling wave group velocity is well known (v_g = 0.0065) and the ratio of shunt impedance to Q, r/Q, is readily scalable from SLAC performance, the relationship between average accelerating gradient, E_a, and power is conveniently expressed as:

$$E_a = [(\omega \, p_o/v_g) \, (r/Q)]^{1/2} \tag{2}$$

The scaled value of r/Q is 546 ohms. The ratio of peak surface field to average accelerating gradient is 1.95.[21] This, with equation (2), predicts a maximum surface gradient of 1.52 GV/m at a power level of 100 MW.

Figure 20 shows the HGS after being dimple-tuned to a frequency of 34.6 GHz to match the frequency of the FEL magnetron driver. Figure 21 shows the test arrangement at ELF. The vacuum in the HGS was only marginal, typically in the mid 10^{-5} torr range. Photomultipliers viewing the input coupling aperture and along the HGS axis served as spark detectors.

Normally, new accelerator structures are preconditioned with pulsed rf whose power level is slowly increased to the rated value. This may take several days or more at repetition rates up to 360 pps. Because of the lack of other high power 35 GHz sources and the ELF repetition rate of 0.5 Hz, there was no opportunity to precondition the HGS. During ELF testing, the power to the HGS was increased until there was first evidence of (partial) breakdown, as indicated by photomultiplier signals. Also, the shape of the microwave pulse transmitted through the HGS started to show some chopping or rolloff at the tail end of the pulse. Staying at this power level for 1/2 hour or more showed some, but small, evidence of improvement.

Our testing time on ELF was limited to about 3 days by other priorities so there was no opportunity for in-depth testing. The best HGS performance is represented by the detected microwave pulse shown in Figure 22. The peak received power level was 3.1 MW. Allowing for a measured 3.0 dB total HGS insertion loss, the calculated corresponding peak field in the input cavity is ~380 MV/m. This is equivalent to an average accelerating field of ~190 MV/m. Considering the marginal vacuum conditions and routine HGS metallurgy, this result is very encouraging.

An attempt was made to speed the conditioning process by increasing the power level, in a few steps, by more than an order of magnitude. This resulted in more energetic discharges but no appreciable gain in the received power level and conditioning. When later sectioned and examined microscopically, the surfaces of the HGS disks and iris apertures appeared clean and smooth, free of any visible arc marks or sputtering.

New accelerator sections are being constructed using the best copper and fabrication techniques. These sections will undergo a thorough high temperature bakeout and will be tested at a vacuum level of 10^{-7} to 10^{-8} torr. We expect to

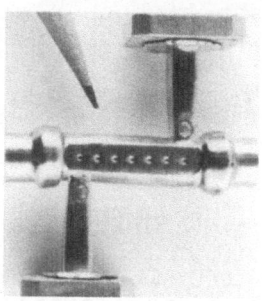

Fig. 20. Dimple-tuned HGS.

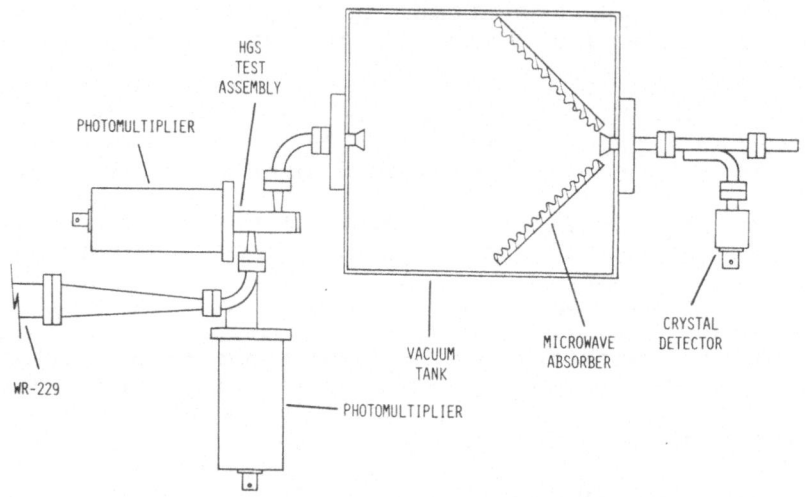

Fig. 21. HGS test arrangement at ELF.

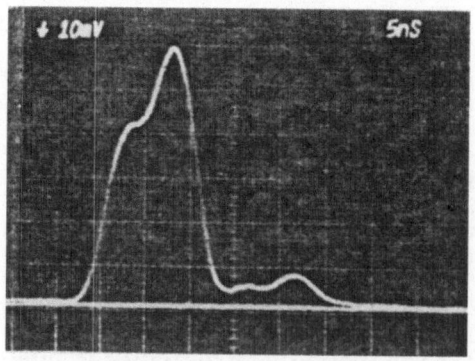

Fig. 22. Maximum gradient achieved in HGS.

demonstrate maximum surface gradients significantly higher than the value quoted above.

Microwave output coupling[19]

A significant fraction of the FEL microwave power must be periodically extracted and coupled to the adjacent accelerator structure in a TBA. This must be done in such a manner that the FEL modal power distribution is not disturbed; i.e., essentially all of the power should continually exist in the desired TE_{01} mode in the FEL. Any power converted to other modes by coupling discontinuties represents an undesirable reduction in overall efficiency.

It does not appear possible to achieve the necessary output coupling in the FEL's oversized interaction waveguide with negligible mode conversion using directional coupling. Our proposed solution is to introduce angled septa into the guide which will function as "scoops" for gracefully removing a fraction of the flowing microwave power. This scheme is shown conceptually in Figure 23. The scoops are gradually tapered to fundamental waveguide size so that power can be transported to the accelerator HGS without mode conversion. The waveguide size and septum locations are chosen so as to accommodate the wiggled beam. The FEL is designed so that the microwave power increase per unit length from FEL action is equal to the average power extracted per unit length.

Figure 24 shows a cross-sectional view of a septum coupler test section being constructed. The tapering and septum angles have been designed to minimize mode conversion. The test section is being made of non-magnetic 310S stainless steel. It will be tested at ELF in the summer of 1986 to

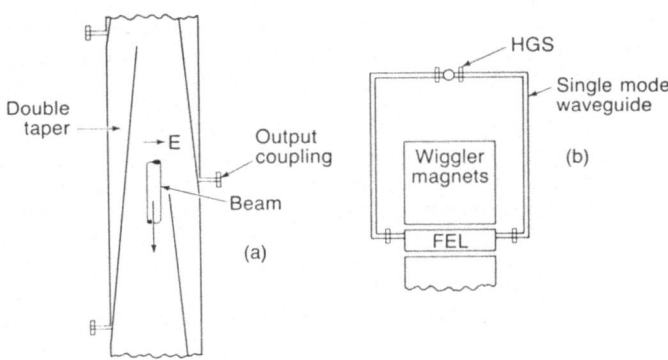

Fig. 23. Septum coupling technique for extracting FEL microwave power.

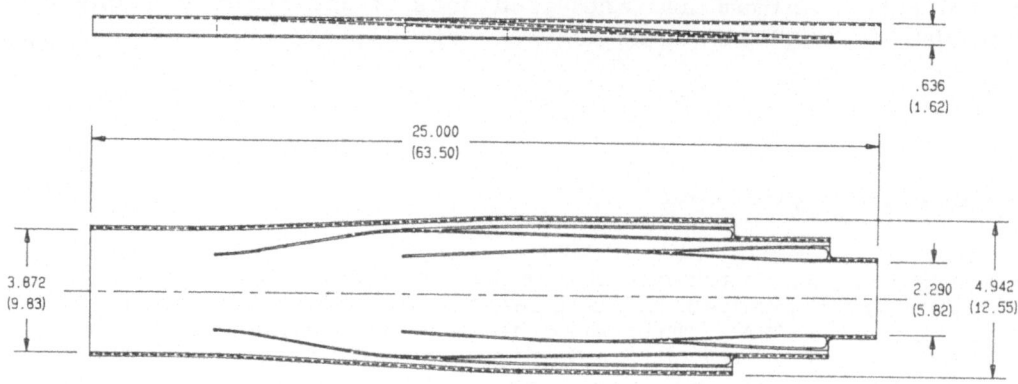

Fig. 24. Septum coupler test section.

investigate any possible ill effects induced by the coupler on microwave mode and harmonic content, and electron beam stability.

TBA problem areas

In addition to the above concerns, several other problem areas exist. An optimized design for an FEL beam reacceleration cavity has yet to be carried out. Its overall beamline insertion length must probably be held to 2-3 inches in order not to seriously degrade the TBA's high average accelerating gradient. Moreover, the microwave power loss incurred in crossing a reacceleration gap should be only a few percent. Initial gap-loss measurements indicate that special microwave focusing or guiding will be required to achieve an adequately low loss.

The lack of a suitable rep-rateable high power microwave source will preclude normal rf preconditioning of accelerator structures for some time. Other conditioning methods are being investigated.

Transverse wakefield effects, originally thought to be a serious limitation in such small accelerators, now appear to be acceptably small if sufficiently large beam holes are provided in the accelerator disks.[22] This needs experimental verification.

Referring to Table I, the desired high-energy beam luminosity is difficult to achieve when accelerating single electron bunches.The situation is mitigated by the acceleration of bursts of multiple bunches. This is because where N ≡ number of particles bunch and n ≡ number of bunches.

$$\pounds \sim N^2 n$$

This mode of operation has yet to be fully theoretically analyzed and optimized.

Perhaps the largest outstanding TBA challenge is in the area of phase stability and control. Analytical studies of the sensitivity of microwave phase to errors in operating parameters are in progress.[23] With no correction, the phase errors resulting from very small, but realistic, deviations from ideal operating conditions are unacceptably large. Two examples are discussed in the next paragraph.

For characterizing two key FEL parameters, we define the normalized wiggler magnetic field intensity, A_w:

$$A_w = 0.093 \ \lambda_w \ (cm) \ B_w (kG), \qquad\qquad (3)$$

and the beam plasma frequency, ω_p, dependent on beam current, I:

$$\omega_p^2 = \frac{1.3 \times 10^{21} \ I(kA)}{a(cm) \ b(cm)}. \qquad\qquad (4)$$

We choose 0.05 radians as the maximum allowable error in the phase of rf delivered to the high gradient accelerator. Figure 25 shows phase deviation vs distance along the FEL for two selected values of error in ω_p. The phase deviation is linear with error amplitude and duration. The effects of multiple sources of error add linearly. Errors in A_w produce results of a different character, as shown in Figure 26.

A phase-stabilizing scheme is clearly required which is automatic and nearly instantaneous. A proposed feedback phase-correcting system is presented in reference 21. However, it is admittedly cumbersome and costly. We are continuing the search for a more practical solution.

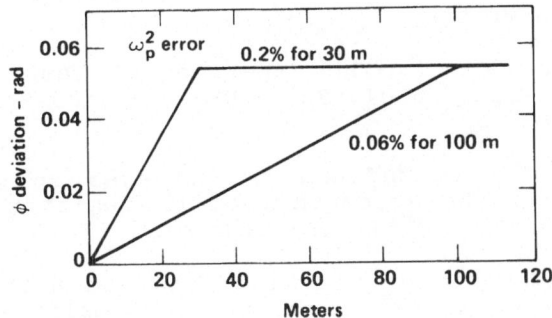

Fig. 25. Phase deviation resulting from FEL beam current error.

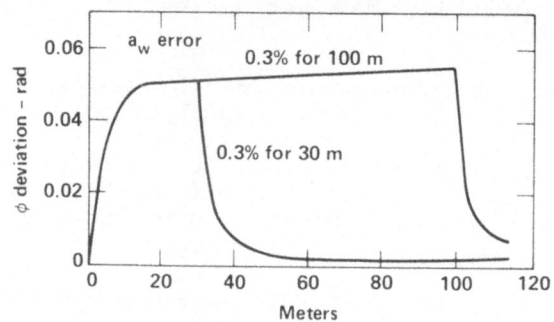

Fig. 26. Phase deviation resulting from wiggler magnetic field error.

Acknowledgments

We appreciate the many helpful discussions with Jonathan Wurtele of MIT, and Gregory Loew, Philip Morton, and Perry Wilson of SLAC. For support during ELF testing, we thank James Dunlap, Scott Hawley, Roy Swafford, David Trimble and the ELF crew headed by James Crawford. We are grateful for the data acquisition routines implemented by Frank Chambers. We also thank David Rogers for his coordination efforts and James Bryan, Michael Burns, G. Robert Gray, John Kinney, Gary Koehler, Hans Krapf and Ted Yokota for their mechanical support. Lastly, we acknowledge with gratitude the excellence of Carolyn Wong in the preparation of this manuscript.

References

1. T.J. Orzechowski, et al, in: Free Electron Generators of Coherent Radiation, C.A. Brau, S.F. Jacobs and M.O. Scully, eds., SPIE, Bellingham, WA, p. 65 (1983).

2. A fourth meter of wiggler is being added to ELF at the time of this writing.

3. R.E. Hester, et al, The Experimental Test Accelerator (ETA), IEEE Trans. Nucl. Sci., NS-26:3, (1979).

4. Edge emittance is defined as the area in phase space which matches the acceptance of the emittance selector and contains 100% of the electron beam.

5. T. J. Orzechowski, et al, High Gain and High Extraction Efficiency From a Free Electron Laser Amplifier Operating in the Millimeter Wave Regime, Proc. 7th Int. Free Electron Laser Conf., Tahoe City, CA, UCRL-93262, (1985).

6. W.B. Colson and A.M. Sessler, Free Electron Lasers, Ann. Rev. Nucl. Part. Sci., 35:25, (1985).

7. N.M. Kroll, P.L. Morton and M.R. Rosenbluth, IEEE J. Quant. Elect., 17, 1436 (1981); also see: W.M. Fawley, D. Prosnitz, and E.T. Scharlemann, Phys. Rev. A30:2472 (1984).

8. T.J. Orzechowski, A Novel Probe for Determining the Size and Position of a Rela- tivistic Electron Beam, in Fiber Optics in Adverse Environments II, R.A. Greenwell, ed., SPIE Vol. 506, p. 36 (1984).

9. H.A. Haus, Noise in Free Electron Laser Amplifier, IEEE J. Quant. Elect., QE-17:1427 (1981).

10. T.J. Orzechowski, et al, High Gain Free-Electron Lasers Using Induction Linear Accelerators, IEEE J. Quant. Elect., QE-21:7, 831 (1985).

11. W.D. Kilpatrick, Rev. Sci. Inst., 28:824 (1957).

12. R.S. Symons and H.R. Jory, Cyclotron Resonance Devices, in Advances in Electronics and Electron Physics, Vol. 55, L. Marton and C. Marton, Eds. Academic Press, Inc. (1981).

13. A.M. Sessler, The Free Electron Laser as a Power Source for a High-Gradient Accelerating Structure, in: Laser Acceleration of Particles, P.J. Channel ed. AIP Conf. Proc. No. 91, American Institute of Physics, New York, p. 154 (1982).

14. J.E. Leiss, Induction Linear Accelerators and Their Applications, IEEE Trans. Nucl. Sci., NS-26: 3, 3870 (1979).

15 E. Sternbach and A.M. Sessler, A Steady-State FEL: Particle Dynamics in the FEL Portion of a Two-Beam Accelerator, Proc. of 7th Int. Free-electron Laser Conf., Tahoe City, CA, LBL-19939, (1985).

16. J.S. Wurtele, On Acceleration by the Transfer of Energy Between Two Beams," in Laser Acceleration of Particles, C. Joshi and T. Katsouleas ed. AIP Conf. Proc. No. 130, American Institute of Physics, New York, p.305 (1985).

17. D.B. Hopkins, A.M. Sessler and J.S. Wurtele, "The Two-Beam Accelerator, LBL-17800, Nuc. Inst. & Meth. for Phys. Res., 228:15 (1984).

18. F.B. Selph, The Two-Beam Accelerator, LBL-18403 (1984), in Proc. of Third Summer School on High Energy Particles, Brookhaven National Laboratory/State University of New York, Stony Brook, NY (1983).

19. D.B. Hopkins and R.W. Kuenning, The Two-Beam Accelerator: Structure Studies and 35 GHz Experiments, IEEE Trans. Nucl. Sci., NS-32:3476 (1985).

20. R.B. Neal, W.A. Benjamin, Inc., Chapter 6 (1968).

21. J.S. Wang and G.S. Loew, Measurements of Ultimate Accelerating Gradients in the SLAC Disk-Loaded Structure (Part I), SLAC/AP-26 (1985).

22. F.B. Selph and A.M. Sessler, Transverse Wakefield Effects in the Two-Beam Accelerator, Submitted to Nuclear Inst. and Meth., LBL-20083 (1985).

23 R.W. Kuenning and A.M. Sessler, Phase and Amplitude Control of the Radio Frequency Wave in the Two-Beam Accelerator, Submitted to Nuclear Inst. and Meth., LBL-20235 (1985).

DISCUSSION

BOSCOLO:
Since your RF extractors are pencil-like, I guess that the wavelength you are using is at certain extent at the limit for exploiting already established -wave technique.

SCHEMPP:
If you produce your pencil-like linac out of thousands pieces per meter, the errors may be comparable to the phase stability requirements you have talked about.
There would probably be advantages in having a bright linac beam for FEL instead of the Induction Linac beam.
Can you compare the RF injection with Schnell scheme?

ACCELERATION OF ELECTRONS BY THE WAKE FIELD

OF PROTON BUNCHES

Alessandro G. Ruggiero

Argonne National Laboratory
Argonne, Illinois 60439, USA

In this paper we re-present and discuss more details of a novel idea to accelerate low-intensity bunches of electrons (or positrons) by the wake field of intense proton bunches travelling along the axis of a cylindrical rf structure. Accelerating gradients in excess of 100 MeV/m and large "transformer ratios", which allow for acceleration of electrons to energies in the TeV range, are calculated. A possible application of the method is an electron-positron linear collider with luminosity of 10^{33} cm^{-2} s^{-1}. The relatively low cost and power consumption of the method is emphasized.

Introduction

Recently a considerable amount of attention has been given to the conceptual design of a linear collider for electrons and positrons in the TeV energy range and luminosity of 10^{33} cm^{-2} sec^{-1}. A list of possible parameters for the collider is given in Table 1. It is obvious that the large luminosity can be obtained mainly because of the assumed beam dimensions at the collision point. These dimensions are considerably smaller than those obtained with present technology.

A crucial parameter in the design of the linear collider is the power in the beam to be accelerated, given by the number of particles per bunch, the final energy and the repetition rate. It is important to keep this number reasonably small, possibly around few MWatts, which corresponds to the example given in Table 1. In fact new methods of acceleration are expected to have

Table 1. Electron-Positron Linear Collider

Energy per Beam	1	TeV
No. of Particles / Bunch	5×10^9	
Frequency of Encounter	5	kHz
Normalized Rms Emittance, σ^2/β^*	1×10^{-6}	m-rad/γ
β^*	5	mm
Rms Beam Spot Size, σ	500	A$^\circ$
Rms Bunch Length	2	mm
Disruption Parameter, D	3	
Luminosity Enhancement Factor, H	3	
Energy Spread from Beamstrahlung, $\Delta E/E$	4	%
Luminosity	1×10^{33}	cm^{-2} sec^{-1}
Beam Power	2×4	MWatt

relatively low efficiency, and the total power required depends on the acceleration efficiency. This fact is so important and overriding that it is the main reason why large luminosities have been proposed with very small beam dimensions.

In order for the linear collider to become a practical reality in the near future, two major considerations are to be taken into account: cost and power efficiency. They will eventually provide the selection for the method to be used. For instance an efficiency of 0.1 % is already too low because it could easily correspond to a total power level of few GWatts.

Linear accelerating structures made of a sequence of metallic rf cavities are certainly among the most promising candidates, since they can support accelerating field gradients in excess of 100 MeV/m. The major problem with these devices is that they have to be operated in the very high frequency range, 30 GHz or more, and it is not easy to find adequate, efficient power sources.

Several ideas have recently been discussed, which involve the use of one or more beams of electrons, tightly bunched, at low energy and high intensity. These beams provide energy to an rf structure which is then used to power the main accelerating linear system where the principal bunches of either electrons or positrons are to be accelerated.[1] It is also possible to stimulate radiation from the beam travelling through wiggler magnets, and to use the radiation at the proper frequency to power the main linear accelerator.[2] In a way, all these methods are reminiscent of the conventional Klystrons; the power of a low-energy, high-intensity electron beam is converted in electro-magnetic power at higher frequency to energize sections of otherwise conventional linear accelerators.

In the past we have proposed a similar concept[3,4] which we wish to expand in this paper, and to correct few misunderstandings at that time. It has been suggested that it is possible to use the same rf structure for both the driver beam and the beam of particles to be accelerated (Beam Transformers),[5] rather then have them travel in separated devices. We concur with this approach, but in our case the primary beam is made of protons, for reasons explained below. We have called the device: the WAKEATRON.

The WAKEATRON

The WAKEATRON is intended to be a linear collider for electrons and positrons with energy of 1 TeV and luminosity of 10^{33} cm^{-2} s^{-1} as shown in Table 1. In this device, electrons and positrons are accelerated on the wake field of intense and relatively long proton bunches. All the beam bunches involved travel along the axis of an rf structure as shown in Fig. 1. The rf structure is made of a sequence of cavities having a gap g, the outer radius b and the inner radius, the region where the beams travel, equal to a.

A possible lay-out of the WAKEATRON is given in Fig. 2. It is made of two parts which are identical to each other but arranged symmetrically around the crossing point where the two beams collide. One part is to accelerate electrons and the other positrons. Each part is made of a proton source which generates tight bunches in a conventional way. There is an electron beam source at one side and a positron beam source at the other. The acceleration of electrons and positrons takes place in the two sections of the WAKEATRON itself which are identical to each other. One proton bunch is extracted from each side and injected into its respective sections of rf structure immediately followed by either an electron or a positron bunch. This will occur at some repetition rate at which all sources are to be adjusted to.

With reference to Fig.1, the driver bunch, made of protons, creates a wake field so that each proton will lose an amount U of energy per unit

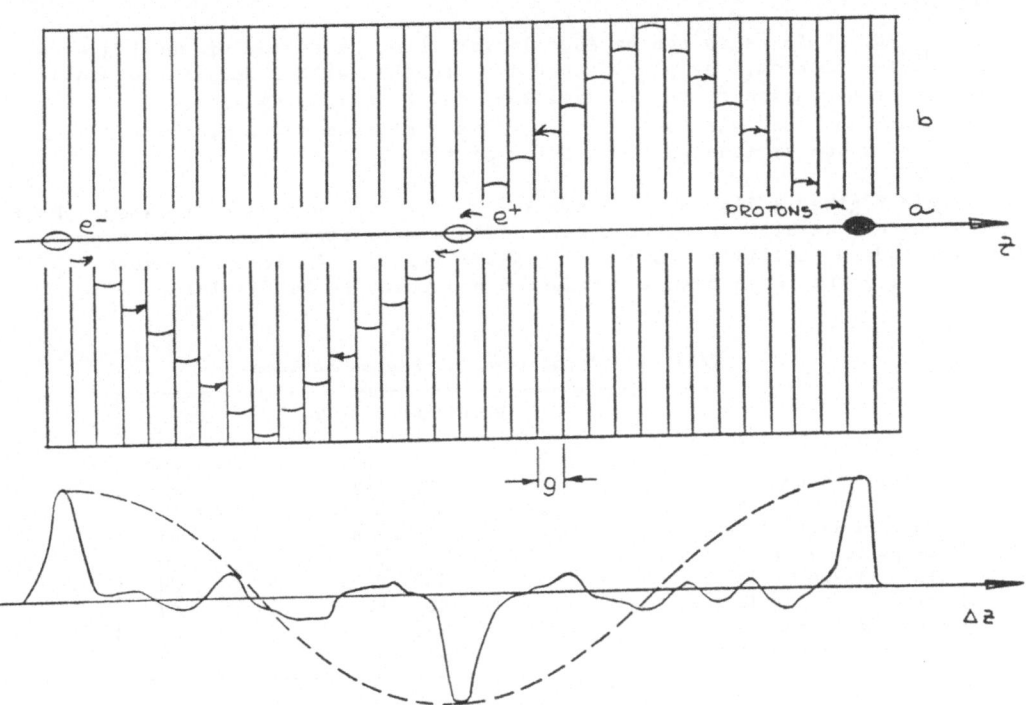

Fig. 1. Section of the WAKEATRON.

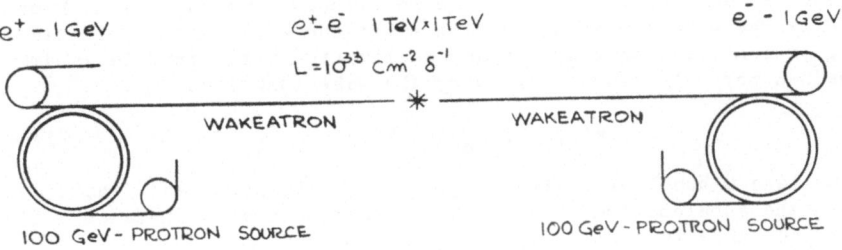

Fig. 2. The Linear Collider based on the principle of the WAKEATRON.

length. This loss is a function of the location of the proton within the bunch. Because protons have a heavier mass than electrons, and because they lose different amount of energy, they move with respect to each other in a process we have called "mixing". Therefore if the mixing frequency is large enough, one is justified to take an average energy loss $\bar{U}$ per unit length, which is the same for all particles. This average value depends on the rms length of the bunch (σ) and on the geometry of the rf cavities (a,b and g).

The wake field behind the proton bunch has an oscillatory behaviour with positive and negative maxima. Very short bunches of electrons or positrons can be located to correspond with the accelerating peaks, where we can define an accelerating gradient U_{max}. This is larger than the average loss $\bar{U}$ per particle in the driver bunch. The "Transformer Ratio" is defined as the ratio $U_{max}/\bar{U}$. This ratio can be made larger than the factor of two suggested by an earlier theorem,[6] by providing driver bunches with a length comparable to the outer dimension of the rf cavities, and by requiring particles in the driver bunch to "mix" with each other. This justifies and explains our suggestion to use protons as the driving particles.

Table 2 gives a list of parameters for a desired source of intense, short proton bunches at high repetition rate. These parameters are rather close to those of hadron facilities (LAMPF II, TRIUMF II, AGS II, European Hadron Facility,...) that have been investigated and found to be feasible.

Table 2. Parameters for Proton Sources

	WAKEATRON Driver		EHF	
Energy	100	GeV	30	GeV
Average Current	240	µAmp	100	µAmp
No. of Protons / Bunch	3.0×10^{11}		2.8×10^{11}	
Bunch Extraction Rate	5.0	kHz	2.25	kHz
Rms Bunch Length	0.3	cm	20.	cm
Beam Average Power	24	MW	3	MW
Total Power for Source	100	MW	30	MW
Power Efficiency	24	%	10	%
Cost of the Facility	400	M$	300	M$

We can assume that protons can be decelerated from an initial value of 100 GeV down to 10 GeV, before they are disposed of. Thus each proton will lose an average total of 90 GeV. If the transformer ratio were 11, electrons and positrons can be accelerated to about 1 TeV.

Calculation of the Wake Fields

A point charge, travelling along the axis of the rf structure shown in Fig. 1, excites an infinite sequence of longitudinal modes, the n-th of which described by the wave number k_n and the amplitude w_n. A test particle following at a distance z receives an acceleration rate, that is an amount of energy gained per unit length, given by the wake function

$$w(z) = \sum_n w_n \cos(k_n z) \qquad (1)$$

If we take a bunch of finite length with N particles and linear density $f(z)$, then the acceleration rate for a test particle at a distance z from the center of the bunch is

$$U(z) = N \sum_n w_n \int_z^\infty f(z') \cos k_n(z'-z) \, dz' \qquad (2)$$

For a gaussian distribution with rms length σ and a particle following at a distance $z \gg \sigma$

$$U(z) = N \sum_n w_n e^{-k_n^2 \sigma^2/2} \cos(k_n z) \qquad (3)$$

We can also calculate the average energy loss $\bar{U}$ per particle in the driver bunch, that is the average of $U(z)$ given by eq. (2) over the distribution function $f(z)$. We have

$$\bar{U} = \frac{N}{2} \sum_n w_n e^{-k_n^2 \sigma^2} \qquad (4)$$

The transformer ratio is then

244

$$r = 2 \frac{\sum_n w_n e^{-k_n^2 \sigma^2/2} \cos(k_n z)}{\sum_n w_n e^{-k_n^2 \sigma^2}} \tag{5}$$

which is a function of the location of the test particle. In the limit of zero bunch length

$$r = 2 \frac{\sum_n w_n \cos(k_n z)}{\sum_n w_n} \tag{6}$$

It is not possible then for this ratio to reach value in excess of 2.

In the limit of long bunches, which is the case we are interested here, it is possible to retain only the lowest mode in the summations at the r.h. side of eq.s (3 and 4). In this case we have

$$U(z) = N w_o e^{-k_o^2 \sigma^2/2} \cos(k_o z) \tag{7}$$

of which the maximum value is

$$U_{max} = N w_o e^{-k_o^2 \sigma^2/2} \tag{8}$$

which we assume to correspond to the location of the test particle. Also

$$\bar{U} = \frac{N}{2} w_o e^{-k_o^2 \sigma^2} \tag{9}$$

and the transformer ratio is

$$r = U_{max}/\bar{U} = 2 e^{k_o^2 \sigma^2/2} \tag{10}$$

In the approximation of long bunches, it is then possible to obtain large transformer ratios. Of course the larger is the ratio the lower is the accelerating gradient, and one is forced to a compromise.

We can estimate the wake amplitude w_o and wave number k_o for the fundamental resonating mode of the cavities shown in Fig. 1. In the case a,g << b, we have

$$w_o = \frac{4e^2}{\pi \epsilon_o b^2} \tag{11}$$

and

$$k_o = \frac{2\pi}{2.61b} \tag{12}$$

We have tested our equations with several runs with TBCI,[7] and we have indeed found agreement between the code results and our single mode approximation model. It seems the model breaks down for $\sigma/b < 0.5$.

Mixing

The energy gain or loss depends on the particle position within the bunch. Particles will acquire different momenta and, because of their heavy mass, will exchange position with respect to each other. We have called this: "mixing". A high rate of mixing is required so that in average, that is over a "mixing period", all the particles lose the same amount of energy. It is because of this effect that large transformer ratios, as calculated before, can be justified.

If the bunch length is comparable to the wavelength of the mode excited, there will be locations within the bunch that are "fixed", that is they correspond to zero energy gain or loss. It is possible to calculate the motion of the particles in the proximity of the fixed points, and calculate the mixing angular frequency Ω in the limit of small displacements. This is given by

$$\Omega = c \sqrt{\frac{Nw_o}{\gamma^2 E} f(\bar{z})} \qquad (13)$$

where $f(\bar{z})$ is the linear density at the location $\bar{z}$ of the fixed point. For a gaussian distribution with centre at $z = 0$

$$f(z) = \frac{\exp(-z^2/2\sigma^2)}{\sqrt{2\pi}\,\sigma} \qquad (14)$$

We expect the shape of the driver bunch to change during the deceleration. Its average energy E, which appears also in the expression (13) for the mixing frequency, will continuously decrease. Therefore the mixing rate, the transformer ratio and the rate of energy loss and gain will also vary. We have verified this with computer codes that simulate both the driver as well as the test beam motion. To compensate for the changing bunch shapes, it may be possible to restore the initial rates by modifying the dimensions of the following rf cavities accordingly.

A Conceptual Design

To achieve large acceleration gradients, one wants cavities with small outer radius b, as one can see from Eq. (11). We propose the following dimensions:

$$a = 1 \text{ mm}$$
$$b = 4 \text{ mm}$$
$$g = 2 \text{ mm}$$

If we take 3×10^{11} protons in a bunch as proposed in Table 2, then the maximum accelerating gradient is

$$Nw_o = 432 \quad \text{MeV/m}$$

For large transformer ratios one has to adjust the bunch length accordingly. A list of possible combinations is given in Table 3. The bunch length is usually small and this, because of the large number of particles involved, is a matter of concern. For instance a transformer ratio of 10 which would allow a final electron beam energy close to 1 TeV, requires $\sigma = 3$ mm.

We report in Table 2 the performance of two proton sources. The first is the one we require for the WAKEATRON driver, and the second is the proposed European Hadron Facility (EHF). It is possible to start from the latter to which then one adds an extra ring to accelerate protons from 30 to 100 GeV. A possible way to shorten the bunches is to adjust the transition energy of the last ring to about the beam energy at extraction.

Table 3. WAKEATRON Performance versus Transformer Ratio

r	σ/b	U_{max} MeV/m	$\bar{U}$ MeV/m	ΔE_{final} TeV	L km	P_e MW	P_e/P_p %	P_e/P_{total} %
2	0.000	432.0	216.00	0.18	0.4	0.72	3.	0.7
5	0.562	172.8	34.56	0.45	2.6	1.80	7.5	1.8
8	0.691	108.0	13.50	0.72	6.7	2.88	12.	2.9
10	0.745	86.4	8.64	0.90	10.4	3.60	15.	3.6
13	0.803	66.5	5.11	1.17	17.6	4.68	19.5	4.7
16	0.847	54.0	3.38	1.44	26.7	5.76	24.	5.8
20	0.891	43.2	2.16	1.80	41.7	7.20	30.	7.2
25	0.933	34.6	1.38	2.25	65.0	9.00	37.5	9.0
30	0.966	28.8	0.96	2.70	93.8	10.80	45.	10.8
36.35	1.000	23.8	0.65	3.27	137.4	13.08	54.5	13.1

The power efficiency of the proton source is the ratio of the final, average power in the proton beam to the total power required to operate the facility. This efficiency is already about 10 %, a relatively large figure, for the EHF and we expect an even larger number for the WAKEATRON driver. This is an interesting and useful feature: the larger the proton beam current the more efficient is the source. Indeed, one can reach a situation where most of the power involved is in the beam and in the rf system that provides acceleration, usually in a one-to-one ratio. The power of the magnet system does not depend on the beam current and will become only a relatively small fraction of the total.

The cost of the source can also be estimated. It is relatively modest compared to larger enterprises like, for instance, the SSC, the 20×20 TEV^2 proton-proton super collider.

Table 3 summarizes our results, based on the single mode approximation. For a given transformer ratio, we have estimated the required rms bunch length, the maximum accelerating gradient U_{max} and the average loss rate $\bar{U}$ per particle in the driver bunch. Assuming that protons are decelerated from 100 down to 10 GeV, we have also calculated the total energy gain ΔE_{final} for beams of electrons and positrons as well the distance L that all the beams have to travel, which then gives roughly the length of the WAKEATRON. We show the average power P_e of the electron (positron) beam for 5×10^9 particles in a bunch at the repetition rate of 5×10^3 bunches per second as shown in Table 1. The energy transfer efficiency from one beam to the other is given by the ratio P_e/P_p, where P_p is the average power in the proton beam which we have taken to be 24 MW, as also shown in Table 2. The transfer efficiency is quite large and useful. This can be thought as the equivalent of the conventional klystron efficiency, and it is actually the energy recovery efficiency of the proton beam. The overall efficiency is obtained by multiplying the transfer efficiency with the efficiency to operate the proton source, which, as also shown in Table 2, we assume to be 24 %. The overall efficiency is given on the last column of Table 3. For a transformer ratio of 11, it is an interesting and useful 4 %.

We have chosen the energy of the proton beam large enough so that it is possible to accelerate electrons to 1 TeV with a low transformer ratio of 10-12. Also we expect that the length of the proton bunches decreases with the energy, and one requires short bunches, as we have seen. On the other hand, the mixing frequency decreases with the beam energy; and if one wants to take advantage of the mixing process, then he is forced to a compromise. We have calculated the mixing period T at the initial energy of 100 GeV for an rms bunch length of 3 mm and the parameters described above. We have found a reasonable 3 μsec, which corresponds to a travel length of about 1 km. Preliminary results of computer simulations of the particle motion show that indeed this mixing period is adequate.

Acknowledgements

The author of this paper is greatly indebted to P. Schoessow and J. Simpson for their assistance in the numerical calculations of details like wake field and mixing of the WAKEATRON concept. The author is also grateful to T. Weiland for conversations he had with him on the nature and parametric dependence of the wake fields. Because of that conversation, it has been possible to correct an original mistake.

References

1. W. Schnell, "Consideration of a Two-Beam RF Scheme for powering an RF Linear Collider. CLIC Note #7, CERN, Nov. 11, 1985.

2. J. S. Wurtele, "On Acceleration by the Transfer of Energy between two Beams", Laser Acceleration of Particles, Malibu, California, 1985, AIP Conf. Proceed., No. 130, p. 305.

3. A.G. Ruggiero, "The Wakeatron: Accleration of Electrons on the Wake Field of a Proton Bunch", Laser Acceleration of Particles, Malibu, Californai, 1985, AIP Conf. Proceed. No. 130, p. 458.

4. A.G. Ruggiero, "The WAKEATRON: Acceleration of Electrons on the Wake Field of a Proton Bunch", The Generation of High Fields, Frascati, Italy 1984, Workshop Proceedings, p. 128.

5. G. Voss and T. Weiland, DESY M-82-10, April 1982 and DESY 82-074, Nov. 1982.

6. R.D. Ruth et al. "A Plasma Wake Field Accelerator", SLAC-PUB-3374, July, 1984.

7. T. Weiland, "On the Numberical Solution of Maxwell's Equations and Applications in the Field of Accelerator Physics", Particle Accelerators, 1984, Vol. 15, p. 245.

†This work was supported by the U.S. Department of Energy, Division of High Energy Physics, Contract W-31-109-ENG-38.

ANALYTIC TOOLS FOR SOLVING NONLINEAR PROBLEMS

IN PARTICLE ACCELERATORS : A REVIEW AND AN EXAMPLE **

Stefania P. Petracca* and Innocenzo M. Pinto[+]

* Dip. di Matematica ed Applicazioni, Università di Napoli
v. Mezzocannone, 8 - 80134 Napoli (Italia)

[+] Dip. di Elettronica, Università di Napoli
v. Claudio, 21 - 80125 Napoli (Italia)

SUMMARY

   Motivations for using analytical tools in studying particle accelerators are suggested, available techniques are reviewed, and a computational example is presented.

INTRODUCTION

   The free-evolution and interaction of charged bunches and beams in particle accelerators is ruled by *nonlinear* partial integrodifferential equations. Representative examples include standard,[1] as well as novel designs (e.g., beat-wave,[2] and plasma-wake,[3] accelerators) and proposals (e.g., rectified-light,[4] devices).

   The present-day standard tools for attacking these problems are numerical codes,[5] , producing huge bulks of dull numbers, after several runs, to be carefully examined before one can draw any conclusion about the role played by the various design parameters. Such a knowledge is of paramount importance, prior to construction, in view of the tremendous cost of the hardware. Furthermore, numerical methods are inherently plagued by finite lattice/mesh size effects, yielding, e.g., unreliable results in the asymptotic (high-energy) spectrum.

   As a complementary or alternative approach, we shall briefly review in Sect.1 some *general* and *systematic* techniques, based on the methods of Functional Analysis, for obtaining accurate, highly readable *formal* (i.e., non-numerical) solutions to nonlinear problems. As an application, a simple one-dimensional space-charge nonlinear evolution equation will be derived and solved in Sect.2

** Work sponsored in part by MPI.

1. ANALYTIC TOOLS : A REVIEW

Solving the nonlinear (functional) problem:

$$N\{f,\phi\} = 0 \tag{1}$$

wherein $\phi(\cdot)$ is a prescribed forcing term,* means finding and efficiently representing (or approximating) the *unknown* nonlinear functional $f\{\phi\}$.

While several well-known *ad-hoc* techniques do exist for solving *special classes* of nonlinear equations (including, e.g., Bäklund transforms,[6], IST,[7], multiple-scale methods,[8], etc.) here we shall focus on *general* and *systematic* approaches.

The basic *function* representation/approximation tools are the (*local*) Taylor-power-series and the (*global*) orthogonal-Fourier expansion. Their *functional* counterparts are the Volterra,[9,10] and Wiener-Cameron-Martin,[11] expansions.

The Volterra series (VS) expansion of an *analytic* functional is:

$$f\{\phi\} - f\{\phi_s\} = \sum_{n=1}^{\infty} f^{(n)}\{\phi-\phi_s\} \tag{2}$$

wherein ϕ_s, $f\{\phi_s\}$ is a known exact solution pair, viz., $N\{f\{\phi_s\},\phi_s\} = 0$ (the usual case is $\phi_s = f\{\phi_s\} = 0$.). The n-term in (2) is the functional counterpart of the n-term of the standard Taylor-power-series:

$$f^{(n)}\{\phi-\phi_s\} = \frac{1}{n!} \cdot \frac{\delta^n f}{\delta\phi(u_1)...\delta\phi(u_n)} \Bigg|_{\phi=\phi_s} \overset{n}{\underset{i=1}{*}} \{\phi(u_i)-\phi_s(u_i)\} \tag{3}$$

wherein the n-order Frèchet derivative is called n-order Volterra Green's function, and $*$ denotes convolution. As a practical recipe for computing the kernels, one lets:

$$\phi(x) - \phi_s(x) = A_1\delta(x-x_1) + ... + A_n\delta(x-x_n) + ... \tag{4}$$

into (2,3 and 1), then differentiates once with respect to $A_1, ... , A_n$, and and finally sets $A_1=A_2=...=A_n = 0$, $n=1,2,...$, so as to obtain a *hierarchy of linear equations* in the unknown kernels,[9,10].

A natural extension of the VS is the (functional) Lagrange-Bürmann expansion,[12] viz., a VS whose argument is a properly chosen functional of $\phi-\phi_s$, instead of $\phi-\phi_s$ itself. A skillful choice of the latter functional may lead to a dramatic convergency improvement,[13].

The Wiener-Cameron-Martin (WCM) expansion,[11], on the other hand, can be heuristically introduced as follows. One replaces the forcing function $\phi(x)$, $a\leq x\leq b$, with the n-*tuple* $\{\phi_1,...,\phi_n\}$, $\phi_k = \phi[a + (k-1)(b-a)/(n-1)]$, $k = 1,2,...,n$, and views the WCM expansion as a suitable limit, as $n \to \infty$ of the n-dimensional Hermite orthogonal expansion of the *function* $f(\phi_1,...,\phi_n)$.

---

* Extension to the free-evolution case is straightforward,[9,10].

250

Solution techniques should be obviously judged on the basis of the achievable trade-off between *generality* and *accuracy*, on one hand, and *computational ease* on the other.

The WCM expansion has been hitherto used for solving only a limited number of problems,[14,15], mainly because one is usually led to the (untrivial,[16]) formal evaluation of Wiener integrals,[17].

The VS, on the other hand does show the following nice features: *i*) extends the Green's function philosophy of *linear* problems to the *nonlinear realm*; *ii*) requires only standard *linear tools* to be worked out to all orders; *iii*) can be *fully* implemented using symbolic manipulation languages (e.g., MACSYMA©, SMP©, etc.); *iv*) can easily accomodate *stochastic* problems, where, e.g., according to (2 and 3), averages and correlations of f{φ} are represented as *known* functionals of the (known) statistical cumulants of φ(·).

Almost obvious drawbacks of the VS approach, however, stem from its (functional) Taylor-series nature, namely: *i*) *truncated* VS expansions may exhibit *secular* behaviour, i.e., growing unbounded as x or t tend to infinity; *ii*) VS may be *poorly* (if not at all) convergent, when *large* nonlinearities and/or forcing terms are involved. Both these shortcomings can be circumvented as follows: *i*) *systematic*,[18] procedures are available for getting rid of secular terms either before (Rytov's scheme) or mean-while (Lighthill's scheme) or after (Pritulo's scheme) solving the problem into a VS; *ii*) *analytic continuation* is needed to exploit the *global* properties of the sought solution starting from its *local* VS representation (analytic functional element). This can be conveniently accomplished,[19] by recasting the VS solution into a functional *rational (Padé) approximant*,[20].

We conclude this short survey by addressing the interested Reader to the References, where he will find all formal details and outfits fully developed.

2. A COMPUTATIONAL EXAMPLE

Here we shall be concerned with the possibly simplest model of space-charge nonlinear evolution equation, describing the longitudinal dynamics of a charged line-bunch traveling along the axis of a straight perfectly conducting pipe with uniform cross section. Space-charge phenomena do play a key-role in high beam-current machines.

Let $\{\zeta,\tau\}$ the bunch-frame, coincident at $t = 0$ with the lab-frame $\{z,t\}$, and drifting thereafter with constant velocity v_o; and let $\beta_o = v_o/c$, $\gamma_o = (1-\beta_o^2)^{-\frac{1}{2}}$, c being the velocity of light *in vacuo*.

The bunch dynamics is most conveniently described in terms of the following quantities:

$$\lambda(z,t) = \textit{line density} \tag{5}$$

$$m = \frac{v(z,t) - v_o}{v_o} = \textit{relative momentum spread} \tag{6}$$

The initial data of the free-evolution problem are:

$$\lambda(z,t=0) = f(z) \quad ; \quad v(z,t=0) = v_o \tag{7}$$

Starting from first principles, the following equations are readily established:

$$\frac{dm}{d\tau} = -\frac{q}{m} v_o^{-1} \frac{\partial \phi}{\partial \zeta} \tag{8}$$

Lorentz-force equation (transverse forces neglected; quasi static approximation, viz., sub-luminal intra-bunch velocities; collisions ignored; q, m ≡ electron charge and mass);

$$\frac{\partial \lambda}{\partial \tau} = -\frac{\partial}{\partial \zeta} (\lambda \frac{d\zeta}{d\tau}) \tag{9}$$

Liouville-continuity equation;

$$\frac{d\zeta}{d\tau} = \beta_o c (1 - \alpha_1 \gamma_o^2) m \tag{10}$$

relative intra-bunch velocity equation,[21],*;

$$\phi(\zeta,\tau) = \gamma_o^{-1} \phi_o \lambda(\zeta,\tau) \tag{11}$$

space-charge potential approximation, accurate provided (bunch-length) ≫ ≫ (pipe diameter),[22].

From eq.s (8 to 11) one easily derives the following set of coupled nonlinear partial differential equations:

$$\frac{\partial m}{\partial \tau} = -m \frac{\partial m}{\partial \zeta} \beta_o c (1 - \alpha_1 \gamma_o^2) - \frac{q}{m} \frac{\phi_o}{v_o \gamma_o} \frac{\partial \lambda}{\partial \zeta} \tag{12}$$

$$\frac{\partial \lambda}{\partial \tau} = -\beta_o c (1 - \alpha_1 \gamma_o^2) \frac{\partial}{\partial \zeta} (m \lambda) \tag{13}$$

Equations (12 and 13) can be conveniently re-scaled, so as to cast them into an *universal* form, leaving all machine-dependent parameters in the scaling equations, and involving only dimensionless quantities, by letting:

$$M = \ell^{\frac{1}{2}} \left\{ \frac{q|\phi_o|}{m v_o \beta_o c \gamma_o |1 - \alpha_1 \gamma_o^2|} \right\}^{-\frac{1}{2}} \cdot \text{sgn}(1 - \alpha_1 \gamma_o^2) \cdot m \quad ; \quad \Lambda = \ell \lambda \tag{14}$$

$$Z = \ell^{-1} \zeta \quad ; \quad T = \ell^{-3/2} \left\{ \frac{m v_o \gamma_o}{q \beta_o c |\phi_o (1 - \alpha_1 \gamma_o^2)|} \right\}^{-\frac{1}{2}} \tau \tag{15}$$

---

*
The *momentum compaction factor*, α_1, is defined according to,[22]: $R/R_o = 1 + \alpha_1 m$, R, R_o being the (betatron-oscillation averaged) orbit radii corresponding to the velocities v, v_o respectively, in storage rings.

Hence :

$$\frac{\partial M}{\partial T} = -\text{sgn}(\phi_o)\,\text{sgn}(1-\alpha_1\gamma_o^2)\,\frac{\partial \Lambda}{\partial Z} - M\,\frac{\partial M}{\partial Z} \qquad (16)$$

$$\frac{\partial \Lambda}{\partial T} = -\frac{\partial}{\partial Z}\,(M\Lambda) \qquad (17)$$

Note that, according to whether:

$$\text{sgn}(\phi_o)\,\text{sgn}(1-\alpha_1\gamma_o^2) \gtrless 0 \qquad (18)$$

eq.s (16 and 17) make an *hyperbolic* or *elliptic* problem, describing, re-spectively, *debunching* and *selfbunching*.

Letting further,[*]

$$M = \frac{\partial \Gamma}{\partial Z} \qquad (19)$$

into (16 and 17) one finally obtains:

$$\Lambda = -\text{sgn}(\phi_o)\,\text{sgn}(1-\alpha_1\gamma_o^2)\,(\frac{\partial \Gamma}{\partial T}+\Gamma\,\frac{\partial \Gamma}{\partial Z}) \qquad (20)$$

$$\frac{\partial}{\partial T}\,(\frac{\partial \Gamma}{\partial T}+\Gamma\,\frac{\partial \Gamma}{\partial Z}) = -\frac{\partial}{\partial Z}\,\{\,\frac{\partial \Gamma}{\partial Z}\,(\frac{\partial \Gamma}{\partial T}+\Gamma\,\frac{\partial \Gamma}{\partial Z})\,\} \qquad (21)$$

with initial conditions:

$$\Gamma(Z,T=0) = 0 \quad ; \quad \frac{\partial \Gamma}{\partial T}\bigg|_{T=0} = -\text{sgn}(\phi_o)\,\text{sgn}(1-\alpha_1\gamma_o^2)\cdot\ell\cdot f(Z) \quad =: F(Z) \qquad (22)$$

The VS solution of (21 with 22), is obtained by first letting (*à la Rytov*),[24] :

$$\Gamma(Z,T) = \exp\left[\psi(Z,T)\right] - 1 \qquad (23)$$

and then expanding $\psi(Z,T)$ into a VS of argument $F(Z)$, viz.:

$$\psi(Z,T) = \sum_{n=1}^{\infty} \psi^{(n)}(Z,T) \qquad (24)$$

After a few lengthy but straightforward manipulations,[24] the follow-ing solution is obtained:

$$\psi^{(1)}(Z,T) = T\,F \qquad (25)$$

$$\psi^{(2)}(Z,T) = -T^2F^2/2 - T^3\left[F\,\ddot{F} + 2\,\dot{F}^2\right]/6 \qquad (26)$$

---

[*] It can be shown that Γ is the problem's scaled dimensionless action function,[23].

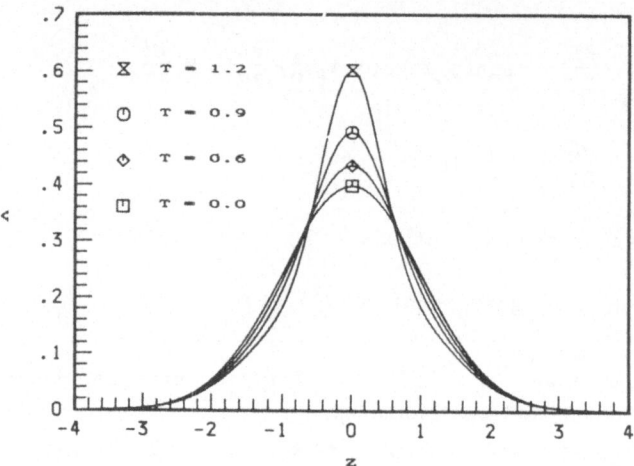

Fig.1 – Scaled density. Self-bunching case.

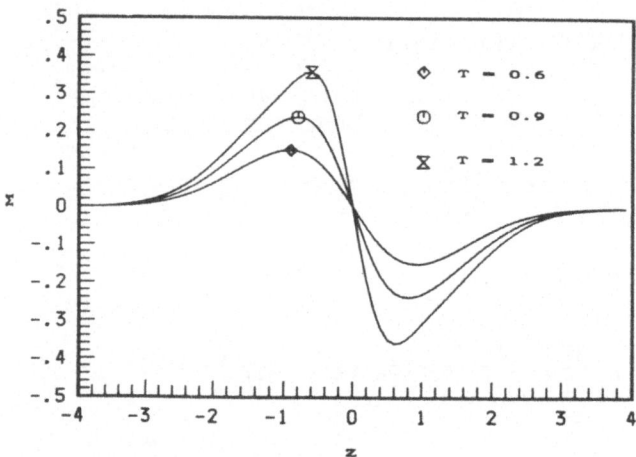

Fig.2 – Scaled momentum. Self-bunching case.

$$\psi^{(3)}(Z,T) = T^3 F^3/3 + T^4 \left[2F\dot{F}^2 + F^2\ddot{F} \right]/4 +$$
$$+ T^5 \left[8F\dot{F}^2 + 14F\dot{F}\dddot{F} + 37\dot{F}^2\ddot{F} + F\ddddot{F} \right]/120 \qquad (27)$$

Note that (25 to 27, etc.) are *completely general*, i.e., valid for an *arbitrary* initial profile F(Z), albeit analytic.

The accuracy of the VS solution has been favourably compared,[24] with available numerical results,[23].

As an illustration, the scaled density and momentum spread for an initial condition:

$$F(Z) = (2\pi)^{-\frac{1}{2}} \exp(-Z^2/2) \qquad (28)$$

(gaussian bunch) have been computed using (25 to 28), and are shown in 5 Fig.s 1 to 4, for the selfbunching and the debunching case.

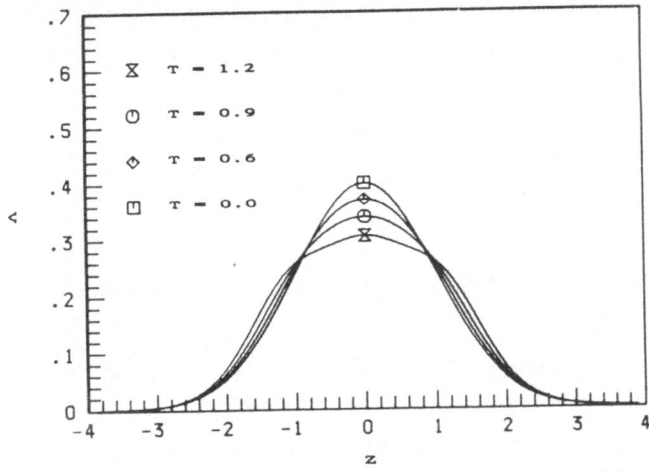

Fig. 3 – Scaled density. De-bunching case.

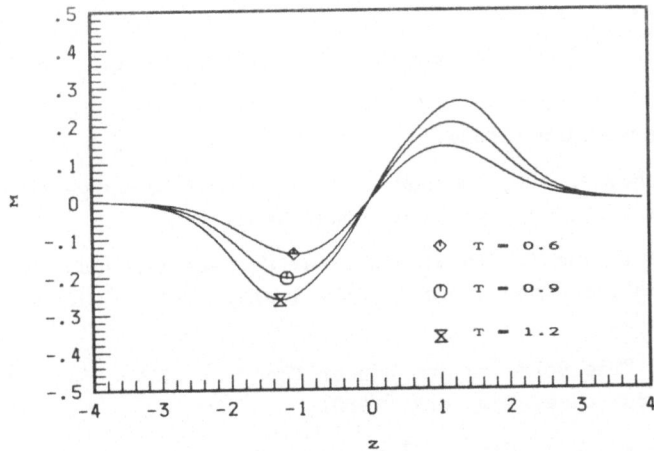

Fig. 4 – Scaled momentum. De-bunching case.

The above should be intended as an extremely simplified preliminary test-problem, though it does show both the applicability of the method, and the way to more or less straightforward extensions, including, e.g., longitudinal and transverse bunch-bunch interactions, collisions (intra-bunch interactions), more general geometries, etc.

CONCLUSIONS

We believe that analytic functional techniques could provide an extremely valuable general and systematic approach to gain physical insight into a number of Particle Accelerator problems where nonlinear and/or stochastic effects are relevant. Accordingly, we do recommend them to be further probed by the Researchers' Community.

ACKNOWLEDGEMENTS

     Prof.s G. Franceschetti and V.G. Vaccaro are kindly acknowledged for a number of stimulating discussions and suggestions.

REFERENCES

1. Proceedings of the US-CERN School on Nonlinear Dynamics, S. Margherita di Pula (Italy), 1985.

2. C.M. Tang and P. Sprangle, Space-Charge Waves in Laser Beat Wave Accelerators, Phys. Fluids (USA) 28:1974 (1985).

3. P. Chen et al., Acceleration of Electrons by Interaction of a Bunched Electron Beam with a Plasma, Phys. Rev. Lett. 54:693 (1985).

4. A.F. Pisarev, Particle Focusing and Deflection in a Ring Accelerator with Rectified Light, Sov. Phys. Tech. Phys. 24:697 (1977).

5. T. Weiland, On the Numerical Solution of Maxwell Equations and Applications in the Field of Particle Accelerators, Particle Acc. (GB) 15:245 (1984).

6. R.M. Miura (ed.), "Bäklund Transforms", Springer Verlag, Berlin (1976).

7. F. Calogero and A. de Gasperis, "Spectral Transforms and Solitons", North Holland, Amsterdam (1982).

8. N.N. Bogolijubov et al., "Methods of Accelerated Convergence in Nonlinear Mechanics", Springer-Hindustan, Dehli (1976).

9. G. Franceschetti and I. Pinto, Nonlinear Propagation and Scattering : Analytic Solution and Symbolic Code Implementation, J. Opt. Soc. Am. A2:997 (1985).

10. M. Schetzen, "The Volterra and Wiener Theories of Nonlinear Systems", Wiley-Interscience, New York (1980).

11. R.H. Cameron and W.T. Martin, The Orthogonal Development of Nonlinear Functionals in Series of Fourier Hermite Functionals, Ann. Math. (USA) 48:385 (1947).

12. D. Gorman and J. Zaborsky, Functional Lagrange Expansion in State Space and the s-Domain, IEEE Trans. Automatic Control AC-11:498 (1966).

13. R. Bellman, Lagrange Expansion for Operators, Bull. Am. Math. Soc. 71:496 (1965).

14. H. Ogura and J. Nakayama, Initial Value Problem of the One Dimensional Wave Propagation in a Homogeneous Random Medium, Phys. Rev. A 11:957 (1975).

15. H. Nakazawa, Nonlinear Stochastic Differential Equations in Statistical Physics and Integral Representation, Progr. Th. Phys (Japan) 56:1411 (1976).

16. A.M. Arthur (ed.), "Functional Integration and its Applications", Oxford Univ. Press, Oxford (1975).

17. I.M. Gel'fand and A.M. Yaglom, Integration in Functional Space and its Applications in Quantum Physics, J. Math. Phys. 1:48 (1960).

18. A. Nayfeh, "Perturbation Techniques", Wiley-Interscience, New York (1973).

19. A. Cuyt, "Padé Approximants for Operators", Springer-Verlag, Berlin (1984).

20. G. Franceschetti and I. Pinto, Padé Approximants in Functional Electromagnetics, Proc. VII Nat.1 Meeting on Applied Electromagnetics, Trieste (Italy), 1986.

21. E. Ciapala et al., The Variation of γ_t with $\Delta p/p$ in the CERN-ISR, IEEE Trans. Nucl. Sci. NS-26:3571 (1979).

22. M.H. Blewett et al. (ed.s), Theoretical Aspects of the Behaviour of Beams in Accelerators and Storage Rings, CERN Rep.t 77-13 (1977).

23. U. Funk et al., Mathematical and Numerical Methods for the Nonlinear Hyperbolic Propagation Problem: A Preliminary Treatment of Collective Space Charge Effects in Accelerator Physics, Kernforschungsanslage Jülich Rep. 1772 (1982).

24. S. Petracca and I. Pinto, Volterra Series Solution of Particle de-Bunching Equation, Lett. Nuovo Cimento 44:573 (1985).

APPROXIMATE SOLUTIONS OF THE EQUATION
FOR THE LONGITUDINAL MOTION OF PARTICLES
IN AN RFQ ACCELERATOR

M.Leo[+], R.A.Leo[+], M.Puglisi[*], C.Rossi[*], G.Soliani[+] and
G.Torelli[o]

+Dipartimento di Fisica dell'Universita, 73100 Lecce, Italy
and INFN - Gruppo di Lecce, Sezione di Bari, Italy
*Dipartimento di Fisica Nucleare e Teorica dell'Universitá
27100 Pavia, Italy and INFN - Sexione di Pavia, Italy
oDipartimento di Fisica dell'Universitá, 56100 Pisa, Italy
and Lab. di San Piero (INFN), Pisa, Italy

ABSTRACT

We obtain approximate analytical solutions of the equation for the
longitudinal motion of a particle in an RFQ accelerator. Our procedure is
exploited in the case in which the accelerated particles are protons. Our
analytical results are in good agreement with those arising from the
numerical integration of the above mentioned equation.

1. INTRODUCTION

As is well know, an important idea in the study of a linear accelerator
structure is due to Kapchinskii and Teplyakov (K–T), who proposed in 1970
a radio frequency quadrupole (RFQ) device where the RF field can be used
both for acceleration and transverse focusing /1,2,3/. A schematic drawing
of an RFQ four vanes resonator is shown in Fig.1.

To the best of our knowledge, so far the equation of motion governing
the beam dynamics of a particle in the K–T framework has been investigated
by means of numerical techniques only. Here we outline an analytical
procedure to obtain an explicit approximate solution of the equation for
the longitudinal motion, which reads

$$\frac{d^2z}{dt^2} = AC \sin(kz)\sin(\omega t + \phi), \tag{1}$$

where

$$AC = (qVAk)/(2M), \tag{2}$$

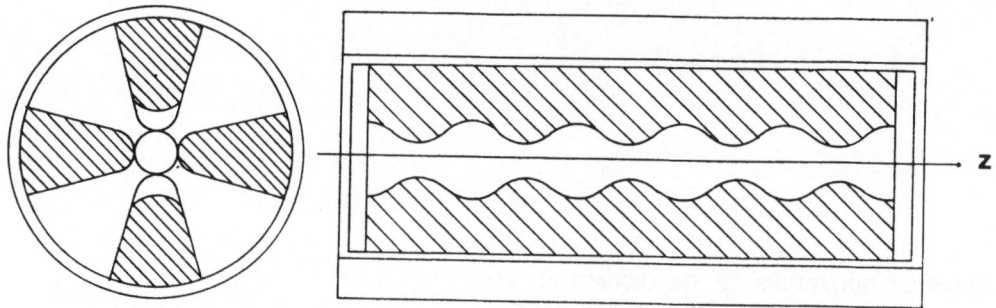

Fig. 1. Schematic (front and side) view of an RFQ resonator.

V is the intervane voltage, and k, ω, φ, q and M are the wave number, the angular frequency of the machine, the phase, the charge and the rest mass of the injected particle, respectively.

We observe that from the synchronism condition between the particle and the RF field, it follows that the integration of Eq.(1) for each value of k yields the cell length z in such a way that kz = π, which is just the half period of the RF field itself /4/.

2. APPROXIMATE SOLUTIONS OF THE EQUATION OF MOTION

Eq.(1) is a nonlinear second order ordinary differential equation with a periodic coefficient, which can be suitably studied resorting to the averaging method of Bogoliubov, Krylov and Mitropolsky (BKM) /5,6/. This procedure can be used to find approximate solutions of equations in the standard form

$$\frac{dx_j}{dt} = \epsilon X_j (t, x_1, x_2, \ldots, x_n) \quad (j = 1,2,\ldots,n), \tag{3}$$

in which ε is a small parameter and X_j can be expessed by

$$X_j (t, x_1, x_2, \ldots, x_n) = \Sigma_\nu e^{i\nu t} X_{j\nu}(x_1, x_2, \ldots, x_n) \tag{4}$$

where ν are constant frequencies.

In order to put Eq.(1) in the form (3), let us introduce the variables

$$\tau = \omega t, \qquad \eta = kz - \tau. \tag{5}$$

Then Eq.(1) can be written as

$$\frac{d^2\eta}{d\tau^2} = \frac{\epsilon^2}{2} [\cos (\eta-\phi) - \cos(\eta + 2\tau +\phi)], \tag{6}$$

where the (adimensional)parameter ε is defined by

$$\epsilon = \left(\frac{kAC}{\omega^2} \right)^{\frac{1}{2}}, \tag{7}$$

which is required to be less than one.

260

Using a vectorial notation, Eq.(6) becomes

$$\frac{dX}{d\tau} = \epsilon F(\tau, X),\tag{8}$$

where X and F are the vectors

$$X = (X_1, X_2)^T,\tag{9}$$

and

$$F(\tau,X) = (X_2, \tfrac{1}{2}[\cos(X_1-\phi)-\cos(X_1+2\tau+\phi)])^T,\tag{10}$$

with

$$X_1 = \eta, \quad X_2 = \frac{dX_1}{\epsilon d\tau},\tag{11}$$

where T denotes the transposed operation. As explicit function of τ, F is of period π.

Following the BKM method, now we make the averaging with respect to the variable τ over the interval $(0,\pi)$. We denote by

$$\xi = (\xi_1, \xi_2)^T\tag{12}$$

the solution of the resulting equation, namely

$$\frac{d\xi}{d\tau} = \epsilon F_0(\xi),\tag{13}$$

where

$$F_0(\xi) = (\xi_2, \tfrac{1}{2}\cos(\xi_1 - \phi))^T.\tag{14}$$

Assuming the initial conditions

$$X_1(0) = \xi_1(0), \quad X_2(0) = \xi_2(0),\tag{15}$$

the function ξ can be considered as the first approximation solution of Eq.(8). Solving (13), we get

$$\xi_1 = 2\arcsin[m \operatorname{sn}(\frac{\epsilon}{\sqrt{2}}\tau + \alpha; m^2)] + \phi + \frac{\pi}{2},\tag{16}$$

where sn stands for the Jacobi elliptic function of parameter m, and

$$m^2 = \sin^2 \tfrac{1}{2}(\phi + \frac{\pi}{2}) + \frac{1}{2\epsilon^2}(k\frac{v_0}{\omega} - 1)^2,\tag{17}$$

$$\alpha = -\operatorname{sn}^{-1}[\frac{1}{m}\sin\tfrac{1}{2}(\phi + \frac{\pi}{2}); m^2],\tag{18}$$

being v_0 the initial velocity of the particle.

261

An improved approximate solution of Eq.(8) can be written as

$$X = \xi + \sum_n \epsilon^n \chi_n (\tau,\xi),$$ (19)

where the vectors χ_n have to be determined in such a way that Eq.(8) is satisfied at any order in the parameter ϵ, with the initial conditions $\chi_n(0, \xi) = 0$. As it is shown in Ref. 7, we derived the following higher order contributions

$$\chi_1 = (0, -\frac{1}{4} \sin(\xi_1 + 2\tau + \phi) + \frac{1}{4} \sin (\xi_1 + \phi))^T,$$ (20)

$$\chi_2 = (\frac{1}{8} \cos(\xi_1+2\tau+\phi) - \frac{1}{8} \cos(\xi_1+\phi) + \frac{1}{4}\tau \sin(\xi_1+\phi),$$
$$\frac{\xi_2}{4} [\frac{1}{2} \sin(\xi_1+2\tau+\phi) - \frac{1}{2} \sin(\xi_1+\phi) - \tau\cos(\xi_1+\phi)])^T,$$ (21)

$$\chi_3 = (\frac{\xi_2}{8}[-\cos(\xi_1+2\tau+\phi)+\cos(\xi_1+\phi)-2\tau\sin(\xi_1+\phi)-2\tau^2\cos(\xi_1+\phi)],$$

$$\frac{1}{8} \{\frac{1}{16} \cos2(\xi_1+\phi) - \frac{1}{16} \cos2(\xi_1+2\tau+\phi) + \frac{1}{4} \cos(2\xi_1+2\tau) - \frac{1}{4}\cos2\xi_1$$

$$+ \frac{1}{4} \cos(\xi_1+\phi)\cos(\xi_1+2\tau+\phi) - \frac{1}{4} \cos^2(\xi_1+\phi) + \frac{\tau}{2} \sin2\xi_1$$ (22)

$$+ \sin(\xi_1+\phi)[-\frac{\tau}{2} \cos(\xi_1+2\tau+\phi) + \frac{1}{4} \sin(\xi_1+2\tau+\phi) - \frac{1}{4}\sin(\xi_1+\phi)]$$

$$+ \frac{\tau^2}{2} \cos2\xi_1 - \xi_2^2[-\frac{1}{2} \sin(\xi_1+\phi) + \frac{1}{2}\sin(\xi_1+2\tau+\phi) - \tau\cos(\xi_1+\phi)$$

$$+ \tau^2\sin(\xi_1+\phi)]\})^T.$$

Inserting (20), (21) and (22) in(19), we get a third order approximate solution of Eq.(8).

3. CONCLUSION

The axial design of an RFQ depends upon the solution of the equation of particles motion and "normally" this solution is obtained by means of numerical methods. It should be pointed out that these may be very sofisticated and time consuming as, for instance, the Runge-Kutta "predictor-corrector" method.
Nevertheless, in spite of the high degree of complexity of the procedures, still troubles arise concerning the accuracy of the solution. In fact, since the parameters that define the motion should change from cell to cell, it follows that the program should restart each time with initial conditions which cannot coincide with the final status previously defined. Moreover as it is non predicable a priori where a non synchronous particle will reach the end (or the beginning for retrograde motion) of each cell, then the definition of the final time may request extremely small steps of integration in contrast with the nature of the program, which tends to find

out the integrating step in accordance with the smoothness of the function. Last but not least is the integration time that plays an important role in the cost of the whole design. For the above reasons the availability of an analytical solution was always deemed very important, since it directly leads to the design as soon as all the input parameters are given.

The third order analytical approximate solution found in Sect.2 was tested from two different point of view. Precisely, the solution always converges with the desired accuracy when the parameter ϵ remains less than one; practically this condition is always obeyed and a very simple calculation would show that, in a nonrelativistic approximation, $\epsilon = 0.5(AV/V_0)^{1/2}$ where V_0 is the injection voltage in a cell (eventually the first) which exibits a static gain equal to AV. It follows also that the accuracy of the solution does not depend on the mass of the accelerated particle. Both for testing the validity of the analytical solution and checking its computer formulation, many solutions have been tested with the usual algebraic test where each cell is divided into a suitable number (40) of intervals and the acceleration in each interval is held constant.

It is also important to note that only an analytical solution offers the possibility of optimizing a structure in a reasonably short computation time. Not to be ignored is the possibility of showing the different behavior of structures which differ very little from one another.

The solution obtained in Sect.2 can be compared with the corresponding one coming from the numerical integration of Eq.(8). The last has been integrated by means of a four order Runge–Kutta predictor–corrector method/8/. We remark that the stability of the numerical algorythm exploited by us has been subjected to several checks. From among these we mention the comparison between the increasing of the energy per cell, calculated numerically, and the corresponding value coming from the analytical solution.

From (7) and (2) we see that the parameters varying along the accelerator are the wave number $k = 2\pi/\beta\lambda$ and the efficiency A. On the other hand, since the particle velocity (in the synchronism condition) is growing, then the value of k tends to decrease accordingly.

As particles to be accelerated we have taken protons, dealing with an RFQ accelerator with N = 150 cells in which A and ϕ are given by

$$A = 0.05 \{1 + (9/2) [1 + \cos(\pi + (N-1)(\pi/119))]\},$$

$$\phi = -85^\circ + (55^\circ/119) (N-1) ,$$

for $1 \leq N \leq 120$, and

$$A = 0.5, \quad \phi = -30^\circ, \quad \text{for} \quad 120 \leq N \leq 150.$$

Starting from an injection energy of 10 KeV, a final energy of about 1.5 MeV is achieved. The length of the whole machine turns out to be of 3.51 meters.

The value of the physical quantities calculated analytically agree with the corresponding ones determined via the numerical integration of Eq.(8) within a few per thousand. To conclude, we remark that the procedure sketched in Sct.2 may be applied as well to investigate the equation for the transversal oscillations.

REFERENCES

1. I.M. Kapchinskii and V.A. Teplyakov, "Linear Ion Accelerator with Spatially Homogeneous Strong Focusing", Pribory i. Tekhnika Experimenta, 119, No.2, pp.19–22, March–April 1970
2. I.M. Kapchinskii and N.V. Lazarev, "The Linear Accelerator Structures with Space–Uniform Quadrupole Focusing", IEEE Trans. on Nuclear Science, 26, 3462 (1979)
3. R.H. Stokes, K.R. Crandall, J.E. Stovall and D.A. Swenson, "RF Quadrupole Beam Dynamics Design Studies", 1979 Linac Accelerator Conference
4. J. Le Duff, "Dynamics and Acceleration in Linear Structures", in Proceedings of CERN Accel. School, CERN 85–19, 1985; M.Puglisi, "Introduction to the RFQ Vane Tips Shaping", Brookhaven National Laboratory, Polarized Proton Technical Note No. 28, 1983
5. N. Krylov and N.N. Bogoliubov, "On Nonlinear Mechanics", Acad. Sci. USSR (1945)
6. N.N. Bogoliubov and Y.A. Mitropolsky, Asymptotic Methods in the Nonlinear Oscillations, Gordon and Breach, N.Y., 1961
7. M. Leo, R.A. Leo, M. Puglisi, C. Rossi, G. Soliani and G. Torelli, Preprint–Dipartimento di Fisica dell'Università, Lecce, Italy, 1986
8. IBM Scientific Subroutine Package H20–0205–3, 1969, p.337.

A MODEL OF FOUR RODS R.F.Q.

A. Fabris, A. Massarotti, and M. Vretenar

INFN - Sezione di Trieste
Trieste, Italy

Fig.1 shows one of the R.F.Q. structures proposed from Frankfurt. It can be thought as built from a chain of modules made with a line of length h, short-circuited at one end and loaded at the other end from the parallel of two lines, each of them of length l/2, open-ended.

If Z_{01} is the characteristic impedance of the first line, its input impedance Z(h) is:

$$Z(h) = jZ_{01} \, \mathrm{tg}\,\beta h$$

with h $<\lambda/2$ and $\beta = 2\pi/\lambda = \omega/c$;
ω angular frequency of excitation.
c speed of light.

If Z_{02} is the characteristic impedance of the quadrifilar line, the impedance at the input is:

$$Z_{\parallel}(l/2) = -j(Z_{02}/2)\,\mathrm{ctg}\,\beta\,l/2$$

with l/2 $<\lambda/4$

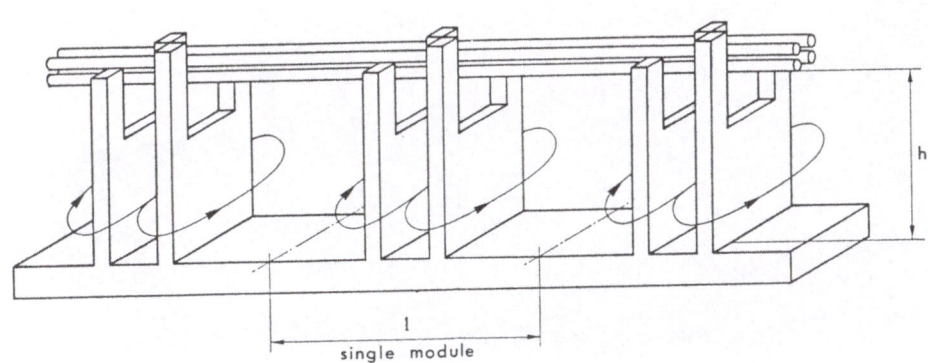

single module

Fig. 1. R.F.Q. structure proposed by Frankfurt.

The resonance condition is given from:

$$|Z(h)| = |Z_{\parallel}(1/2)|$$

This structure exhibits two drawbacks:
a) because of the different paths for the currents percurring homologous rods there is an asymmetry in the quadrupolar voltages;
b) the voltage along the rods changes with $\cos\beta x$, in the range $0 \leq x \leq 1/2$.

The structure proposed from Trieste solves the first problem imposing to the currents equal paths (Fig. 2) and the second imposing a longitudinal symmetry (Fig. 3) with electrodes equally spaced at such distances that one can consider equipotential the single rods.

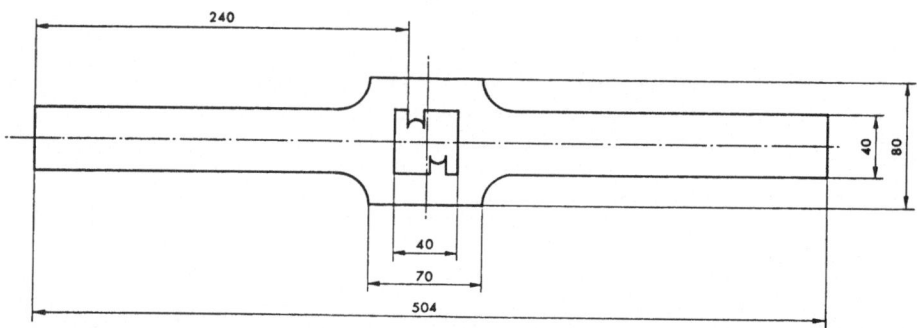

Fig. 2. Electrode proposed by Trieste.

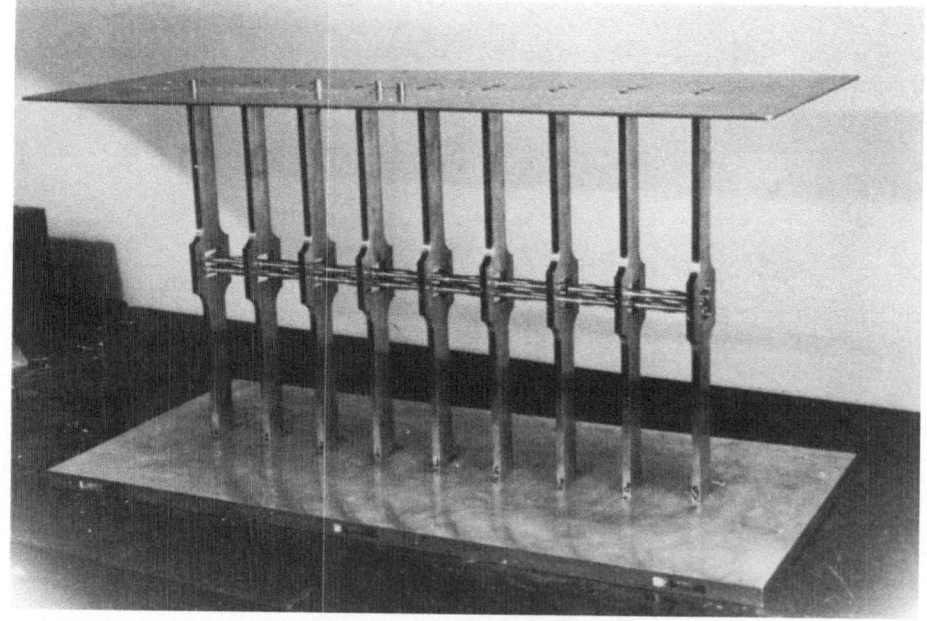

Fig. 3. Structure proposed by Trieste.

The choice of the parameters for the constructed models has been done calculating in first approximation, the characteristic impedances with the usual formulae from the literature[2], obtaining errors lower than 10%. Then models have been built with two and four cells to measure the capacitance per unit length of the lines corresponding to external and internal cells. In this way one can calculate the characteristic impedance of a structure with any number of cells and its resonance frequency within 1%.

A computer code has been used to calculate at each frequency the shunt resistance of the whole structure as a function of the number of cells (this is of the free geometrical parameters of the cell, height and length), giving the capacitance between the rods (in our case $C = 76$ pF), the overall length of the structure and the width of the electrodes. From Fig. 4, the maximum for R_{sh} depends from the frequency and from the number of cells, keeping constant the width of the electrodes ($L = 40$ mm in the figure).

Fig.5 shows for a certain number of frequencies the maximums obtained for R_{sh}, as a function of the width B, and the corresponding number of cells giving the maximum for the given width. One can see that at each frequency there is a combination of B,L,h for which R_{sh} is maximum.

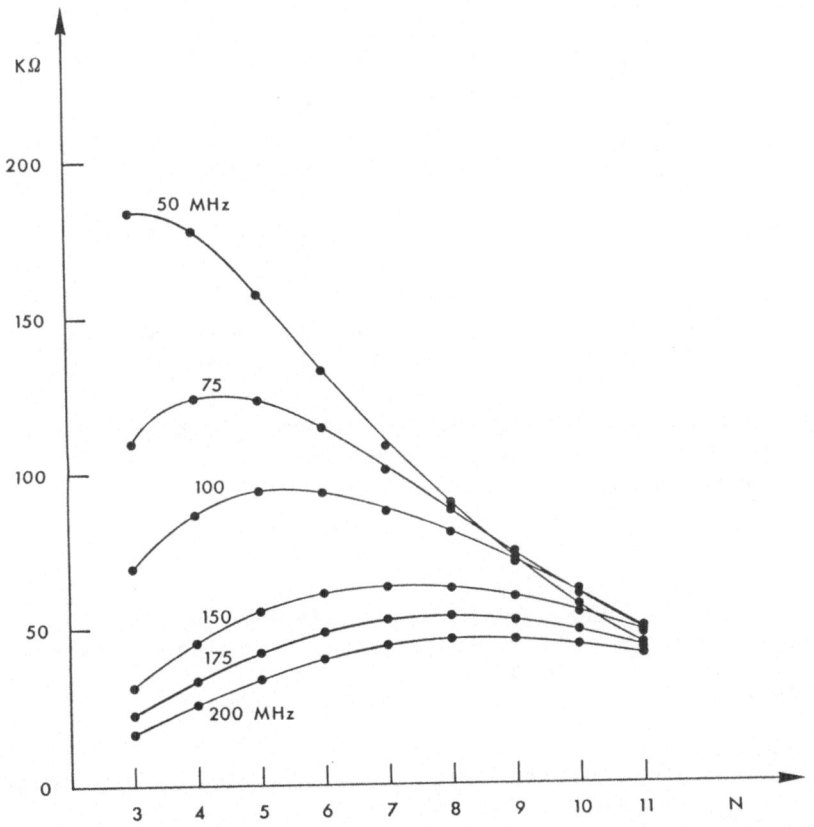

Fig. 4. Maxima of R_{sh} for given rods capacitance (C=76 pF), width of the electrodes (L=40 mm) and overall length of the structure.

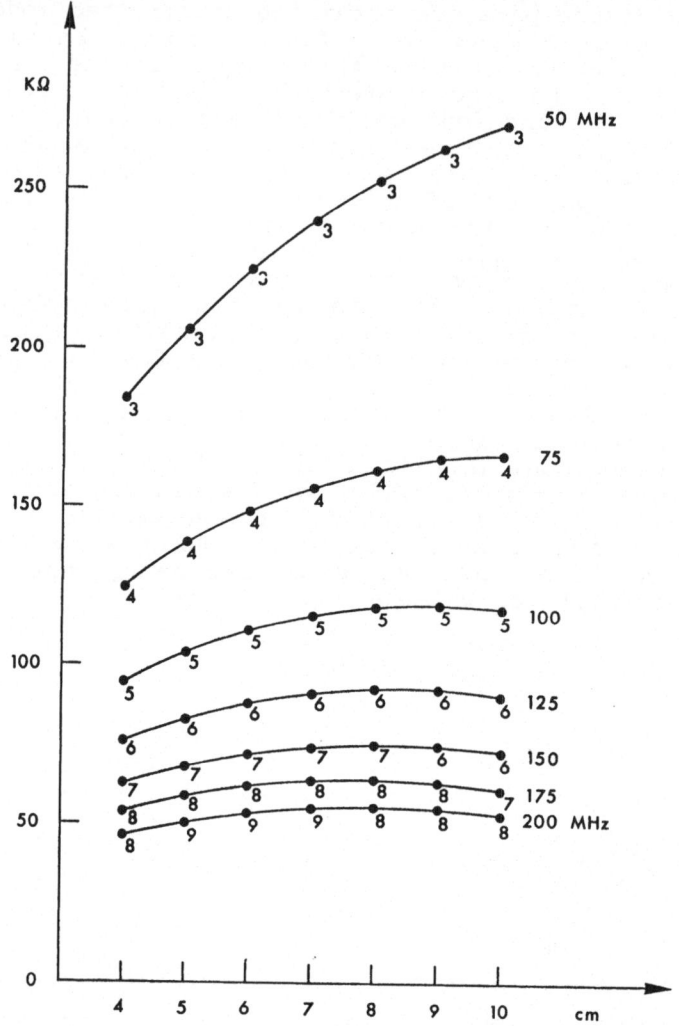

Fig. 5. Maxima of R_{sh} for a given capacitance (C=76 pF) as function of the width of the electrodes with the corresponding number of cells.

Several models have been built, with electrodes of different shape. The data are referred to the aluminium (without welding) model of Fig.2 and 3.

The characteristic impedance of the structure is:

Z_{om} = 23.25 Ω (measured $\pm 0.5\ \Omega$)
Z_{oc} = 22.97 Ω (calculated from the measurements with 2 and 4 cells, $\pm 0.5\ \Omega$)
$\Delta Z_{oc}\ /Z_{om}$ = 1.2%

With cylindrical rods one has, at the frequency f_o= (123.413$\pm$.001) Mhz:

$R_{shm} = (52380 \pm 950) \; \Omega$ (measured value)
$R_{shc} = 70600 \; \Omega$ (calculated value)
$Q_m = 3085 \pm 50$
$Q_c = 4800$

For the same structure with modulated rods, at the frequency $f_o = 138.061$ MHz:

$R_{shm} = (49210 \pm 950) \; \Omega$
$Q_m = 2875 \pm 50$

This is due to the decrease of the capacitance between the rods and to the increase of the paths for the current.

For a structure with 4 cells, and unmodulated rods, at the frequency $f_o = (77.046 \pm .001)$ MHz:

$R_{shm} = (77100 \pm 1300) \; \Omega$
$R_{shc} = 118800 \; \Omega$
$Q_m = 2750 \pm 50$
$Q_c = 4400$

Using copper instead of aluminium, one can extimate an increase in the shunt resistances by nearly a factor 1.5.

It is then justified to think that structures of this kind can be used in the range between 50 MHz and 150 MHz, with excitation power between 10 and 25 kW, giving nearly 50 kV between the rods.

REFERENCES

1. A. Schempp, H. Deitinghoff, M. Ferch, P. Junior and H. Klein, Four Rod - $\lambda/2$ - R.F.Q. for Light Ion Acceleration, Institut fur Angewandte Physik, Frankfurt am Main, (1984).
2. See for instance: Meinke-Gundlach, Taschenbuch der Hochfrequenz Technik, Springer-Verlag, Berlin (1962).

RFQ FIELD STABILIZATION USING DIPOLE SUPPRESSORS

M. Vretenar

INFN - Sezione di Trieste
Trieste, Italy

In the frame of the CERN RFQ2 project, a certain effort has been devoted to problems of stabilization of electromagnetic fields in the RFQ. In particular one tried to consider schemes which would reduce distorsions of the quadrupolar field due to transverse asymmetries in the structure.

The RFQ2 is a high intensity accelerator (200mA of protons) and it is intended to replace the present injection system into the CERN Linac2. One wishes to have in RFQ2 a very good field quality and, in addition, have a machine very little sensitive to mechanical and tuning errors.

Following a proposal by W.Pirkl[1], a stabilization scheme "Dipole suppressor" has been analized. It consists of four rectangular loops closed with tunable capacitances (small trimmer capacitors in the test model, eventually parallel plates in a real cavity) built inside one end cover of the RFQ, see Fig. 1. At each end of the RFQ the quadrupole magnetic flux splits into two parts and returns in the adjacent quadrants, and the loops are placed in the plains of separation for these two fluxes, in such a way that they don't

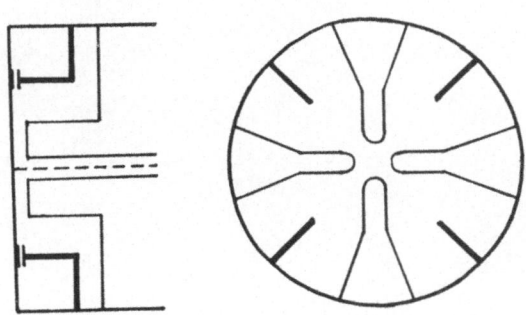

Fig. 1. Dipole suppressors.

interact with the quadrupole mode. Instead, they intercept nearly all the flux relative to the dipole modes always present in the cavity and responsable for the instability. The dipole suppressors are then a tunable oscillating circuit magnetically coupled only to the dipole mode, and their effect will be to split this dipole into two different resonances, two dipole modes. Adjusting the capacitances it is possible to arrange the two new dipole modes to an exact symmetry in position around the quadrupole frequency. Generally, a mechanical perturbation to the cavity results in the introduction, at the quadrupole frequency, of a dipole component inversely proportional to the difference between the dipole and quadrupole frequencies. This dipole component can be positive or negative, depending if the dipole frequency is higher or lower than the quadrupole one. If the perturbing modes are two and equally spaced around the quadrupole, one can expect that their respective perturbative components cancel out giving a cavity virtually insensitive to transverse perturbations.

This behaviour has been verified on a full scale model cavity (Fig. 2 shows its end plate with the dipole suppressors). The outputs from calibrated monitoring loops in two opposite quadrants of the model were connected to a 90° hybrid, which gave either their sum or their difference. The modal symmetry is such that the sum is proportional to the pure quadrupolar field, the difference to the pure dipolar field. Dipolar and quadrupolar resonances then can be observed separately on the screen of a network analyzer.

For the unstabilized model, the situation was as in Fig. 3, for the two pairs of opposite quadrants: there is a strong dipole mode in the pair of quadrants containing the feeder loop, a very low dipole in the other pair (whose level is proportional to the asymmetries in the structure), and the quadrupole. Inserting the dipole suppressors (Fig. 4) the

Fig. 2. Test Model end plate with the dipole suppressors loops.

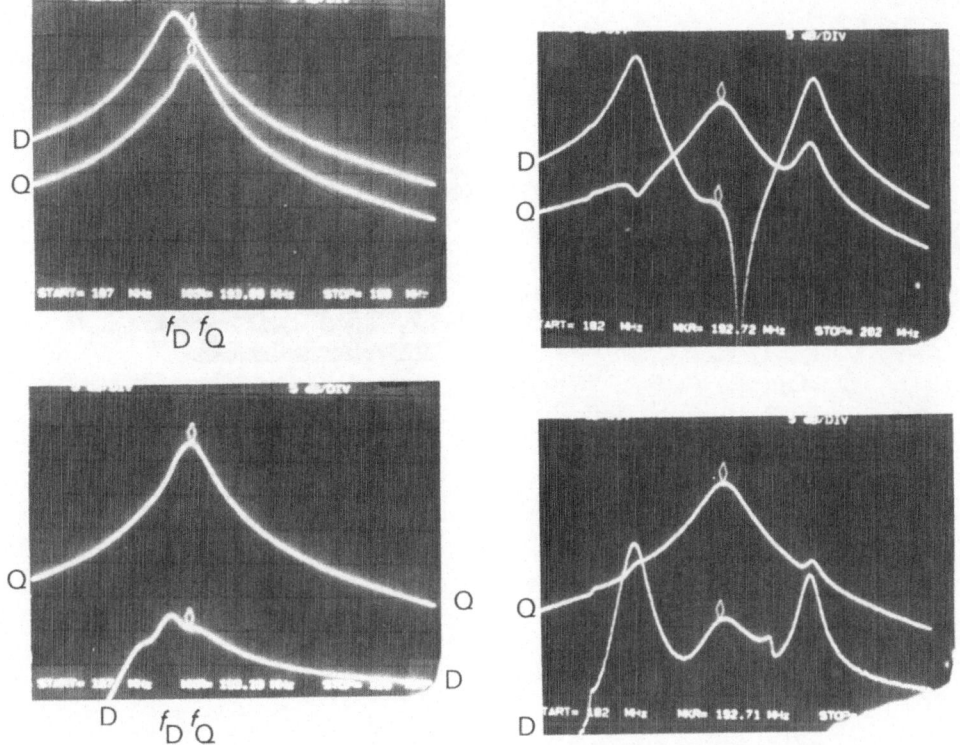

Fig. 3. Modal responses in the unsta- Fig. 4. Modal responses in the RFQ
      bilized RFQ.   stabilized with dipole sup-
                                         pressors.

quadrupole is unchanged while the dipole, as foreseen, is splitted into two peaks.

This arrangement allows also to see directly how the cavity reacts to perturbations: a mechanical perturbation is simulated by pushing two piston tuners in one quadrant progressively in. Fig. 5 shows the effect on the modal responses for the unstabilized RFQ (pair of quadrants without the feeder) of this gradually increasing perturbation: it increases all the modes frequencies, causes the dipole level, at the dipole frequency, to rise, but mainly causes a second perturbative peak of dipolar field to rise at the quadrupole frequency. To stabilize the RFQ is to avoid this peak to rise, and Fig. 6 shows that perturbing the model after the insertion of the dipole suppressors the frequencies and the levels go up, but there is no dipole peak rising at the quadrupole frequency.

Fig. 7 shows the experimentally obtained "sensitivity curves" for three different RFQ structures, unstabilized, stabilized with "$\lambda/2$ lines", (another stabilizing scheme, proposed by W.Pirkl[2], which moves the dipole mode to a frequency 8 MHz higher than the quadrupole) and stabilized with dipole suppressors. The curves are obtained plotting the ratio D/Q between dipole and quadrupole fields measured at the operating quadrupole frequency for increasing perturbations, indicated with their induced relative shift of the quadrupole frequency from its imperturbed value. The asymmetry in the

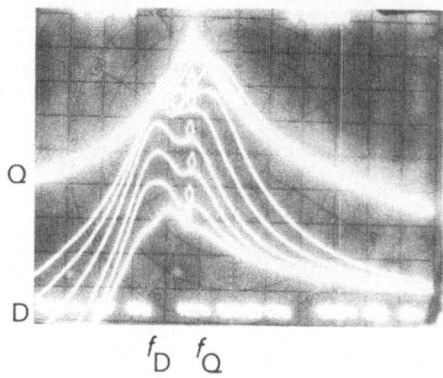

Fig. 5. Reaction to perturbations of the unstabilized RFQ.

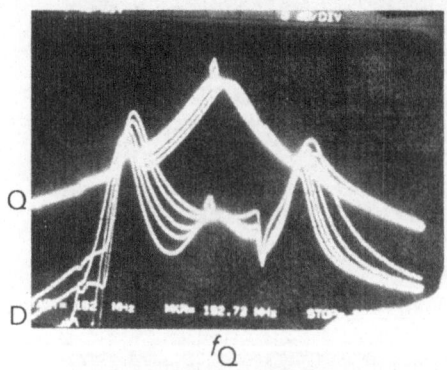

Fig. 6. Reaction to perturbations of the RFQ stabilized with dipole suppressors.

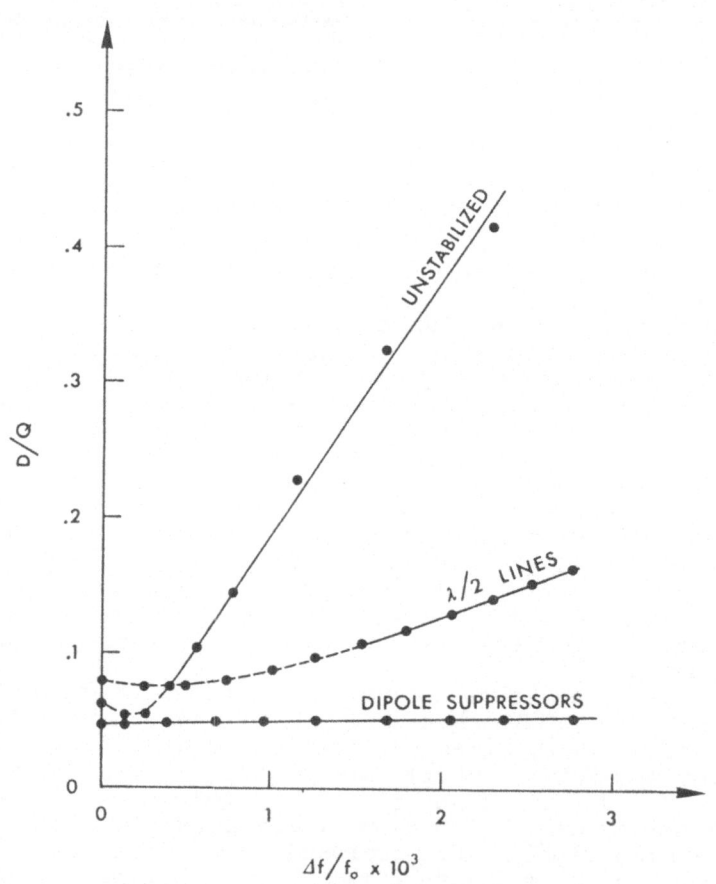

Fig. 7. Induced fields asymmetry versus increasing per‑turbations for unstabilized and stabilized RFQs.

fields induced from the perturbation is very high in the unstabilized case, it is reduced with the $\lambda/2$ lines, while the dipole suppressors curve is nearly flat. The slope of the curves in the linear region is taken as measure for the transverse sensitivity of the structure. The measured values for this test show that using dipole suppressors it's possible to improve the transverse stability of the RFQ cavity by nearly a factor 100.

REFERENCES

1. W.Pirkl, One more RFQ stabilization method (dipole shifter), CERN PS/RF/Note 85-6 (1985).
2. W.Pirkl, Another RFQ stabilization scheme, CERN PS/RF/Note 85-5 (1985).

PARTICIPANTS

BIZZARRI, U.

ENEA, Dip. TIB-FIS, Centro Ricerche Energia,
P.O.Box 65,
00044 Frascati, Italy

BOSCOLO, I.

INFN - Sezione di Bari, Gruppo di Lecce, and
Dip. di Fisica dell'Università, Via Arnesano,
73100 Lecce, Italy

BRESSAN, M.

Dip. di Elettronica dell'Università,
Via Abbiate Grasso 209,
27100 Pavia, Italy

CALABRETTA, L.

INFN - Sezione di Catania, and Dip. di Fisica
dell'Università, Corso Italia 57,
95129 Catania, Italy

CAVENAGO, M.

INFN - Sezione di Pisa, and Scuola Normale
Superiore, Via Livornese
56010 S. Piero a Grado, Italy

CAVALLINI, M.

Proel Tecnologie, V.le Machiavelli 29,
50125 Firenze, Italy

CIRRI, G.

Proel Tecnologie, V.le Machiavelli 29,
50125 Firenze, Italy

CLAUS, J.

Brookhaven National Laboratory,
Upton, Long Island, NY 11973, USA

COURANT, E.

Brookhaven National Laboratory,
Upton, Long Island, NY 11973, USA

DE ANGELIS, U.

INFN - Sezione di Napoli, and Dip. di Fisica
dell'Università, Mostra d'Oltremare, pad. 20,
80138 Napoli, Italy

DELLE CAVE, G.

Istituto di Cibernetica del CNR, Via Toiano,
80072 Arco Felice, Italy

FEDELE, R.

INFN - Sezione di Napoli, and Dip. di Fisica
dell'Università, Mostra d'Oltremare, pad. 20,
80138 Napoli, Italy

GNEPF, S.

ETH, Inst. für Quantum Electronics, HPF,
8092 Zurich, Switzerland

GUSTAFSSON, S. INFN - Sezione di Catania, C.so Italia 57,
95129 Catania, Italy

HOPKINS, D.B. Lawrence Berkeley Laboratory, Building 52B,
Berkeley, CA 94720, USA

JOHNSEN, K. CERN,
1211 Geneva 23, Switzerland

JOHO, W. SIN,
5234 Villigen, Switzerland

KLEIN, H. University of Frankfurt, R. Mayer Str. 2/4,
6000 Frankfurt am Main, Germany

LUCHES, A. Dip. di Fisica dell'Università, Via Arnesano,
73100 Lecce, Italy

MASSAROTTI, A. INFN - Sezione di Trieste, and Dip. di Fisica
dell'Università, Via Valerio 2,
34127 Trieste, Italy

NASSISI, V. Dip. di Fisica dell'Università, Via Arnesano,
73100 Lecce, Italy

NEUMAN, W. ETH, Inst. für Mittelenergie-Physik, HPF,
8093 Zurich, Switzerland

PAGANI, C. INFN - Sezione di Milano, Via Celoria 16,
20133 Milano, Italy

PALMER, R.B. Stanford Linear Accelerator Center,
Stanford, CA 94305, USA

PALUMBO, L. Dip. di Energetica dell'Università "La
Sapienza", Via Scarpa 14,
00161 Roma, Italy

PEGORARO, F. Scuola Normale Superiore, P.zza dei Cavalieri,
56100 Pisa, Italy

PELLEGRINI, C; CERN,
1211 Geneva 23, Switzerland

PETRACCA, S.P. Dip. di Matematica ed Applicazioni della
Università, Via Mezzocannone 8,
80134 Napoli, Italy

PICARDI, L. ENEA, Dip. TIB-FIS, Centro Ricerche Energia,
P.O.Box, 65
00044 Frascati, Italy

PINTO, I.M. Dip. di Elettronica dell'Università,
Via Claudio 21,
80125 Napoli, Italy

POGGIANI, R. Dip. di Fisica dell'Università,
P.zza Torricelli 2,
56100 Pisa, Italy

PUGLISI, M.   INFN - Sezione di Pavia, and Dip. di Fisica
dell'Università, Via Bassi 6,
27100 Pavia, Italy

RAINO', A.   INFN - Sezione di Bari, and Dip. di Fisica
dell'Università, Via Amendola 173,
70126 Bari, Italy

RATTI, A.   Dip. di Fisica dell'Università, Via Bassi 6,
27100 Pavia, Italy

ROSSI, C.   INFN - Sezione di Pavia, and Dip. di Fisica
dell'Università, Via Bassi 6,
27100 Pavia, Italy

ROSSI, L.   INFN - Sezione di Milano, Lab. Ciclotrone,
Via Celoria 16,
20133 Milano, Italy

ROSATELLI, C;   Dip. di Ingegneria Elettr. dell'Università,
Via Opera Pia 11/A,
16145 Genova, Italy

RUGGIERO, A.G.   Argonne National Laboratory,
Argonne, IL 60439, USA

SABIA, E.   ENEA, Dip. TIB-FIS, Centro Ricerche Energia,
P.O.Box 65
00044 Frascati, Italy

SCHEMPP, A.   Inst. für Angenwandte Physik der Universität,
6000 Frankfurt am Main, Germany

SCHNELL, W.   CERN,
1211 Geneva 23, Switzerland

SERAFINI, L.   INFN - Sezione di Milano, Via Celoria 16,
20133 Milano, Italy

SOLIANI, G.   INFN - Sezione di Bari, Gruppo di Lecce, and
Dip. di Fisica dell'Università, Via Arnesano,
73100 Lecce, Italy

SOUTHWORTH, B.   CERN,
1211 Geneva 23, Switzerland

STAGNO, V.   INFN - Sezione di Bari, and Dip. di Fisica
dell'Università, Via Amendola 173,
70126 Bari, Italy

TAZZARI, S.   INFN - Laboratori Nazionali di Frascati,
P.O. Box 13,
00044 Frascati, Italy

TAZZIOLI, F.   INFN - Laboratori Nazionali di Frascati,
P.O. Box 13,
00044 Frascati, Italy

TORELLI, G.   INFN - Sezione di Pisa, and Dip. di Fisica
                               dell'Università, Via Livornese,
                               56010 S. Piero a Grado, Italy

VACCARO, V.G.   INFN - Sezione di Napoli, and Dip. di Fisica
                               dell'Università, Mostra d'Oltremare, pad. 20,
                               80138 Napoli, Italy

VAN DER MEER, S.   CERN,
                               1211 Geneva 23, Switzerland

VIGNATI, A.   ENEA, Dip. TIB-FIS, Centro Ricerche Energia,
                               P.O. Box 65
                               00044 Frascati, Italy

VRETENAR, M.   INFN - Sezione di Trieste, and Dip. di Fisica
                               dell'Università, Via Valerio 2,
                               34127 Trieste, Italy

WANGLER, T.P.   Los Alamos National Laboratory,
                               Los Alamos, NM 87545, USA

WEILAND, T.   DESY, Notkestrasse 85,
                               2000 Hamburg 52, Germany

WEISS, M.   CERN,
                               1211 Geneva 23, Switzerland

WILSON, E.   CERN
                               1211 Geneva 23, Switzerland

INDEX

Accelerating structures, 8,9,58
Acceleration
  operating mode for, 50
Acceptance, 131,132,133
Alvarez proton linac, 59

Beam diagnostic, 35
Beamstrahlung, 142
  description, 114,115,116
  limited collider, 121
Beat wave accelerator, see Plasma
BKM, see Method

Cavity
  for micro lasertron, 101,102
  indipendent, 51
Cerenkov monitor, 35, 37
Colonnate π modes, 64,65
  2π modes (Palmer), 62,63
Compensated multibunch
  description, 72,73,74,76
  efficiency, 85
Compton scattering, see FEL Compton
Cooling rate, 122
  rings, 128,129,130,135,136,137
  for SLC, 124,125,129,133,134
Cost estimation
  for high energy, 45,105,106
  for a linear collider, 105,106,
    107,108,118,119
  for a peak power, 108,109
  of a SSC, 109,118
Cost optimization, 112,113
CW klystron, 76,85

Damping rings, 156,157,158
Debunching problem, 253
DESY, 31,42
Dipole suppressor scheme, 271

Drive beam, 79
Drive linac, 77,78

Electron laser facility, 220-222
  features, 223
  gain measurement, 223
  high power diagnostics, 224
  performance, 226
Electron-positrons
  beam parameters, 141,142,143
  cost estimation, 105,118
  linear collider, 6,7,140,241
  storage rings, 139
Emittance, 183,184,197,198
  equilibrium, 123
  growth equations, 186,188
  growth parameters, 188,190
  longitudinal, 127,128
  RMS: numerical studies, 190-194
    real problem and
    experiments, 194,195
European hadron facility, EHF,
  202,244,247
Experimental test accelerator,ETA,
  220,223

Free-electron laser, 75,220,230
  acceleration techniques, 148,149
  compton mechanism, 163-166
  driven linac, 146,147
  low technology, 170-178
  microwave power, 235
  process, 166-170
  properties, 152,153,154,159
Focusing optics, 116
  conventional, 116,117
  laser, 117
  super disruption, 117,118

FODO
  numerical simulation studies, 196,197
  structure, 79
Foxhole structure
  2π mode, 58,59,60
  , $2\pi/3$ mode, 60,61
Four rods
  RFQ model, 266,267,269
  structure, 211,212,215,216

Gap monitors, 36
Gyrotrons, 230

Hadron colliders, 3
  Eloisatron project, 6
  Lep tunnel, 4
  SSC, 5 (see also Super super-
    conducting collider SSC)
HERA, 6
  RFQ project, 207-211,215
  RFQ resonant rings, 209
High energy buncher, 42
High frequency linac, 149-152
  SCC, FEL system, 155,156,160
Hollow beam spectrometer, 37

IFEL, see FEL
Intermediate energy storage, 84
Intrabeam scattering, 126,127

Jungle gym structure, 71

Kapchinskiy, 202
Kapchinskiy-Vladimirskii
  distribution, 182
Kapchinskiy-Teplyarov, 259
Kilpatrick limit, 230
Klystron, 225,229
  cost, 112
  (see also C.W.Klystron)

Lasertron, 89,90
  cost, 109
Linear accelerator, 29,31-33,40
  accelerating cavities, 33
  Alvarez proton linac, 59
  high energy buncher, 42
  prebuncher cavities, 33
  wake field, 42
Linear collider, 7,67,139,140,
    219,229

Linear collider (continued)
  cost, 106,107,108,119
  (see also Wakeatron)
Linear ion accelerator, 181-183
Liouville continuity equation, 252
Lorentz force, 252
Luminosity, 46-48,53,54,73-75,141,
    241

Main linac, 68,69,70
  advantages, 85,86
  energy, 85
Methods
  of BKM, 260,261
  of Runge-Kutta, 262,263
Microlasertron, 91,92,102,103
  efficiency, 94,96
  equivalent circuit, 93,94
  low-rf field, 95
  one-dimensional simulation,
    96,97,98,99,100
  short transit time, 95
  parameter, 101,102
Multibunch operation, 49

Non-linear equations, 249,250
Non-linear evolution equation,
    251,252,253
Non-linear field energy, 183-185
Numerical simulation
  FODO, 196,197
  microlasertron one-dimensional
    simulation, 96-100
  RMS, 190,191,192,193,194
  waktrak, 40

Optoelectric switches, 10

Palmer's colonnate 2π modes,
    62,63
Penning effect, 32
Plasma, 12
  BWA, 12,13,15,18,21,22
  chamber, 25
  cold, 17,19
  decay instability, 23
  low-frequency response, 18
  wake field, 12,15
Ponderomotive force, 16
Pirkl, 271,273

Radiation matter interaction, 163

RFQ accelerator, 201,262,263
  analytical solution, 259-262
  application, 205
  beam dynamics, 201,202
  bunch system, 203,204
  for high ions, 213
  HERA project, 207-211,215
  prototype, 214
  resonant rings, 209
  resonator, 212,213,259
  structures, 206,265,266
RFQ2 accelerator, 271,272
Runge-Kutta method, 262,263

Self-bunching problem, 253
Single long bunch, 47
Superconducting
  accelerating structures, 9
  cavities, 9,10,149,151
  cavities UHF, 86
  RF drive linacs, 75,85

Super-superconducting collider,
  5,247
  cost estimation, 109,118
Switched power linac, 89

Textronix digitizer, 226
Transfert structure, 80
Two beam accelerator, 118,160,
  219,231,232,233,236,237
  scheme, 75,76

Volterra series, 250,251

Wake field, 109,110
  transformer, 29,42
  see also Plasma wake field
Wakeatron, 242,243,244,245 (see
  also linear collider)
Waktrack computer simulation, 40
Wiener-Cameron-Martin, 250